MOLECULAR BIOLOGY INTELLIGENCE UNIT

PROTEIN TOXIN STRUCTURE

Michael W. Parker, D.Phil.

St. Vincent's Institute of Medical Research
Melbourne, Australia

CHAPMAN & HALL
ITP An International Thomson Publishing Company

New York • Albany • Bonn • Boston • Cincinnati • Detroit • London • Madrid • Melbourne •
Mexico City • Pacific Grove • Paris • San Francisco • Singapore • Tokyo • Toronto • Washington

R.G. LANDES COMPANY
AUSTIN

MOLECULAR BIOLOGY INTELLIGENCE UNIT
PROTEIN TOXIN STRUCTURE

R.G. LANDES COMPANY
Austin, Texas, U.S.A.

Please address all inquiries to the Publishers:
R.G. Landes Company, 909 Pine Street, Georgetown, Texas, U.S.A. 78626
Phone: 512/ 863 7762; FAX: 512/ 863 0081

North American distributor:
Chapman & Hall, 115 Fifth Avenue, New York, New York, U.S.A. 10003

CHAPMAN & HALL

U.S. and Canada ISBN: 0-412-11081-4

While the authors, editors and publisher believe that drug selection and dosage and the specifications and usage of equipment and devices, as set forth in this book, are in accord with current recommendations and practice at the time of publication, they make no warranty, expressed or implied, with respect to material described in this book. In view of the ongoing research, equipment development, changes in governmental regulations and the rapid accumulation of information relating to the biomedical sciences, the reader is urged to carefully review and evaluate the information provided herein.

Library of Congress Cataloging-in-Publication Data

Protein toxin structure / [edited by] Michael W. Parker.
    p. cm. — (Molecular biology intelligence unit)
  Includes bibliographical references and index.
  ISBN 1-57059-368-X (RGL : alk. paper). —ISBN 3-540-61191-6 (SV : alk. paper)
  1. Toxins—Structure-activity relationships. I. Parker, Michael W., 1959- . II. Series.
QP631.P76 1996
615'.373—dc20

                        96-21770
                        CIP

Publisher's Note

R.G. Landes Company publishes six book series: *Medical Intelligence Unit, Molecular Biology Intelligence Unit, Neuroscience Intelligence Unit, Tissue Engineering Intelligence Unit, Biotechnology Intelligence Unit and Environmental Intelligence Unit*. The authors of our books are acknowledged leaders in their fields and the topics are unique. Almost without exception, no other similar books exist on these topics.

Our goal is to publish books in important and rapidly changing areas of bioscience and environment for sophisticated researchers and clinicians. To achieve this goal, we have accelerated our publishing program to conform to the fast pace in which information grows in bioscience. Most of our books are published within 90 to 120 days of receipt of the manuscript. We would like to thank our readers for their continuing interest and welcome any comments or suggestions they may have for future books.

Deborah Muir Molsberry
Publications Director
R.G. Landes Company

CONTENTS

EDITOR

Michael W. Parker, D.Phil.
The Ian Potter Foundation Protein Crystallography Laboratory
St. Vincent's Institute of Medical Research
Melbourne, Australia
Chapter 2, Chapter 5

CONTRIBUTORS

Joseph E. Alouf
Institut Pasteur
Unité des Toxines Microbiennes
Paris, France
Chapter 1

Charles E. Bell
Molecular Biology Institute
    and UCLA DOE Laboratory
    of Structural Biology and Molecular
    Medicine
University of California at Los
    Angeles
Los Angeles, California, USA
Chapter 3

Melanie J. Bennett
Department of Biochemistry
    and Biophysics
School of Medicine, University
    of Pennsylvania
Philadelphia, Pennsylvania, USA
Chapter 3

Robert F. Bozarth
Department of Life Sciences
Indiana State University
Terre Haute, Indiana, USA
Chapter 15

J. Thomas Buckley
Department of Biochemistry
    and Microbiology
University of Victoria
British Columbia, Canada
Chapter 5

Maia M. Chernaia
Medical Research Council of Canada
    Group in Protein Structure
    and Function
Department of Biochemistry
University of Alberta
Edmonton, Alberta, Canada
Chapter 9

R. John Collier
Department of Microbiology
    and Molecular Genetics
Harvard Medical School, Shipley
    Institute of Medicine
Boston, Massachusetts, USA
Chapter 3

Cathleen A. Earhart
Department of Biochemistry
University of Minnesota Medical
    School
Minneapolis, Minnesota, USA
Chapter 11

David Eisenberg
Molecular Biology Institute
  and UCLA DOE Laboratory
  of Structural Biology and
  Molecular Medicine
University of California at Los
  Angeles
Los Angeles, California, USA
Chapter 3

William H. Flurkey
Department of Chemistry
Indiana State University
Terre Haute, Indiana, USA
Chapter 15

Juan C. Fontecilla-Camps
Institut de Biologie Structurale Jean-
  Pierre Ebel, CEA-CNRS
Grenoble, France
Chapter 14

Marie E. Fraser
Medical Research Council of Canada
  Group in Protein Structure
  and Function
Department of Biochemistry
University of Alberta
Edmonton, Alberta, Canada
Chapter 9

William Furey
Biocrystallography Laboratory
VA Medical Center
Pittsburgh, Pennsylvania, USA
Chapter 12

Fei Gu
Department of Biological Sciences
Purdue University
West Lafayette, Indiana, USA
Chapter 15

Bart Hazes
Department of Medical Microbiology
  and Immunology
University of Alberta
Edmonton, Alberta, Canada
Chapter 10

Wim G.J. Hol
Biomolecular Structure Center
Howard Hughes Medical Institute
University of Washington School of
  Medicine
Seattle, Washington, USA
Chapter 8

Dominique Housset
Institut de Biologie Structurale Jean-
  Pierre Ebel, CEA-CNRS
Grenoble, France
Chapter 14

Michael N.G. James
Medical Research Council of Canada
  Group in Protein Structure and
  Function
Department of Biochemistry
University of Alberta
Edmonton, Alberta, Canada
Chapter 9

Anis Khimani
Department of Life Sciences
Indiana State University
Terre Haute, Indiana, USA
Chapter 15

Yuri V. Kozlov
Englehardt Institute of Molecular
  Biology, Academy of Sciences
  of Russia
University of Oslo Center for Medical
  Studies
Moscow, Russia
Chapter 9

Jade Li
Molecular Research Council
  Laboratory of Molecular Biology
Cambridge, England
Chapter 4

Robert C. Liddington
Dana-Farber Cancer Institute
Boston, Massachusetts, USA
Chapter 6

Ethan A. Merritt
Department of Biological Structure
Biomolecular Structure Center
University of Washington School
  of Medicine
Seattle, Washington, USA
Chapter 8

David T. Mitchell
Department of Biochemistry
University of Minnesota Medical
  School
Minneapolis, Minnesota, USA
Chapter 11

Arthur F. Monzingo
Department of Chemistry
  and Biochemistry
University of Texas
Austin, Texas, USA
Chapter 13

Douglas H. Ohlendorf
Department of Biochemistry
University of Minnesota Medical
  School
Minneapolis, Minnesota, USA
Chapter 11

Franc Pattus
Departement Recepteurs et Proteines
  Membranaires
CNRS, Ecole Superieure
  de Biotechnologie de Strasbourg
Illkirch, France
Chapter 2

Carlo Petosa
Dana-Farber Cancer Institute
Boston, Massachusetts, USA
Chapter 6

G. Sridhar Prasad
Molecular Biology Department
Scripps Research Institute
La Jolla, California, USA
Chapter 11

R. Radhakrishnan
Center of Macromolecular
  Crystallography
Department of Pharmacology
University of Alabama
  at Birmingham
Birmingham, Alabama, USA
Chapter 11

Stan Rane
Department of Biological Sciences
Purdue University
West Lafayette, Indiana, USA
Chapter 15

Randy J. Read
Department of Medical
    Microbiology and Immunology
University of Alberta
Edmonton, Alberta, Canada
Chapter 10

Jon D. Robertus
Department of Chemistry
    and Biochemistry
University of Texas
Austin, Texas, USA
Chapter 13

Martin Sax
Biocrystallography Laboratory
VA Medical Center
Pittsburgh, Pennsylvania, USA
Chapter 12

Patrick M. Schlievert
Department of Microbiology
University of Minnesota Medical
    School
Minneapolis, Minnesota, USA
Chapter 11

Michael P. Schlunegger
Molecular Biology Institute
    and UCLA DOE Laboratory
    of Structural Biology and
    Molecular Medicine
University of California at Los
    Angeles
Los Angeles, California, USA
Chapter 3

David L. Scott
Medical Services
Massachusetts General Hospital
Boston, Massachusetts, USA
Chapter 7

Thomas J. Smith
Department of Biological Sciences
Purdue University
West Lafayette, Indiana, USA
Chapter 15

Boyd A. Steere
Molecular Biology Institute
    and UCLA DOE Laboratory
    of Structural Biology
    and Molecular Medicine
University of California at Los
    Angeles
Los Angeles, California, USA
Chapter 3

Penelope E. Stein
Department of Medical
    Microbiology and Immunology
University of Alberta
Edmonton, Alberta, Canada
Chapter 10

S. Swaminathan
Biocrystallography Laboratory
VA Medical Center
Pittsburgh, Pennsylvania, USA
Chapter 12

Demetrius Tsernoglou
Department of Biochemical Sciences
Università di Roma "La Sapienza"
Rome, Italy
Chapter 2, Chapter 5

Focco van den Akker
Department of Biological Structure
Biomolecular Structure Center
University of Washington School
    of Medicine
Seattle, Washington, USA
Chapter 8

F. Gisou van der Goot
Département de Biochimie
Université de Genève
Geneva, Switzerland
Chapter 5

Ingrid R. Vetter
Max-Planck-Institute for Molecular
  Physiology
Dortmund, Germany
Chapter 2

Manfred S. Weiss
Molecular Biology Institute
  and UCLA DOE Laboratory
  of Structural Biology and
  Molecular Medicine
University of California at Los
  Angeles
Los Angeles, California, USA
Chapter 3

Edwin M. Westbrook
Center for Mechanistic Biology
  and Biotechnology
Argonne National Laboratory
Argonne, Illinois, USA
Department of Biochemistry,
  Molecular Biology and Cell Biology
Northwestern University
Evanston, Illinois, USA
Chapter 7

Rong-Guang Zhang
Center for Mechanistic Biology
  and Biotechnology
Argonne National Laboratory
Argonne, Illinois, USA
Chapter 7

ACKNOWLEDGMENTS

I wish to thank Dr. Colin Blake for introducing me to the world of protein crystallography and Prof. Demetrius Tsernoglou in whose lab my interest in toxin structures developed. I thank the Wellcome Trust for their support of my work through the award of a Wellcome Australian Senior Research Fellowship in Medical Science. Finally, I wish to thank Marisa Bertocchi for her assistance in preparing the manuscript of this book.

Michael W. Parker
May 1996

========= CHAPTER 1 =========

INTRODUCTION

Joseph E. Alouf

"The empirical basis of objective science has thus nothing 'absolute' about it. Science does not rest upon solid bedrock. The bold structure of its theories rises, as it were, above swamp. It is like a building erected on piles. The piles are driven down from above into the swamp, but not down to any natural or 'given' base; and if we stop driving the piles deeper, it is not because we have reached firm ground. We simply stop when we are satisfied that the piles are firm enough to carry the structure, at least for the time being."

-*Karl Popper (The Logic of Scientific Discovery)*

When I was invited to write this introductory chapter I was honored and pleased to have the opportunity to briefly highlight some important issues relevant to structural aspects of bacterial protein toxins. The present book attempts for the first time to provide into one volume a series of chapters prepared by invited experts on the structure of these fascinating macromolecules. A particular emphasis is placed on the molecular features of these highly active biomolecules in relation to their biological properties and mechanisms of action.

The researchers in bacterial protein toxinology of my generation have had the privilege of witnessing over 40 years the tremendous development of this field of biological sciences, particularly in the past decade. The application to this field of the methodologies and concepts of the widely expanding domains of molecular genetics, protein biochemistry and engineering, cell biology, immunology, pharmacology and neurobiology allowed tremendous progress in our knowledge of the structure and mode of action of bacterial protein toxins at the cellular, subcellular and molecular levels. Furthermore, the achievements realized in this domain greatly contributed to the use of many toxins as exquisite and highly specific tools to probe numerous functions and architectural components of eukaryotic cells, in addition to their use for prophylactic and therapeutic strategies in medicine

Protein Toxin Structure, edited by Michael W. Parker. © 1996 R.G. Landes Company.

such as vaccination, anti-cancer and anti-viral therapy, management of severe muscular disorders (botulinum toxins), adjuvants and other applications.

Since the seminal discovery of the first bacterial toxin (diphtheria toxin) in 1888 by Emile Roux and Alexandre Yersin at Pasteur Institute (Paris), about three hundred peptide and protein toxins have been discovered to date. More than 100 were characterized within the past 15 years. These molecules are quite diverse as concerns molecular features, physico-chemical, and biological properties. Their molecular size vary from less than 30 amino acid residues (cyanobacteria cyclic microcystins, the 26-amino acid residue *Staphylococcus aureus* delta-toxin) to as large as 2,710 residues (*Clostridium difficile* toxin A, the largest single chain bacterial protein hitherto identified). Bacterial toxins may be single-chain polypeptides or oligomeric components constituted from 2 up to 8 protomers, such in the case of the A-B type ADP-ribosylating protein toxins. The B moieties of these macromolecules are either identical (cholera toxin, Shiga and Shiga-like toxins) or different (*Bordetella pertussis* toxin).

The outstanding achievements in our present knowledge of the molecular features and mechanisms of action of bacterial protein toxins closely depended on the progress during the past four decades in the concepts, methodology and technology of the elucidation of the primary and three-dimensional (3-D) structure of proteins. Three major landmarks in the biochemistry of these macromolecules characterize this progress:

(i)   The establishment of the first definitive order of amino acids within a protein molecule (insulin) in the early 1950s by F. Sanger;

(ii)   The first determination in 1959 of the 3-D structure of a protein (myoglobin) by X-ray crystallography by J.C. Kendrew;

(iii) The application in the 1980s of the methodologies of molecular genetics which enabled the easy establishment of the primary structure of proteins deduced from the nucleotide sequences of their encoding structural genes.

Currently the complete primary structures of thousands of prokaryotic and eukaryotic proteins essentially deduced from gene sequencing has been established. Similarly, the 3-D structure of several hundreds of proteins has been also elucidated to date.

As concerns the primary structure of the bacterial protein toxins, the complete amino acid sequence of not more than 20 different molecules arduously determined by the classical Edman degradation was known at the dawn of the 1980s. Since then, the successful sequencing of the cloned structural genes of more than 120 protein toxins, particularly those of major basic and clinical importance, has allowed us to deduce their amino acid sequences. The determination of the primary structure of this considerable number of protein toxins led to the finding that a great number of toxins constitute structurally and in most cases, functionally-related families or superfamilies: the ADP-ribosylating toxins, the thiol-activated toxins, the cholera-like and Shiga-like toxins, the membrane damaging RTX toxin group, the staphylococcal and streptococcal shock-inducing superantigenic toxins, etc.

The achievements of the past ten years have paved the way for considerable progress in the elucidation of the respective mechanisms of action and a better understanding of structure-activity relationship particularly through the modification of the corresponding structural genes (site-directed/oligonucleotide-directed mutagenesis, deletion mapping, insertional mutagenesis, gene fusion etc.) and the expression of the recombinant toxins in appropriate cells followed by their purification and the evaluation of their biological properties.

Furthermore, the comparative structural and functional studies have enabled us to discover related toxin families or groups not only between prokaryotic ones but also between prokaryotic and eukaryotic protein toxins. This is particularly the case of the plant toxin ricin (and other related plant toxins) on the one hand and *Shigella dysenteriae* serotype-1 Shiga toxin and *E. coli* Shiga-like I and Shiga-II toxins. These plant and enterotropic bacterial toxins act on target cells as specific N-glycosidases exhibiting identical mechanism of action by cleaving a single residue from near the 3' end of the 28S rRNA component of the ribosomal complex of eukaryotic cells.

The establishment of the 3-D structure of bacterial protein toxins remained for a long time the dream or the Holy Grail of many researchers. The dream came true in 1986 with the publication of the crystal structure of *Pseudomonas aeruginosa* exotoxin A by V.S. Allured and coworkers.[1] Since then, the 3-D structures of the following bacterial toxins were established to our knowledge: *Escherichia coli* colicin,[2] *E. coli* heat-labile enterotoxin,[3] *Bacillus thuringiensis* δ-endotoxin,[4] *E. coli* verotoxin-1 B oligomer,[5] diphtheria toxin,[6] *Staphylococcus aureus* enterotoxins B[7] and C,[8] *S. aureus* toxic shock syndrome toxin-1,[9,10] *Aeromonas hydrophila* proaerolysin,[11] pertussis toxin,[12] *Shigella* holotoxin,[13] and cholera toxin.[14-16]

The stage is now set for exploiting all outcomes of structural studies on bacterial toxins for a deeper understanding of their properties and their precise role in the pathogenesis of infectious diseases elicited by the toxin producing Gram-positive and Gram-negative bacteria.

> *"I have yet to see any problem, however complicated, which, when you looked at it in the right way, did not become still more complicated."*
>
> *-Poul Anderson*

It is hoped that this book will be of interest to a wide audience of scientists including microbiologists, physicians, cell and molecular biologists, biochemists, immunologists and many others. One may expect that the various contributions prove useful and serve to arouse the interest of senior students of the biological sciences to pursue specific problems which remain to be unraveled in the rapidly growing areas of bacterial toxinology and more widely basic and clinical microbiology.

REFERENCES

1. Allured VS, Collier RJ, Carroll SF et al. Structure of exotoxin A of *Pseudomonas aeruginosa* at 3.0 Å resolution. Proc Natl Acad Sci USA 1986; 83:1320-24.
2. Parker, MW, Pattus F, Tucker AD et al. Structure of the membrane-pore-forming fragment of colicin A. Nature 1989; 337:93-96.
3. Sixma TK, Pronk SE, Kalk KH et al. Crystal structure of a cholera toxin-related heat-labile enterotoxin from *E. coli*. Nature 1991; 351:371-378.
4. Li J, Carroll J, Ellar DJ. Crystal structure of insecticidal d-endotoxin from *Bacillus thuringiensis* at 2.5Å resolution. Nature 1991; 353:815-821.
5. Stein PE, Boodhoo A, Tyrell GJ et al. Crystal structure of the cell-binding B oligomer of verotoxin-1 from *E. coli*. Nature 1992; 355:748-50.
6. Choe S, Bennett MJ, Fujii G et al. The crystal structure of diphtheria toxin. Nature 1992; 357:216-22.
7. Swaminathan S, Furey W, Pletcher J et al. Crystal structure of staphylococcal enterotoxin B, a superantigen. Nature 1992; 359:801-806.
8. Hoffmann ML, Jablonski LM, Crum KK et al. Predictions of T cell receptor and major histocompatibility complex binding sites on staphylococcal enterotoxin C1. Infection and Immunity 1994; 62:3396-407.
9. Prasad GS, Earhart CA, Murray DL et al. Structure of toxic shock syndrome toxin-1. Biochemistry 1993; 32:13761-13766.
10. Acharya KR, Passalacqua EF, Jones EY et al. Structural basis of superantigen action inferred from crystal structure of toxic-shock syndrome toxin-1. Nature 1994; 367:94-97.
11. Parker MW, Buckley JT, Postma JPM et al. Structure of the *Aeromonas* toxin proaerolysin in its water-soluble and membrane-channel states. Nature 1994; 367:292-295.
12. Stein PE, Boodhoo A, Armstrong GD et al. The crystal structure of pertussis toxin. Structure 1994; 2:45-57.
13. Fraser ME, Chernaia MM, Kozlov YV et al. Crystal structure of the holotoxin from *Shigella dysenteriae* at 2.5 Å resolution. Nature Structural Biology 1994; 1:59-64.
14. Merritt EA, Sarfaty S, van den Akker F et al. Crystal structure of cholera toxin B-pentamer bound to receptor GM1 pentasaccharide. Protein Science 1994; 3:166-175.
15. Zhang R-G, Westbrook ML, Westbrook EM et al. The three-dimensional structure of cholera toxin. J Mol Biol 1995; 251:563-573.
16. Zhang R-G, Maulik PR, Westbrook EM et al. The 2.4 Å crystal structure of the cholera toxin B subunit pentamer: choleragenoid. J Mol Biol 1995; 251:550-562.

INSIGHTS INTO MEMBRANE INSERTION BASED ON STUDIES OF COLICINS

Ingrid R. Vetter, Michael W. Parker, Franc Pattus
and Demetrius Tsernoglou

Colicins are a distinctive class of antibiotics produced by various strains of *E. coli*. The name "colicin" was first used by Gratia and Fredericq[1] to describe antibiotics acting only on the same or closely related species. Analogous relationships exist between the related cloacins and *Enterobacter cloacae* and the klebicins and *Klebsiella* species. When induced, colicins are usually expressed in large amounts and secreted into the extracellular medium with the aid of a plasmid-encoded lysis-gene product. The cell commits suicide, making it at first sight difficult to explain how the colicin plasmid could have possibly survived. The answer is that each plasmid encodes a so-called "immunity protein" which protects the other plasmid-containing (but not colicin producing) cells, guaranteeing the survival of the plasmid (ca. 30% to 50% of Gram-negative bacteria isolated from natural sources have colicin plasmids).[2] The induction functions through the SOS-system for the vast majority of colicins, colicin V being the only exception. The release mechanism is probably not a specific export since none of the colicins has a signal sequence and, furthermore, it can be observed that the cell envelope of the secreting cell is severely disturbed by lipoproteins which are products of the lysis-genes mentioned above.[3] Following the secretion, the molecules bind to metabolite receptors on the outer membrane of sensitive cells, thereby parasitizing these bacterial transport systems and using them as entry ports. The translocation across the periplasm then takes place with the aid of the Tol or Ton protein systems which have been shown to be associated with adhesion sites between the inner and outer membrane of the bacterium.[4] The subsequent killing of the bacterial cell can be achieved

Protein Toxin Structure, edited by Michael W. Parker. © 1996 R.G. Landes Company.

through several processes, including inhibition of protein, murein and DNA synthesis, enzymatic cleavage of DNA or ribosomal RNA and deenergization of the cytoplasmic membrane by pore formation.[5,6]

Colicins have emerged not only as powerful model systems for investigation of the mechanism and energetics of protein unfolding and transfer across membranes, but also are relevant for the mechanism of toxic activity and the targeting, insertion and translocation of macromolecules in general.[5,6,74] Furthermore, they form well-defined voltage-gated ion channels, reminiscent of ionic channels in excitable membranes which are of particular interest since they undergo structural changes in the transition from the water-soluble to the membrane bound and inserted form.

DOMAIN ORGANIZATION OF THE COLICINS

Colicins have been grouped as A or B on the basis of the proteins used in their translocation and uptake. Group A colicins, including colicins A, N, E1 and K, use the Tol QRAB gene cluster pathway, whereas group B colicins use the TonB pathway (Table 2.1). These pathways are shared with some types of bacteriophages, e.g., T1 and filamentous bacteriophages.[7]

Table 2.1. Colicins, their receptors and translocation pathways

Group A ID	mol.wt.	receptor	function	translocation	killing
A	63000	BtuB/OmpF	see below	tolQRABC	pore
N	42000	OmpF	div. < 600	tolQA	pore
E1	57000	BtuB	Vit. B12	tolQRAC	pore
K	69000	TsX	nucleosides	tolQRAB	pore
E2	61629	BtuB/OmpF	see above	tolQRAB	DNAse
E3	57963	BtuB/OmpF	see above	tolQRAB	RNAse
E4–E9	var.	BtuB	Vit. B12	tolQRAB	DNAse/RNAse

Group B ID	mol.wt.	receptor	function	translocation	killing
B	54700	FepA	enterochelin	tonB	pore
Ia	69400	cir	iron accumul.	tonB	pore
Ib pore	69900	cir	iron accumul.	tonB	
D	74688	FepA	enterochelin	tonB	protein synth. inhibitor
M	29482	FhuA	iron accumul.	tonB	murein synth. inhibitor
V	5800	?	?	tonB	pore
10	53342	TsX	nucleosides	tonB/tolC	pore
5	56000	TsX	nucleosides	tonB	?

Characteristic for colicin molecules is the organization into domains which are associated with each of the functional steps mentioned above, namely, receptor binding, translocation and killing (Fig. 2.1). These functional domains have been defined initially by limited proteolysis and deletion analysis and later confirmed for some colicins by structural analysis. As shown in Figure 2.1, the colicins follow the same arrangement independent of the specific colicin type.

It has been proposed that these domains function completely independently, leading to the hypothesis that colicins evolved by recombination of DNA fragments which encode the uptake and activity domains.[8] Corroborating this, the receptor binding and translocation domain of colicins Ia and Ib (which both use the same receptor and translocation system) have sequences with 98% identical amino acids, whereas the channel-forming domains share only 43% identical residues. Furthermore, the other pore-forming colicins A, B, N and E1 share a highly related channel-forming domain, but there are no detectable homologies between the other domains, although some of them bind to a common receptor, e.g., colicin A and colicin N both can attach to the OmpF porin. Even more diverse is the secondary structure and function of the domains of the colicins that do not form a pore.

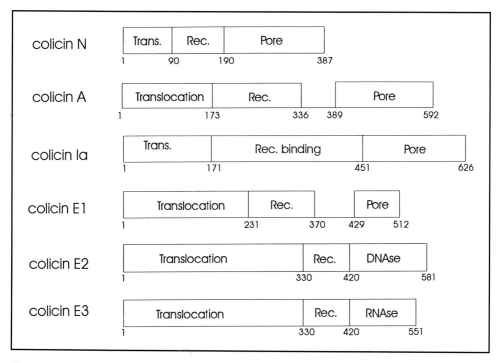

Fig. 2.1. Domain organization of the colicins.

RECEPTOR BINDING

The first step of the colicin entry into the cell is the binding to an accessible molecule in the outer membrane of the target cell. As mentioned above, colicins can use metabolic receptors like the OmpF porin, the vitamin B12 receptor (BtuB) or iron transport systems like cir and FepA as entry points into the cell. The structure of these molecules has been predicted as being composed of membrane-crossing β-strands.[6] The interaction with the colicins is supposed to take place at the external surface of the receptors and is energy-independent. Mutational studies on some receptors proved that single mutations can abolish binding activity, leading to the conclusion that the binding process is a true molecular recognition event.[9] It has been shown that the purified OmpF porin channel by itself is sufficient for binding to colicin N, so, at least in case of colicin N, there are no additional membrane components or interactions with the membrane necessary to provide the binding sites.[10]

TRANSLOCATION

The above-mentioned Tol and Ton proteins, respectively, are an absolute requirement to accomplish the transport of the pore-forming domain from the extracellular space to the inner membrane where it forms its channel. The Tol proteins (TolQ, TolR, TolA and TolB) are involved in maintaining the structure of the *E. coli* cell envelope and have no known physiological transport function, whereas the TonB protein participates in an energy-coupled uptake system together with the ExbB and ExbD proteins. The two transport systems are related since the ExbB and ExbD proteins share 25% sequence identity with the TolQ and TolR proteins, respectively.[11]

The putative arrangement of the Tol and Ton proteins in the cell envelope is depicted in Figure 2.2. TonB forms a link between the inner and outer membrane, spanning the periplasmic space, whereas TolA is anchored only in the inner membrane and presumably reaches out towards the outer membrane with one enormously long α-helix comprising 223 residues.[11,12] The C-Terminus of TolA is proposed to be in the vicinity of the outer membrane where it might interact with components of the outer membrane or the periplasm.[13] In this way TolA could also participate in contact sites between the inner and outer cell membrane, which is supported by the fact that TolA cofractionates with these adhesion sites.[4]

The proteins TolQ and ExbB span the cytoplasmic membrane with three α-helical segments, whereas TolR and ExbD have only one transmembrane segment. TonB probably forms a complex with ExbB and ExbD, and there is evidence for a direct interaction between TolQ and TolR via their transmembrane segments.[11] However, the detailed structures of the Tol and Ton transport complexes remain undetermined.

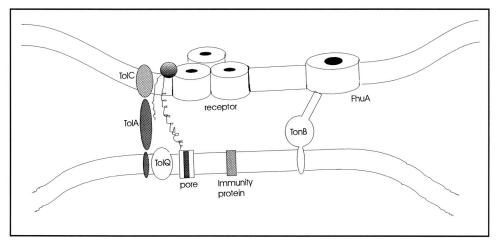

Fig. 2.2. Putative arrangement of the Tol and Ton proteins in the inner and outer bacterial cell envelope including a pore-forming colicin and its receptor.

Little is known about the translocation event itself, although the sequences of the colicins and their receptors encourage some hypotheses. Interestingly, a sequence motif near the N-terminus of the translocation domain (D/E - T/I - X - V/X - V), called the TonB-box, is shared by the TonB-dependent colicins with some receptor proteins, including BtuB and FepA.[14-16] It is thus tempting to assume that these colicins may compete with the receptors for interaction with TonB, initiating the transport across the periplasmic space. This is supported by the finding that TolA and TonB determine the specificity for the different groups of phages and colicins, and, furthermore, a direct interaction between TonB and colicin M and between TonB and the FhuA and cir receptors has been shown.[17,18] The interactions between TolA and the group A colicins seem to follow an equivalent pattern. Supporting this, binding of colicin A to TolA was detected by in vitro binding studies.[19] A "TolA box" near the N-terminus of the translocation domain of the group A colicins has been proposed—in analogy to the TonB box—with the sequence DGTGW (residues 11 to 15 in colicin A), but this hypothesis still needs more experimental evidence.[20]

THE BACTERIOCIDAL EFFECT

The killing of *E. coli* cells by pore-forming colicins follows single-hit inactivation kinetics, i.e., one molecule is able to kill a bacterial cell via membrane depolarization.[21,22] This implies that a single molecule must be able to form a complete ion channel, a hypothesis that is confirmed by studies on colicin E1 insertion into phospholipid vesicles.[23] The primary killing effect of the pore-forming colicins is the efflux of cytoplasmic potassium and phosphate, leading to a breakdown of the cell potential and, subsequently, to a rapid depletion of the ATP pool of the cell.[24] The kinetics of the efflux of cytoplasmic potassium induced by colicin A shows a lag time of approximately 30 seconds between addition of the colicin and the beginning of the potassium efflux, reflecting the time the colicin molecule needs to bind to the receptor and cross the periplasmic space.[25] The channel-forming event itself is fast compared to the binding/translocation time[26] and, in contrast to the binding and translocation events, requires a membrane potential.[25]

The most detailed investigations on the properties of the colicin ion channel have been conducted on colicin E1. Generally, the channels formed by the colicins seem to be quite unspecific, in the case of colicin E1 the pore is permeable to monovalent organic and inorganic cations and to mono- and polyvalent organic anions having diameters of up to 9 Å.[27-29] The ion selectivity is pH-dependent (colicin E1 is strongly anion-selective at pH values below 6.0) and can be altered by site-directed mutagenesis of charged residues.[27,30,31] The colicin E1 channel exhibits multiple states with conductivities between 10 and 60 pS as determined by patch clamp methods, and the pore can open and close on a timescale faster than 2 kHz ("flickering").[5] These observations suggest that there is not a single membrane-inserted conformation but several different intermediates.

After the pore has formed in the membrane, the receptor-binding and translocation domains still have an influence on the gating properties, although the pore-forming fragment can form functional channels by itself.[32] In vivo these domains might be even more important than in vitro as deduced by so-called "trypsin-rescue" experiments. These studies indicate that channels formed by colicin A or colicin E1 in vivo can be closed by adding trypsin to the extracellular medium. Control experiments confirmed that this is not an artifact due to the presence of trypsin in the periplasmic space, leading to the conclusion that the colicin is accessible from the outside even though its pore has already formed. Therefore, one has to assume that the colicin in its active form spans the distance between the outer and the inner membrane.[33,34]

Details of the channel formation will be discussed later after a closer look at the structural data available for the pore-forming domain.

STRUCTURE OF THE PORE-FORMING DOMAIN

The crystal structure of the pore-forming domain of colicin A was first reported in 1989[37] and an improved model subsequently published in 1992.[38] The final model was refined to a resolution of 2.4 Å with a conventional R-factor of 0.18 for all data between 6.0 and 2.4 Å resolution. The polypeptide chain of 204 amino acid residues was found to fold into 10 α-helices arranged in a three-layer structure (Figs. 2.3 and 2.4). Although the fold was considered unique at the time, other

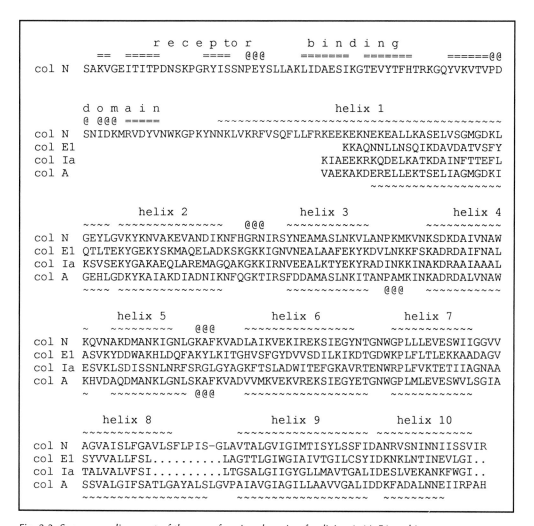

```
                      r e c e p t o r          b i n d i n g
            ==  =====          ====  @@@     =======  =======    ======@@
    col N   SAKVGEITITPDNSKPGRYISSNPEYSLLAKLIDAESIKGTEVYTFHTRKGQYVKVTVPD


            d o m a i n                              helix 1
            @ @@@ =====          ~~~~~~~~~~~~~~~~~~~~~~~~~~~~~~~~~~~~~~~~~~
    col N   SNIDKMRVDYVNWKGPKYNNKLVKRFVSQFLLFRKEEKEKNEKEALLKASELVSGMGDKL
    col E1                            KKAQNNLLNSQIKDAVDATVSFY
    col Ia                            KIAEEKRKQDELKATKDAINFTTEFL
    col A                             VAEKAKDERELLEKTSELIAGMGDKI
                                              ~~~~~~~~~~~~~~~~~


                 helix 2                  helix 3              helix 4
            ~~~~ ~~~~~~~~~~~~~~~     @@@  ~~~~~~~~~~~        ~~~~~~~~~~~~
    col N   GEYLGVKYKNVAKEVANDIKNFHGRNIRSYNEAMASLNKVLANPKMKVNKSDKDAIVNAW
    col E1  QTLTEKYGEKYSKMAQELADKSKGKKIGNVNEALAAFEKYKDVLNKKFSKADRDAIFNAL
    col Ia  KSVSEKYGAKAEQLAREMAGQAKGKKIRNVEEALKTYEKYRADINKKINAKDRAAIAAAL
    col A   GEHLGDKYKAIAKDIADNIKNFQGKTIRSFDDAMASLNKITANPAMKINKADRDALVNAW
            ~~~~ ~~~~~~~~~~~~~~~          ~~~~~~~~~~~~    @@@  ~~~~~~~~~~~


                 helix 5                helix 6              helix 7
            ~   ~~~~~~~~~~     @@@    ~~~~~~~~~~~~~~~      ~~~~~~~~~~~~
    col N   KQVNAKDMANKIGNLGKAFKVADLAIKVEKIREKSIEGYNTGNWGPLLLEVESWIIGGVV
    col E1  ASVKYDDWAKHLDQFAKYLKITGHVSFGYDVVSDILKIKDTGDWKPLFLTLEKKAADAGV
    col Ia  ESVKLSDISSNLNRFSRGLGYAGKFTSLADWITEFGKAVRTENWRPLFVKTETIIAGNAA
    col A   KHVDAQDMANKLGNLSKAFKVADVVMKVEKVREKSIEGYETGNWGPLMLEVESWVLSGIA
            ~   ~~~~~~~~~~   @@@    ~~~~~~~~~~~~~~~~~     ~~~~~~~~~~~~


                 helix 8                helix 9            helix 10
            ~~~~~~~~~~~~~~        ~~~~~~~~~~~~~~~~~~~~   ~~~~~~~~~~~~~~~
    col N   AGVAISLFGAVLSFLPIS-GLAVTALGVIGIMTISYLSSFIDANRVSNINNIISSVIR
    col E1  SYVVALLFSL.........LAGTTLGIWGIAIVTGILCSYIDKNKLNTINEVLGI..
    col Ia  TALVALVFSI.........LTGSALGIIGYGLLMAVTGALIDESLVEKANKFWGI..
    col A   SSVALGIFSATLGAYALSLGVPAIAVGIAGILLAAVVGALIDDKFADALNNEIIRPAH
            ~~~~~~~~~~~~~~~~~     ~~~~~~~~~~~~~~~~~~~~   ~~~~~~~~~~
```

Fig. 2.3. *Sequence alignment of the pore-forming domain of colicins A, N, E1 and Ia.*

workers have since demonstrated a plausible resemblance to the globin fold.[39] Similar α-helical bundles to colicin were subsequently discovered in the membrane insertion domains of insecticidal δ-endotoxin[40] and diphtheria toxin,[41] despite there being no detectable sequence similarities between the toxins. The similar polypeptide folds suggested there may be a common mechanism of insertion into biological membranes.[42] A key feature of the colicin fold is the presence of a helical hairpin sandwiched in the middle layer of the structure. The hairpin is completely buried and consists of hydrophobic residues only. This feature led to the "umbrella" model of membrane insertion in which interaction of the hydrophobic loop of the helical hairpin with the lipid bilayer

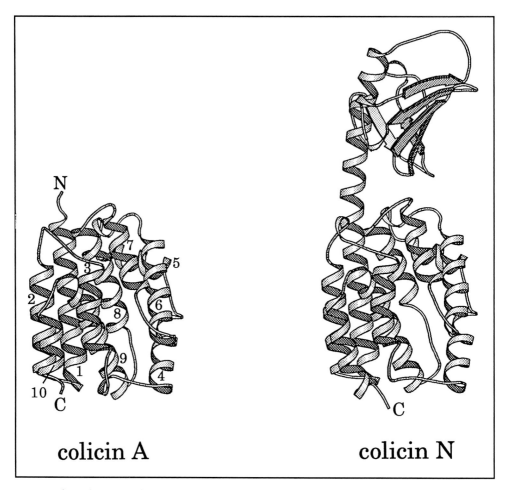

N

7

3   5

2   8   6

8

9

1   4

10   C

C

colicin A   colicin N

Fig. 2.4. Three-dimensional structures of colicin A and colicin N. The pore-forming domain is shown in the same orientation for colicin A and colicin N. The figure was made using the program MOLSCRIPT.[73]

would be followed by spontaneous insertion of the entire hairpin into the bilayer (Fig. 2.5).[38] Because the passage of charged residues is unfavorable on energetic grounds, the insertion of the hairpin necessitated the opening up of the colicin structure leaving the two outer helical layers embedded in the surface of the bilayer, in a process rather like the opening of an umbrella. Two other features of the colicin structure were thought relevant to the insertion mechanism. Firstly, the presence of a ring of eight positively charged residues surrounding the hydrophobic hairpin that may play a role in initial electrostatic interactions with the surface of the membrane. Secondly, the presence

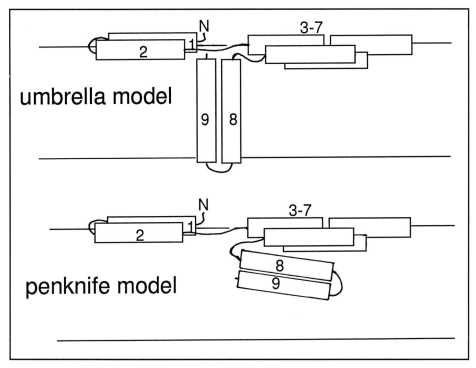

Fig. 2.5. Arrangement of the helices of the pore-forming domain in the umbrella model and the penknife model, respectively.

of a spiral of cavities within the protein core which might promote the unfolding of the toxin.[38]

Solution studies by NMR have established the same overall topology for the pore-forming domains of colicin E1 and colicin A, although there is a slight difference in the helix packing, particularly for helices 1 and 5.[35]

The crystal structure of colicin Ia has been determined to an effective resolution of 4 Å. The putative pore-forming domain of colicin Ia has been assigned according to the structural motif it shares with the corresponding domain of colicin A, although the two domains have only 27% identical amino acids. Due to the limited resolution, the sequence could not be assigned unambiguously, the model was built as a polyalanine chain with some of the loop regions missing.[36] Colicin Ia as well as colicin E1 has a deletion of 10 amino acids in the hydrophobic hairpin formed by helices 8 and 9 as compared to colicin A, leading to a considerable shortening of the hairpin by about 7 to 8 Å. Otherwise the structure of the pore-forming domains are very similar (Fig. 2.4).

The crystal structures of the receptor binding and the pore-forming domain of colicin N have been solved to 3 Å resolution (Vetter, Parker, Tucker, Pattus and Tsernoglou, to be published). Colicin N has 56% identical amino acids in its pore-forming domain as compared to colicin A. This similarity reflects itself in a highly homologous three-dimensional structure of this domain. The major differences can be found in the loop regions and in helix 5 which has been found to be the most flexible helix in colicin A.[38] The deletion of a single leucine residue as compared to colicin A allows for another feature of colicin N that is different from colicin A: namely, a hydrophilic channel leading through the hydrophobic core of the molecule which is filled with water molecules. The adjacent helix bearing the deletion is stretched out, making room for the entrance of the hole. The channel almost meets a large depression on the other side of the molecule whose boundaries are formed by the flexible (as judged from the temperature factors) loop between helices 3 and 4. This hole in the protein might have a function in easing up the formation of the proposed molten globule-state of the molecule (see below), or it even could be a part of the active ion channel itself although this is highly speculative. That the environment in and around the hole is hydrophilic is corroborated not only by the water molecules bound in the inside but also by the presence of heavy atom binding sites (which were identified during the process of structure determination) in the above-mentioned large depression as well as in the entrance of the hole.

CHANNEL FORMATION

The pore-forming domain of colicins associates with lipids even in the absence of a membrane potential as shown by various techniques.[43,44] This state has been described as an "insertion competent" conformation with some characteristics of a "molten globule" state.[45] For most colicins, the insertion is facilitated by an acidic pH, probably due to a partial destabilization of the protein which allows for the rearrangement of the tertiary structure into the insertion competent state.[44] The secondary structure is largely conserved in this pre-insertion form of the channel peptide as judged by fluorescence and CD spectroscopy studies.[45-47] This "loosening" of the structure coincides with an increased protease susceptibility of the membrane-bound protein. For colicin E1, especially the N-terminal 50 residues, corresponding to helices 1 and 2, are accessible to trypsin, although the peptide is bound tightly to the membrane with a K_d of 10^{-7}M.[48,49]

The 10-helix bundle of the pore-forming domain is not the smallest possible fragment sufficient for channel activity. A fragment of colicin E1 with 88 amino acids, corresponding to a region starting at the end of helix 5 and extending to the end of helix 10, is able to form an active channel.[50] The C-terminus (Helix 10) cannot be shortened too much without affecting colicin activity in vivo and in vitro.[51] The shortest colicin A fragment with channel-forming activity comprises 136 amino acids, lacking helices 1-3 of the complete domain with 204 residues.[52] Together with the evidence that the colicin channel can be formed by a monomer, this leads to the conclusion that the ionophoric pore can be formed by only five helices. However, it is difficult to imagine a channel with 9 Å diameter formed by such a small peptide, leading to the proposal of oligomeric (trimeric) channel models.[38]

Although the channel stoichiometry is still under discussion, mutational studies further elucidated the properties of the channel as well as channel formation. For example, deletion mutants of colicin A were inactive if the deletions were located in the region of helices 5 to 9, whereas deletions in the first four helices of the pore-forming domain yielded active proteins. Furthermore, even an exchange of only two amino acids in the loop between helix 5 and helix 6 (A492D and F493P) produced a protein which still bound lipids but was incapable of forming a channel.[53] This corroborates the importance of helices 5 to 10 for channel formation. Information about the channel properties was obtained from point mutants of colicin E1 positioned at four different sites in helices 9 and 10, altering the ion selectivity of the channel (T501E, G502E in helix 9 and L510M and K512Y in the loop connecting helices 9 and 10).[31] This implies a direct participation of these helices in the pore itself. There is also evidence for a role of helices 5 and 6 as a "gating hairpin" which might insert into the membrane when the channel opens.[54]

Introduction of disulfide bridges into colicin A yielded interesting results concerning the translocation and membrane insertion. The disulfide bridges were introduced between helices 1 and 9, helices 5 and 6 and helices 9 and 10, respectively. All mutants were inactive in vivo and in vitro if the disulfide bridges were in an oxidized state, but activity could be restored by reducing the disulfide bonds with DTT. Analysis of the kinetics of pore formation in vivo suggested that the colicin is able to translocate through the periplasmic space although the unfolding of the pore-forming domain was prevented, at least partially, by the disulfide bonds. Furthermore, the link between helices 1 and 9 prevented the voltage-independent insertion of colicin A into membrane vesicles, whereas the links between helices 5/6 and helices 9/10, respectively, allowed the insertion but prevented the voltage-dependent opening of the channel. These experiments together with the results described above imply that a separation of the hairpin formed by helices 1 and 2 from the remainder of the pore-forming domain is necessary to allow insertion into the inner membrane.[26,55]

These results are consistent with fluorescence energy transfer distance measurements, using the three tryptophans of the colicin A pore-forming domain as fluorescence donors together with a fluorescence probe, acetyl-N'-(5-sulpho-1-naphthyl)ethylenediamine (AEDANS), attached to five newly introduced cysteines. The distances between the tryptophans and the AEDANS moieties were determined by measuring the tryptophan fluorescence intensity in the absence and then in the presence of the AEDANS, leading to a value for the energy transfer efficiency and, in turn, to an estimate of the distance between the two groups. The resulting data provide evidence for a separation of the 1-2 hairpin from the other helices and suggest that the hydrophobic 8-9 hairpin is not inserted into the membrane as proposed originally, but stays in close packing arrangement with helices 3 to 7.[38,56] Use of brominated phospholipids allowed a localization of the tryptophans at close distance to the quenching bromine atoms in membrane-bound colicin A. This implies that helices 8 and 9 are well embedded in the hydrophobic core of the membrane, whereas the amphipathic helices are located at the surface of the membrane, exposing the tryptophans to the surrounding bromine atoms but protecting them from soluble quenchers.[57]

Site-directed mutagenesis of colicin Ia and subsequent biotinylation of cysteine 544, corresponding to the beginning of helix 6, prohibited the channel formation if streptavidin was added to the cis side of the membrane if the channels were initially in the closed state, but not if they were open.[58] This strongly suggests that the beginning of helix 6 is exposed in the closed state of the channel but buried in the open state.

Neutron scattering studies indicated that in the absence of a membrane potential the colicin A channel has an association with the membrane comparable to plasma lipoproteins, i.e., the molecule lies mostly at the surface of the membrane.[59] This finding is corroborated by the increased accessibility of the surface bound molecule to proteases, e.g., trypsin, under these conditions.[60]

In summary, these results suggest a model for the closed channel called the "penknife-model" as depicted in Figure 2.5. In this model, helices 1 and 2 are separated from the rest of the molecule (helices 3 to 10), which remain in a close packing arrangement with conserved secondary structure elements as compared to the soluble form and lies mostly on the surface of the membrane. Helices 8 and 9 are thereby supposed to be immersed in the lipid bilayer. The application of a membrane potential might lead to an insertion of some or even all helices of the 3-10 helix bundle. Readers are encouraged to read other recent reviews for alternative viewpoints on the insertion mechanism.[5,74]

STRUCTURE OF THE RECEPTOR BINDING AND TRANSLOCATION DOMAIN

Although the pore-forming domain has attracted major attention, the recent solution of the complete colicin Ia molecule,[36] as well as the solution of a colicin N fragment containing the receptor binding domain in addition to the pore-forming domain (Vetter, Parker, Tucker, Pattus and Tsernoglou, to be published), has shed new light on the functions of receptor binding and translocation. The colicin Ia molecule is composed entirely of α-helices which are arranged in an elongated "Y" shape. The N-terminal translocation domain is closely packed against the pore-forming domain, whereas the receptor binding domain forms the stalk of the "Y", connecting the other two domains. This shape of the molecule permits the molecule to contact both outer and inner membranes simultaneously and might serve to orient the pore-forming domain in the correct position for membrane insertion.[36] This is consistent with membrane protection experiments showing a protease-accessible region of the molecule (~residues 190-358) that corresponds to the receptor binding domain.[43] The translocation domain is not only important for the entry of the colicins but is also known to modify the voltage-dependence, conductance and open lifetimes of the channel-forming fragment of colicin Ia as compared to intact colicin Ia. The same observation can be made for colicin A and colicin E1, although the differences are less pronounced as compared to colicin Ia.[36]

Colicin N has a receptor binding domain consisting of a six-stranded β-sheet, wrapped around an α-helix which is an extension of helix 1 of the pore-forming domain (Fig. 2.4). The topology is roughly similar to the α1- and α2-domains of the human class I histocompatibility antigen AW 68.1 with an r.m.s.-deviation of the main chain of 2.9 Å, although there is no significant sequence homology. The most prominent feature of the receptor binding domain is a large cleft formed by the twisted β-sheet, and it is tempting to assume this cleft to be involved in binding to the receptor of colicin N.

BINDING AND TRANSPORT OF COLICIN N

The uptake mechanism of colicin N is in several respects the simplest among the colicins since there is only one receptor molecule (OmpF porin) which is necessary for binding and translocation and, furthermore, colicin N is the shortest colicin that comprises all the functionalities of the colicins.[61] Mutational studies of the OmpF receptor indicated that colicin N binds in the vicinity of the L7-loop since the point mutations E284K and G285D each abolish any binding activity as judged by the binding of iodinated colicin N to the surface of intact cells bound to nitrocellulose.[9] This loop is located at the outer rim of the trimeric porin channel. Hybrids of OmpF and the related porin OmpC, which is not active in colicin N uptake, have been used to map the region of the porin required for translocation of colicin N (between residues 143 and 262).[62] This central region of the porin is situated at the outside of the porin barrel directly below the loop L7, corroborating the importance of this porin region. Somewhat puzzling was the finding that the mutant G119D is resistant to colicin N as well, the mutation being buried in the porin channel far away from the external surface. The structure of this mutant has been solved and shows no large structural changes in the external loop regions or the outside of the β-barrel, implying that either the mutation exerts long-range effects that are not visible in the X-ray structure or that this site is directly involved in binding. It has been considered as a possibility that translocation of the colicin occurs through the porin channel, as unfolding has been suggested for the passage of colicin A through the outer membrane.[34,63] On the other hand, the disulfide-crosslinked colicin A was still able to proceed to the inner membrane, making a transfer through the porin channel lumen unlikely.[26,55] The exact mechanism of translocation therefore remains a mystery.

IMMUNITY PROTEINS

Each colicin has its specific immunity protein which is constitutively expressed from the colicin plasmid at very low levels (10^2 to 10^3 molecules per cell). These proteins are incorporated into the inner membrane of the bacterial cell and are supposed to sample large areas of the lipid layer in very short times in order to detect and inhibit the incoming colicins. It was suggested that the immunity proteins are targeted to the colicin entry points to explain the high efficiency of the protection, but on the other hand there are more adhesion zones than expressed immunity proteins, so the proteins must still be able to move from one site to another.[5]

Although the function is similar, the topologies of these proteins seem to be quite different. The structure of the colicin E3 immunity protein (ImmE3) has been solved by NMR and consists of a 4-stranded β-sheet,[64] exposing the three residues determining the specificity, whereas the ImmE1 and ImmA proteins are proposed to have three and four membrane spanning helices, respectively.[65,66] Immunity proteins are able to protect the cell against concentrations of the colicin 10^4-10^7 times higher than that required to kill nonimmune cells by interaction with the pore-forming domains of the colicins.[5] There is evidence that this protection mechanism involves a helix-helix interaction in case of colicin E1 and its immunity protein.[67] With the aid of colicin A and B hybrids the helix 8-9 hairpin of the pore-forming domain of these colicins was pinned down as the determinant for immunity specificity, most likely through interactions with the last three helices of the immunity protein.[68] The corresponding construction of hybrids between ImmA and ImmB revealed unexpectedly high structural constraints for these two proteins, i.e. most of the loops were required for ImmA function without carrying any determinants for colicin recognition.[69] In case of the colicin E6 immunity protein, the structure distinguishing between colicin E6 and the related colicin E3 was shown to be a single amino acid sidechain, namely tryptophan 48. When mutated to a cysteine (like in ImmE3), the ImmE6 specificity changed from colicin E6 to colicin E3, suggesting a localized molecular recognition.[70]

CONCLUSIONS

The structures now available for the colicins are a large step forward in our knowledge about the mechanisms of macromolecular translocation through cell membranes, membrane insertion and pore formation. The flexibility and adaptability of these bacteriocins sheds a new light on proteins in general, which are usually viewed as rather rigid structures. The new structural information allows a whole new set of experiments designed to elucidate the exact model of translocation and corroborate or falsify the current models. High resolution structures of colicin Ia and the pore-forming domain of colicin E1 will be available soon.

Experiments introducing colicins as a model system for investigation of cell sorting have been conducted with fusion proteins containing, besides the colicin, a mitochondrial precursor, allowing the construction of molecules which were able to insert into *E. coli* plasma membranes independent from Tol proteins.[71,72]

Thus, in the coming years, the colicins are expected to yield insights into a large number of different phenomena such as transport into and across cell membranes, targeting of proteins, pore formation and protein structure and function in general.

REFERENCES

1. Gratia A, Fredericq CR. Soc Biol 1946; 140:1032.
2. Pugsley AP. The ins and outs of colicins. Part I: Production and translocation across membranes. Microbiol Sci 1984; 1:168-78.
3. Howard SP, Cavard D, Lazdunski C. Amino acid sequence and length requirements for assembly and function of the colicin A lysis protein. J Bacteriol 1989; 171:410-18.
4. Guihard G, Boulanger P, Bénédetti H et al. Colicin A and the Tol Proteins involved in its translocation are preferentially located in the contact sites between the inner and outer membranes of *Escherichia coli* cells. J Biol Chem 1994; 269:5874-80.
5. Cramer WA, Heymann JB, Schendel SL et al. Structure-function of the channel-forming colicins. Annu Rev Biophys Biomol Struct 1995; 24:611-41.
6. Lakey JH, van der Goot FG, Pattus F. All in the family: the toxic activity of pore-forming colicins. Toxicology 1994; 87:85-108.
7. Pattus F, Massotte D, Wilmsen HU et al. Colicins: prokaryotic killer-pores. Experientia 1990; 46:180-92.
8. Riley MA. Molecular mechanisms of colicin evolution. Mol Biol Evol 1993; 10:1380-95.
9. Fourel D, Mizushima S, Bernadac A et al. Specific regions of *Escherichia coli* OmpF protein involved in antigenic and colicin receptor sites and in stable trimerization. J Bacteriol 1993; 175:2754-57.
10. el Kouhen R, Hoenger A, Engel A et al. In vitro approaches to investigation of the early steps of colicin-OmpF interaction. Eur J Biochem 1994; 224:723-28.

11. Lazzaroni JC, Vianney A, Popot JL et al. Transmembrane α-helix interactions are required for the functional assembly of the *Escherichia coli* Tol complex. J Mol Biol 1995; 246:1-7.

12. Levengood SK, Beyer WF Jr, Webster RE. TolA: A membrane protein involved in colicin uptake contains an extended helical region. Proc Natl Acad Sci USA 1991; 88:5939-43.

13. Levengood-Freyermuth SK, Click EM, Webster RE. Role of the carboxyl-terminal domain of TolA in protein import and integrity of the outer membrane. J Bacteriol 1993; 175:222-28.

14. Brewer S, Tolley M, Trayer IP et al. Structure and function of X-Pro dipeptide repeats in the TonB proteins of *Salmonella typhimurium* and *Escherichia coli*. J Mol Biol 1990; 216:883-95.

15. Schramm E, Mende J, Braun V et al. Nucleotide sequence of the colicin B activity gene cba: consensus pentapeptide among TonB-dependent colicin and receptors. J Bacteriol 1987; 169:3350-57.

16. Roos U, Harkness RE, Braun V. Assembly of colicin genes from a few DNA fragments. Nucleotide sequence of colicin D. Mol Microbiol 1989; 3:891-902.

17. Pilsl H, Glaser C, Gross P et al. Domains of colicin M involved in uptake and activity. Mol Gen Genet 1993; 240:103-12.

18. Traub I, Braun V. Energy-coupled colicin transport through the outer membrane of *Escherichia coli* K-12: mutated TonB proteins alter receptor activities and colicin uptake. FEMS Microbiol Lett 1994; 119:65-70.

19. Bénédetti H, Lazdunski C, Lloubes R. Protein import into *Escherichia coli*: colicins A and E1 interact with a component of their translocation system. EMBO J 1991; 10:1989-95.

20. Pilsl H, Braun V. Novel colicin 10: assignment of four domains to TonB- and TolC-dependent uptake via the TsX receptor and to pore formation. Mol Microbiol 1995; 16:57-67.

21. Wendt L. Mechanism of colicin action: early events. J Bacteriol 1970; 104:1236-41.

22. Peterson AA, Cramer, AW. Voltage-dependent, monomeric channel activity of colicin E1 in artificial membrane vesicles. J Membr Biol 1987; 99:197-204.

23. Levinthal F, Todd AP, Hubbell WL et al. A single tryptic fragment of colicin E1 can form an ion channel: stoichiometry confirms kinetics. Proteins 1991; 11:254-62.

24. Guihard G, Bénédetti H, Besnard M et al. Phosphate efflux through the channels formed by colicins and phage T5 in *Escherichia-coli* cells is responsible for the fall in cytoplasmic ATP. J Biol Chem 1993; 268:17775-80.

25. Bourdineaud JP, Boulanger P, Lazdunski C et al. In vivo properties of colicin A: channel activity is voltage dependent but translocation may be voltage independent. Proc Natl Acad Sci USA 1990; 87:1037-41.

26. Duché D, Baty D, Chartier M et al. Unfolding of colicin A during its translocation through the *Escherichia coli* envelope as demonstrated by disulfide bond engineering. J Biol Chem 1994; 269:24820-25.

27. Bullock JO, Kolen ER. Ion selectivity of colicin E1: III. Anion permeability. J Membr Biol 1995; 144:131-45.

28. Bullock JO, Kolen ER, Shear JL. Ion selectivity of colicin E1: II. Permeability to organic cations. J Membr Biol 1992; 128:1-16.

29. Bullock JO. Ion selectivity of colicin E1: modulation by pH and membrane composition. J Membr Biol 1992; 125:255-71.

30. Jakes KS, Abrams CK, Finkelstein A et al. Alteration of the pH-dependent ion selectivity of the colicin E1 channel by site-directed mutagenesis. J Biol Chem 1990; 265:6984-91.

31. Shirabe K, Cohen FS, Xu S et al. Decrease of anion selectivity caused by mutation of Thr501 and Gly502 to Glu in the hydrophobic domain of the colicin E1 channel. J Biol Chem 1989; 264:1951-57.

32. Collarini M, Amblard G, Lazdunski C et al. Gating processes of channels induced by colicin A, its C-terminal fragment and colicin E1 in planar lipid bilayers. Eur Biophys J 1987; 14:147-53.

33. Cramer WA, Zhang Y-L, Schendel S et al. Dynamic properties of the colicin E1 ion channel. FEMS Microbiol Immun 1992; 5:71-81.

34. Bénédetti H, Lloubés, R, Lazdunski, C et al. Colicin unfolds during its translocation in *Escherichia coli* cells and spans the whole cell envelope when its pore has formed. EMBO J 1992; 11:441-47.

35. Wormald MR, Merrill AR, Cramer WA et al. Solution NMR studies of colicin E1 C-terminal peptide. Structural comparison with colicin A and the effects of pH changes. Eur J Biochem 1990; 191:155-61.

36. Ghosh P, Mel SF, Stroud RM. The domain structure of the ion channel-forming protein colicin Ia. Nature Struct Biol 1:597-604.

37. Parker MW, Pattus F, Tucker AD et al. Structure of the membrane-pore-forming fragment of colicin A. Nature 337:93-96.

38. Parker MW, Postma JPM, Pattus F et al. Refined structure of the pore-forming domain of colicin A at 2.4 Å resolution. J Mol Biol 1992; 224:639-57.

39. Holm L, Sander C. Structural alignment of globins, phycocyanins and colicin A. FEBS Lett 1993; 315:301-06.

40. Li J, Carroll J, Ellar DJ. Crystal structure of insecticidal δ-endotoxin from *Bacillus thuringiensis* at 2.5 Å resolution. Nature 1991; 353:815-21.

41. Choe S, Bennett MJ, Fujii G et al. The crystal structure of diphtheria toxin. Nature 1992; 357:216-22.

42. Li J. Bacterial toxins. Curr Opin Struct Biol 1992; 2:545-56.

43. Mel SF, Falick AM, Burlingame AL et al. Mapping a membrane-associated conformation of colicin Ia. Biochemistry 1993; 32:9473-79.

44. Muga A, González-Mañas JM, Lakey JH et al. pH-dependent stability and membrane interaction of the pore-forming domain of colicin A. J Biol Chem 1993; 268:1553-57.

45. van der Goot FG, González-Mañas JM, Lakey JH et al. A "molten-globule" membrane-insertion intermediate of the pore-forming domain of colicin A. Nature 1991; 354:408-10.

46. Mel SF, Stroud RM. Colicin Ia inserts info negatively charged membranes at low pH with a tertiary but little secondary structural change. Biochemistry 1993; 32:2082-89.

47. Lakey JH, Massotte D, Heitz F et al. Membrane insertion of the pore-forming domain of colicin A. A spectroscopic study. Eur J Biochem 1991; 196:599-607.

48. Zhang Y-L, Cramer WA. Constraints imposed by protease accessibility on the trans-membrane and surface topography of the colicin E1 ion channel. Protein Sci 1992; 1:1666-76.

49. Schendel SL, Cramer WA. On the nature of the unfolded intermediate in the in vitro transition of the colicin E1 channel domain from the aqueous to the membrane phase. Protein Sci 1994; 3:2272-79.

50. Liu QR, Crozel V, Levinthal F et al. A very short peptide makes a voltage-dependent ion channel: the critical length of the channel domain of colicin E1. Proteins 1986; 1:218-22.

51. Shiver JW, Cohen FS, Merill AR et al. Site-directed mutagenesis of the charged residues near the carboxy-terminus of the colicin E1 ion channel. Biochemistry 1988; 27:8421-28.

52. Baty D, Lakey J, Pattus F et al. A 136-amino-acid-residue COOH-terminal fragment of colicin A is endowed with ionophoric activity. Eur J Bioch 1990; 189:409-13.

53. Baty D, Kniebiehler M, Verheij H et al. Site-directed mutagenesis of the COOH-terminal region of colicin A: Effect on secretion and voltage-dependent channel activity. Proc Natl Acad Sci USA 1987; 84:1152-56.

54. Abrams CK, Jakes KS, Finkelstein A et al. Identification of a translocated gating charge in a voltage-dependent channel. Colicin E1 channels in planar phospholopid bilayer membranes. J General Physiol 191; 98:77-93.

55. Duché D, Parker MW, González-Mañas JM et al. Uncoupled steps of the colicin A pore formation demonstrated by disulfide bond engineering. J Biol Chem 1994; 269:6332-39.

56. Lakey JH, Duché D, González-Mañas JM et al. Fluorescence energy transfer distance measurements. J Mol Biol 1993; 230:1055-67.

57. González-Mañas JM, Lakey JH, Pattus F. Brominated phospholipids as a tool for monitoring the membrane insertion of colicin A. Biochemistry 1992; 31:7294-300.

58. Qiu X-Q, Jakes KS, Finkelstein A et al. Site-specific biotinylation of colicin Ia: A probe for porotein conformation in the membrane. J Biol Chem 1994; 7483-88.

59. Jeanteur D, Pattus F. Membrane-bound form of the pore-forming domain of colicin A. J Mol Biol 1994; 235:898-907.

60. Massotte D, Yamamoto M, Scianimanico S et al. Structure of the membrane-bound form of the pore-forming domain of colicin A: A partial proteolysis and mass spectrometry study. Biochemistry 1993; 32:13787-94.

61. Bourdineaud J-P, Fierobe H-P, Lazdunski C at al. Involvement of OmpF during reception and translocation steps of colicin N entry. Mol Microbiol 1990; 4:1737-43.

62. Fourel D, Hikita C, Bolla J-M et al. Characterization of OmpF domains involved in *Escherichia coli* K-12: sensitivity to colicins A and N. J Bacteriol 1990; 172:3675-80.

63. Jeanteur D, Schirmer T, Fourel D et al. Structural and functional alterations of a colicin-resistant mutant of OmpF porin from *Escherichia coli*. Proc Natl Acad Sci USA 1994; 91:10675-79.

64. Yajima S, Muto Y, Morikawa S et al. The three-dimensional structure of the colicin E3 immunity protein by distance geometry calculation. FEBS Lett 1993; 333:257-60.

65. Geli V, Baty D, Pattus F et al. Topology and function of the integral membrane protein conferring immunity to colicin A. Mol Microbiol 1989; 3:679-87.

66. Song HY, Cramer WA. Membrane topography of ColE1 gene products: (II) The immunity protein. J Bacteriol 1991; 173:2935-43.

67. Zhang Y-L, Cramer WA. Intramembrane helix-helix interactions as the basis of inhibition of the colicin E1 ion channel by its immunity protein. J Biol Chem 1993; 268:10176-84.

68. Geli V, Lazdunski C. An α-helical hydrophobic hairpin as a specific determinant in protein-protein interaction occurring in *Escherichia coli* colicin A and B immunity systems. J Bacteriol 1992; 174:6432-37.

69. Espesset D, Piet P, Lazdunski C at al. Immunity proteins to pore-forming colicins: Structure-function relationships. Mol Microbiol 1994; 13:1111-20.

70. Masaki H, Akutsu A, Uozumi T et al. Identifiction of a unique specificity determinant of the colicin E3 immunity protein. Gene 1991; 197:133-38.

71. Espesset D, Corda Y, Cunningham K et al. The colicin A pore-forming domain fused to mitochondrial intermembrane space sorting signals can be functionally inserted into the *Escherichia coli* plasma membrane by a mechanism that bypasses the Tol proteins. Mol Microbiol 1994; 13:1121-31.

72. Olschlager T. A colicin M derivative containing the lipoprotein signal sequence is secreted and renders the colicin M target accessible from inside the cell. Archives Microbiol 1991; 156:449-54.

73. Kraulis JP. J Appl Crystallogr 1991; 24:946-50.

74. Lazdunski CJ. Colicin import and pore formation: a system for studying protein transport across membranes? Molec Microbiol 1995; 16:1059-66.

A STRUCTURE-BASED MODEL OF DIPHTHERIA TOXIN ACTION

David Eisenberg, Charles E. Bell, Melanie J. Bennett,
R. John Collier, Michael P. Schlunegger,
Boyd A. Steere and Manfred S. Weiss

Scientific efforts to find the cause of the disease diphtheria date back to at least 1740, when the clergyman Jonathan Dickenson, later the first president of Princeton, wrote his paper "Observations on that terrible Disease vulgarly called the Throat Distemper with advices as to the Method of cure."[1] Dickenson reported his observations during the diphtheria epidemic that swept the northern American colonies in the years 1735-1740, eventually infecting in attenuated form about a fourth of the people in Boston. He wrote:

> "The first Assault was in a Family about ten Miles from me, which proved fatal to eight of the Children in about a Fortnight. Being called to visit the distressed Family, I found upon my arrival, one of the Children newly dead, which gave me the Advantage of a Dissection, and thereby a better Acquaintance with the Nature of the Disease, than I could otherwise have had .... It frequently begins with a slight Indisposition, much resembling an ordinary Cold, with a listless Habit, a slow and scarce discernable Fever, some soreness of the Throat and Tumefaction of the Tonsils: and perhaps a running of the Nose, the Countenance pale, and the eyes dull and heavy. The patient is not confined, nor any Danger apprehended for some Days, till the Fever gradually increases, the whole Throat, and sometimes the Roof of the Mouth and Nostrils are covered with a cankerous Crust.... When the lungs are thus affected, the Patient is first afflicted with a dry hollow Cough, which is quickly succeeded with an extraordinary Hoarseness and total Loss of the Voice, with the most distressing asthmatic Symptoms

Protein Toxin Structure, edited by Michael W. Parker. © 1996 R.G. Landes Company.

and difficulty of Breathing, under which the poor miserable creature struggles, until released by a perfect Suffocation, or Stoppage of the Breath."

Progress towards understanding the microbial and molecular causes of diphtheria had to await the development of medical microbiology in the laboratories of Koch and Pasteur late in the following century. In Koch's laboratory, Friedrich Loeffler, a Prussian army surgeon studied diphtheria. He was able to grow a bacillus in pure cultures, first isolated from tissues of patients who had died from diphtheria. When he inoculated 23 guinea pigs with the bacillus, they all died in two to five days. The lesions at the site of inoculation contained many bacilli. Then Loeffler made another observation and a remarkable inference. He found dense brownish lesions away from the site of the inoculation, in the lungs and elsewhere, but none of these secondary lesions contained bacilli. Loeffler concluded that the bacilli growing at the site of inoculation produce a soluble poison that is transported elsewhere in the body by blood. His work was published in 1884.

Working in the Pasteur Institute in Paris, Emile Roux and Alexandre Yersin[2] were able to isolate the soluble poison that had been inferred by Loeffler. They grew a pure culture of diphtheria bacilli in a broth medium for a week. Then they forced the culture through an unglazed porcelain filter. The clear filtrate contained no bacteria, but when injected into laboratory animals, it caused most of the symptoms of the diphtheria bacillus, including death. They had discovered diphtheria toxin (DT).[1] Sixty-five years later[3] protein crystals were produced from pure, concentrated diphtheria toxin, and in 1982 Collier first grew X-ray grade crystals of DT.[4] From these and other crystal forms,[5] the first atomic model for DT was built[6]—103 years after discovery of the toxin.

The virtual elimination of the disease diphtheria in developed countries followed rapidly from the isolation of the toxin. In 1890 in Koch's lab, researchers found that animals immunized with a vaccine prepared from diphtheria bacilli, and ten days later challenged with a lethal dose of virulent diphtheria bacilli, all survived. In another paper just one week later, Emil von Behring, another Prussian army surgeon in Koch's lab, announced the discovery of diphtheria antitoxin. He found that the serum of animals immunized with the diphtheria vaccine was capable of attenuating the harmful effects of DT. The first Nobel Prize in physiology and medicine was awarded to von Behring for diphtheria antitoxin, later realized to be antibodies against DT. von Behring's next discovery on diphtheria toxin had the most profound impact on public health. In 1913[7] he reported the immunization of children against diphtheria by using a mixture of toxin and antitoxin. The antitoxin protected the subject from the harmful effects of the toxin, the actual vaccine. This mixed toxin-antitoxin vaccine was replaced after 1923 by a vaccine of a formaldehyde-treated DT, called toxoid.

Today, because of near universal vaccination, diphtheria is virtually unknown in most developed countries, although in 1990 a diphtheria epidemic broke out in the former Soviet Union.[8] In 1994, the World Health Organization WHO reported over 80,000 cases since the epidemic began.[9]

MOLECULAR INTRODUCTION

Diphtheria Toxin (DT) is a 535 residue protein.[10] It is secreted from strains of the bacterium *Corynebacterium diphtheriae* which have been infected with a phage carrying the DT gene.[11] It kills human and other eukaryotic cells by inactivating an essential factor of the translation machinery, elongation factor 2 (EF-2).[12,13] Helping to integrate biochemical and structural clues into a coherent model are three recent atomic structures for DT and its components.[6,14-17] In this chapter we present the known structures (Table 3.1) and combine the structural information with a wealth of biochemical data to arrive at a structure based intoxication pathway of DT (Fig. 3.1). The known structures are dimeric DT (dDT) (Fig. 3.2a)[6,15] monomeric DT (mDT) (Fig. 3.2b,c)[14] and the isolated catalytic (C) domain (Fig. 3.2d)[17] each one complexed with the inhibitor adenylyl-3',5'-uridine-monophosphate (ApUp). Currently under investigation are trimeric DT (tDT), nucleotide free dDT and dDT bound to NAD+ (see Table 3.1).

Figure 3.2a displays the secondary structural elements of mDT[6-14] as determined by X-ray crystallography. mDT consists of three distinct structural domains, each of which performs a specific biological function (see Mechanisms for Intoxication by DT). The N-terminal catalytic (C) domain includes residues 1-193 and contains α-helices and β-strands. The active site is located between the two subdomains of this kidney-shaped domain. The middle transmembrane (T) domain comprises residues 205-378 and is entirely α-helical. The receptor-binding (R) domain consists of residues 386-535 and is a flattened β-barrel

---

Table 3.1. The crystal structures of diphtheria toxin and its C domain

DT form	Ligand	Residues	pH	Resolution (Å)	R–factor (%)	Ref.	PDBcode
Dimer	ApUp	1–535	7.5	2.0	19.5	15	1DDT
Monomer	ApUp	1–535	7.5	2.3	20.7	14	1MDT
C domain	ApUp	1–190	5.0	2.5	19.7	17	1DTP
Dimer	NAD+	1–535	7.5	2.3	22.7	65	1TOX

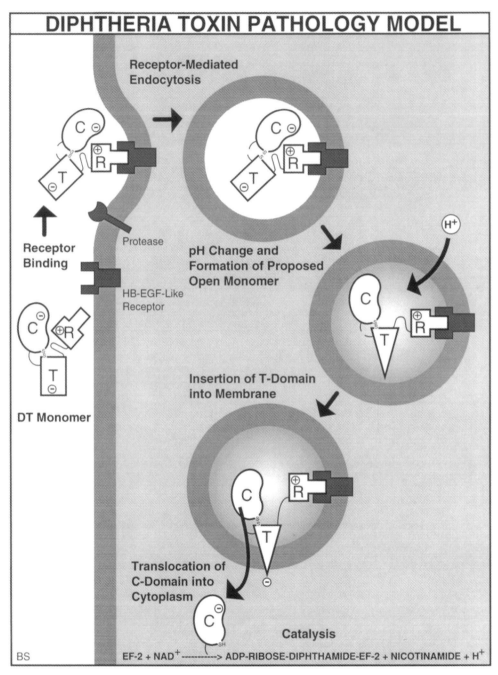

Fig. 3.1. Schematic diagram showing the intoxication pathway of DT. After DT has been secreted by Corynebacterium diphtheriae, a flexible loop between the C and the T domain is cleaved, possibly by a cell surface protease (left side of figure). DT then binds to its cell surface receptor (left top) and undergoes receptor-mediated endocytosis (middle top). After a pH drop in the endosome, an "open" structure forms, and the T domain undergoes a conformational change and inserts into the endosomal membrane (middle center). Then, the C domain gets translocated into the cytosol, and inactivates elongation factor 2 (middle bottom), leading to cell death.

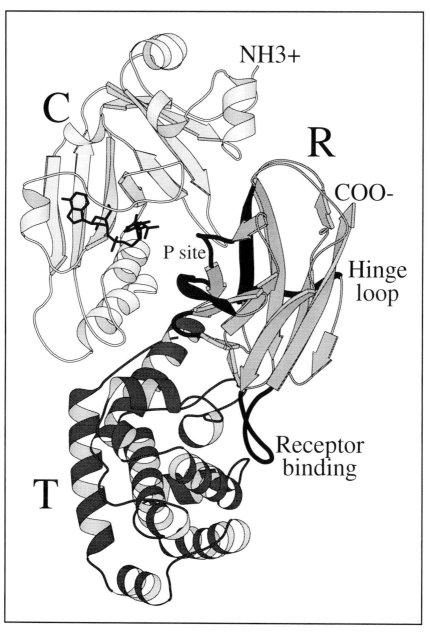

Fig. 3.2. MOLSCRIPT[63] diagrams displaying the X-ray crystal structures of DT. 3.2a.
Monomeric DT ("closed" monomer). mDT is a globular protein that consists of three
domains, the catalytic (C) domain, the transmembrane (T) domain and the receptor-
binding (R) domain. The C domain is a kidney shaped $\alpha + \beta$ structure with the active site
being located in a cleft between two subdomains. The inhibitor ApUp which was used
for crystallization is shown as it is bound in the active site. The receptor binding loop, the
putative P site and the hinge loop are displayed in black (see text). The loop linking the
C and T domain is not shown as it is flexible and no electron density was found in the
refined maps.

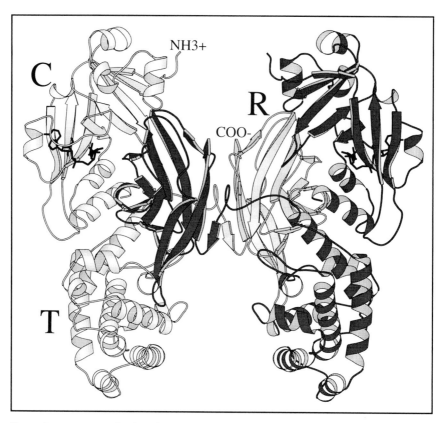

Fig. 3.2b. Dimeric DT displays the same domain structure as mDT, although the R domain is exchanged between the two subunits of the dimer, such that all the interdomain interactions in mDT remain the same. This phenomenon is called "domain swapping" (see text). No polypeptide chain is broken during domain swapping.

with a jelly-roll-like topology, similar to that of an immunoglobulin variable domain[18] or tumor necrosis factor (TNF).[19,20]

Dimeric DT (dDT; Fig. 3.2b) consists of two identical subunits of mDT. A special feature, called "domain swapping" occurs as the R domains of the monomers are exchanged in the dimeric structure. Such a domain swapped dDT consists of two intertwined "open" mDT molecules, a structure depicted in Figure 3.2c. The striking difference between the "open" monomer (Fig. 3.2c) and the "closed" monomer (Fig. 3.2a) is the orientation of the R domain relative to the rest of the molecule. A more detailed description of this phenomenon follows in the section below.

Finally, the structure of the isolated C domain of DT (also called DTA or fragment A) has also been determined and is displayed in Figure 3.2d.[17] Based on this structure a detailed description of ApUp

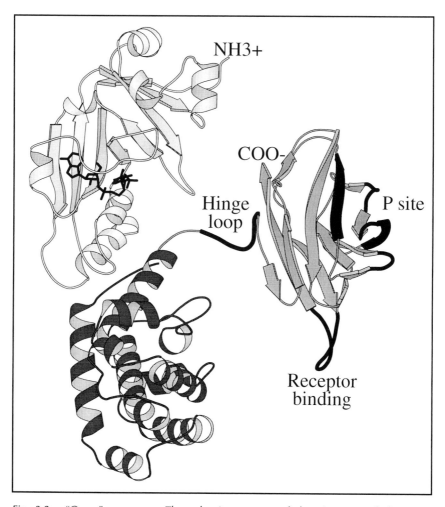

Fig. 3.2c. "Open" monomer. The subunit structure of domain-swapped dDT is a nonglobular structure which is called "open" monomer. The physiological relevance of this structure remains to be established, although it seems likely that a structure like this exists in the endosome at low pH, between the event of binding to the receptor and inserting into the endosomal membrane. The receptor-binding loop, the putative P site and the hinge loop (see text) are highlighted in black and labeled. It is clearly visible that the P site as well as the receptor binding loop are much more accessible in the open monomer as compared to the closed monomer (Fig. 3.2a).

binding was obtained and the implications of NAD⁺-binding could be inferred.

Future work will include the structure determination of DT with its bound substrate NAD⁺ (Table 3.1). These results will give further insight into the catalytic mechanism of DT. As described later, DT also forms higher oligomers such as trimers, tetramers, pentamers,

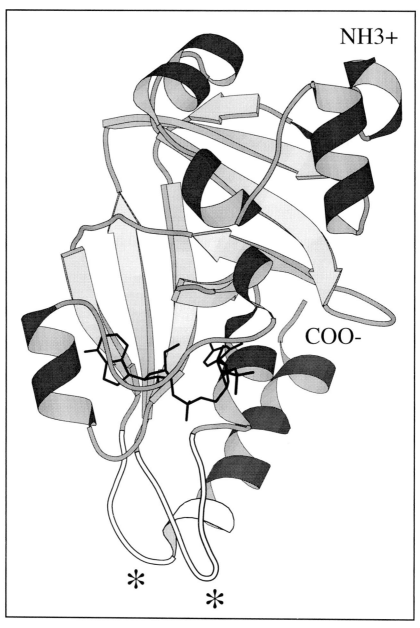

Fig. 3.2d. C domain. The structure of the isolated C domain at pH 5 is essentially identical to the structure of the C domain within the DT molecule at pH 7.5. Two loops, highlighted in white and labeled with a * at the bottom of the diagram, change conformation, however. This change may be important for the translocation event although this is merely speculative at this point.

hexamers or even heptamers as determined by HPLC gel filtration. The biological significance of these oligomers remains to be discovered. Trimeric DT (tDT) is currently being crystallized and it is assumed that its structure will further elucidate the mechanism of domain swapping and formation of intersubunit interfaces.

OLIGOMERIC FORMS OF DT

A long-lived, metastable dimer of DT can be formed by freezing mDT in phosphate buffer, which effectively forms a concentrated, low pH solution of mDT.[21] Upon thawing this solution, a mixture of oligomeric forms of DT is observed. DT dimer obtained by this method displays, in addition to the three domains of each DT molecule, another level of complexity: "domain swapping," a novel mode of protein association.[16]

STRUCTURAL DIFFERENCES BETWEEN MONOMERIC AND DIMERIC DT: "DOMAIN SWAPPING"

The crystal structures of mDT and dDT (Table 3.1)[6,14-15] have revealed essentially identical structures for the three domains (C, T and R). However, there is a small but significant difference in the loop 379-386 which covalently links the T domain to the R domain. Changing the main-chain torsion angles of this hinge loop (Figs. 3.2a, 3.2c) allows the rotation of the entire 15 kDa R domain by 180 degrees. This rotation swings the R-domain away from the rest of the molecule, with atomic movements up to 65 Å. This involves breaking the noncovalent interactions between the R domain and the C and T domains and results in a large difference in the position of the R domain relative to the other two domains of the same molecule. All the noncovalent interactions between the C, T and R domains of a single polypeptide chain in mDT are reproduced exactly in dDT by the C and T domains of one polypeptide chain and the R domain of the neighboring chain in the dimer. We refer to this exchange of R domains in dDT as "domain swapping."[16] One monomer within the dDT crystal structure[15] can be described as an "open" monomer (Fig. 3.2c), in contrast to the "closed" monomer (Fig. 3.2a) observed in the mDT crystal structure.[14]

The domain swapped DT dimer may not be a physiological form of DT since dDT itself is nontoxic.[21] However, its structure reveals a conformation of a single DT molecule that may exist, at least transiently, at the low pH of the endosome (Fig. 3.1), during the process of DT insertion into the endosomal membrane. This open monomer forms part of our model for the pathogenic pathway of DT. At the very least, the domain swapped dimer reveals the extent of conformational change which DT is capable of assuming.

As a result of domain swapping in dDT, two classes of interactions can be defined in the dimer interface. One class includes interdomain interactions that are identical in mDT (C-R, T-R) and domain swapped dDT (C-R', T-R'). This class has been called the "primary interdomain interface."[16] The interfaces between the domains in dDT and mDT are identical in the amounts of solvent-accessible surface area buried (1,860 Å^2 in the C-R'- and T-R'-interface in dDT and 1,910 Å^2 in the C-R- and T-R-interface in mDT, respectively).[22] The other class includes interactions found only in dDT. These have been called the "secondary interdomain interface" (R-R'-interface)[16] and include a buried solvent-accessible surface area of only 410 Å^2 per subunit.[22] The net area buried upon dimerization in the DT dimer consists therefore only of the secondary interdomain interface and is much smaller (as is the hydrophobic association energy estimated from atomic solvation parameters[23]) than is found in dimers of comparable molecular weight (1,600-4,900 Å^2 per subunit).[16] In short, the small R-R' dimer interface, produced not by natural selection but by freezing in phosphate buffer, does not provide enough binding energy to compensate the loss of entropy upon dimerization. It is therefore doubtful that a non domain swapped dimer forms in solution. No DT dimers of any sort are formed at high monomer concentrations (30mg/ml).[21]

Domain swapping seems to be a more common phenomenon than first expected and has been found in a number of other proteins.[16]

ENERGETICS OF DIMERIZATION

A large conformational change during DT dimerization is consistent with biochemical observations. mDT does not spontaneously convert to dDT at neutral pH,[21] indicating a kinetic barrier to dimerization. However, DT can be induced to dimerize by freezing the protein in mixed phosphate buffers, which are known to decrease in pH from 7.0 to 3.6 during freezing.[24] Freezing in buffer which lack this property does not lead to dimerization.[25] Based on the comparison of the mDT and dDT crystal structures we propose that the decrease in pH converts mDT to an open form which then dimerizes by domain swapping at the high concentrations of the eutectic mixture as the pH returns to neutral during thawing.[21]

When freezing DT to produce dimers, we also observe higher oligomers by size-exclusion HPLC. These higher oligomers include not only tetramers, but also trimers and pentamers. It is conceivable that these oligomers are formed by domain swapping, making linear or cyclized aggregates in which each molecule interacts with two neighboring molecules by providing a R domain to one and accepting a R domain from another. This hypothesis is currently being tested by investigating the structure of tDT.

Although the dimer is long lived and stable at high salt, guanidinium hydrochloride and urea, it is thermodynamically unstable: it dissociates to monomers at a rate of 5-10% per several weeks.[21] This shows that the binding energy contributed by the dimer interface is insufficient to overcome the loss of entropy upon dimerization.

Figure 3.3 displays the free-energy relationships in the association pathway from closed monomer to open monomer and finally to domain swapped dimer. The Gibbs free-energy of formation for dimerization (this and the following values are estimated from experiments and calculations), ΔG°_{dim}, must be positive given the slow dissociation of dimer to monomer.[21] We calculate a value of 9 kcal/mol of monomer for ΔG°_{dim} as a sum of the terms $-T\Delta S^{\circ}$ and ΔG°_{sol}, where ΔS° is

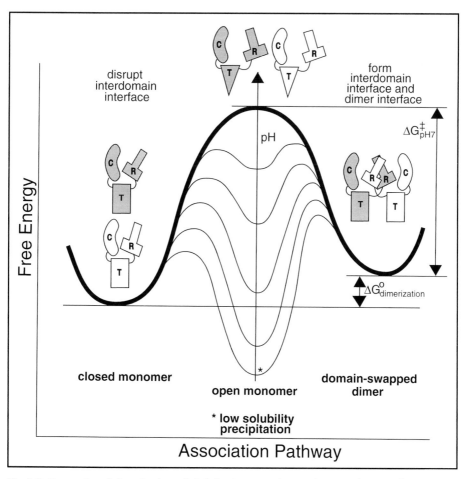

Fig. 3.3. Energetics of dimerization of diphtheria toxin. Thermodynamic diagram showing the free energy relationships between mDT, "domain swapped" dDT and the "open" monomer. At pH 7, the free energy difference between mDT and dDT is about 10 kcal/mol, the free energy of activation for the conversion from mDT to dDT is about 27 kcal/mol. At low pH this barrier has to be considerably lower.

the entropy change of rotation and translation[26] and ΔG°_{sol} is the free-energy change of solvation of the surfaces in the dimer.[23] $T\Delta S^{\circ}$ is estimated to be -10 kcal/mol of monomer at 277K[26] and ΔG°_{sol} to be -1 kcal/mol of monomer based on atomic solvation energies.[23] The free-energy barrier between monomer and dimer can be estimated from experimental rates of dimer dissociation. mΔT converts to dDT only at very high concentration and low pH. At neutral pH, monomer does not convert to dimer and dimers dissociate only slowly, indicating a large activation barrier $\Delta G^{\ddagger}_{pH7}$. $\Delta G^{\ddagger}_{pH7}$ is estimated to be 27 kcal/mol of dimer based on the measured rate of dimer dissociation to monomers[21] and the Eyring rate equation, assuming first-order kinetics and unity in the transmission coefficient. Although the DT dimer is thermodynamically unstable in low concentrations, the large activation barrier endows it with a long lifetime. At low pH, $\Delta G^{\ddagger}$ must be much smaller as shown by formation of DT oligomers. This is indicated by the pH axis in the figure.

MECHANISM OF "OPEN" MONOMER FORMATION AND ITS IMPLICATIONS FOR TOXIN FUNCTION

In order for mDT to form this highly intertwined domain swapped dimer, an open monomer must exist at least transiently (Fig. 3.2c). This structure may resemble the open monomer within dDT. The mDT structure also suggests how low pH can trigger open monomer formation. The interdomain interface between the R and C domains is unusually charged and polarized. There are nine basic and only three acidic residues on the R domain interface surface and seven acidic residues on the C domain interface surface. Three salt bridges stabilize the interface at neutral pH. At low pH, these salt bridges will be disrupted due to protonation of the acidic residues and there will be further destabilization due to isolated, buried positive charges in the interface, favoring the formation of an open-monomer-like structure.

Because low pH triggers not only the formation of the open monomer structure but also membrane insertion[27] it is possible that the open monomer resembles a membrane insertion intermediate.[15] A phosphate-binding site in DT, termed the P site, has been proposed to be involved in membrane insertion by binding to phospholipids (Figs. 3.2a and 3.2c).[28] A cluster of nine charges within 6 Å of the interface between the C and the R domains may constitute the P site of DT (Lys 447, His 449, Arg 455, Lys 456, Arg 458, Arg 460, Lys 474, His 488 and His 492). One of the charged residues in the cluster, Lys 474, was identified by crosslinking with ADP-ribose.[29] Another residue in the cluster, Arg 458, interacts with the 3' terminal phosphate of ApUp in our model.[6] In the closed structure of mDT, this plausible P site is buried in the C-R interface whereas in the open monomer, the P site is accessible. Thus, the open monomer structure may help to explain binding of the P site to phospholipids during membrane insertion.

In closed mDT, the receptor binding portion of DT (residues 482-535, including β hairpin loop 514-525) is tightly packed against the T domain (Fig. 3.2a). This and the possible presence of the DT receptor bound to the R domain might sterically hinder the conformational changes in the T domain and insertion into the membrane might not occur. In contrast, formation of an open monomer brings the receptor binding β-hairpin loop 20 Å away from the T domain, perhaps enabling the T domain to go into action.

The open monomer also exposes apolar surfaces that are buried in the interface between the T and R domains of mDT. Consequently, forming the open monomer costs about 12 kcal/mol in hydrophobic folding energy, as calculated from the atomic solvation parameters.[23] Because the interface between the C and R domains is largely polar, the unfavorable energy is the result of exposing apolar residues at the interface of the T and R domains. Apolar segments in the T-R interface that are exposed in the open monomer are: residues 306-311 (between helix 6 and 7 in the T domain), 316-319 (between helix 7 and 8 in the T domain), 367-371 (helix 9 in the T domain), 426-430 (β-strand 4 in the R domain) and residues 476 and 481 (next to β-strand 8 in the R domain). Because exposure of these apolar segments to aqueous solvent is unfavorable, they might be expected to favor the insertion of the T domain into the membrane.

In summary, the hypothesis of the open monomer as an intermediate in solution helps to interpret the mechanism of domain swapping, provides a plausible P site accessible for phospholipid binding and supports the conformational changes in the T domain needed for membrane insertion.

MECHANISM FOR INTOXICATION BY DT

Each of the three folding domains of DT (Fig. 3.2a) serves a function in the intoxication pathway. The R domain recognizes the target cell, the T domain translocates the C domain, which is the only part of the DT molecule that actually enters the cytosol of the cell to play its deadly role there. The whole intoxication pathway of DT is depicted schematically in Figure 3.1.

After DT is secreted from *Corynebacterium diphtheriae*, a flexible loop (residues 188-199) between the C domain and the T domain has to be proteolyzed ("nicked").[30,31] A cell surface protease named furin has been implicated in this reaction,[31] although it is not clear whether furin is the only protein that nicks DT in vivo.[32,33] It is clear, however, that expression of toxicity of DT depends on cleavage of this loop before or during the toxin's entry into cells.[30] This proteolysis allows the C domain to dissociate from the T domain after translocation to the cytosol, where the disulfide bridge between Cys 186 and Cys 201 is reduced.[34]

Also at the surface of the susceptible cell, a prominent β-hairpin loop in the R domain (residues 514-525) binds to the DT receptor at the surface of the target cell.[35] In this receptor binding loop, replacement of Lys 516, Val 523, Lys 526 or Phe 530 with alanine produced marked reductions in toxicity as assayed on Vero cells.[35] Replacing any of the other residues in this loop caused little or no change. The receptor has been shown to be a 185 residue integral membrane protein, known as heparin-binding epidermal growth factor-like precursor[36,37] which is used opportunistically by DT as its docking site. Binding leads to receptor mediated endocytosis.[38] The cytoplasmic domain of the receptor contains two tyrosine residues that may be involved in receptor mediated endocytosis. However, there is conflicting evidence: another study[39] showed that the cytoplasmic domain of the receptor is not required for receptor-mediated endocytosis.

Once the DT-receptor complex is endocytosed, a drop in the pH of the endosome possibly leads to the formation of an open monomer-like structure (see section above).[27] In addition to the conformational change in the hinge loop (residues 379-386) that leads to the open monomer formation, the T domain itself undergoes a structural transition at low pH and inserts into the endosomal membrane.[40] This event has been studied by mimicking the process at the plasma membrane by exposing cells to low pH. In vitro, DT undergoes a conformational change as the pH is lowered below 5, characterized by increased hydrophobicity.[40,41] Also at low pH, DT forms ion-conducting pores in lipid bilayers.[42-45] The exact nature of this transition is still unclear, but it is quite certain that some parts of the T domain completely span the hydrophobic part of the membrane and extend into the cytosol of the cell.

For the structurally related domain of colicin A, an umbrella-like model has been proposed.[46] The helices are arranged in three layers, and it has been proposed that the outer layers move away from the inner layer and remain on the membrane surface, whereas the inner layer inserts into the membrane, pushing all the way through the membrane. A similar mechanism can operate in DT. The central helices of the T domain (helices 8 and 9) are the ones that reach all the way through the membrane. The loop between helix 8 and 9 contains two acidic amino acid residues.[47] At the low pH of the endosome, these are largely protonated and uncharged, but once this loop reaches the cytosol, they become deprotonated and charged and lock the inserting helices in their transmembrane conformation.

The insertion of the T domain is accompanied by a concomitant appearance of ion-conducting pores in lipid bilayers.[42-45] It is unclear, however, whether these ion channels, which can be observed in vitro and in vivo, have anything to do with the actual translocation step.

The C domain is then somehow translocated from the endosomal lumen to the cytosol. In the structure of the C domain at pH 5.0[17] it has been shown that a conformational change occurs in two loops that make contact with the T domain (Fig. 3.2d). It is conceivable that this change is part of a structural rearrangement that allows the C domain to be inserted into the membrane and ultimately be translocated.

Sometime during or after translocation, the native disulfide bond between residues Cys 186 of the C domain and Cys 201 of the T domain is reduced[34] and the C domain is released into the cytosol. This cell-mediated reduction of the disulfide bond has recently been shown to be the rate determining step of toxin entry into the cell.[48] The transmembrane part of the DT receptor contains one cysteine residue that may be involved in the reduction of the 186-201 disulfide bridge between the T and the C domain.[36] Alternatively, it has been proposed that the reduction of the disulfide bridge occurs at a later stage by cytosolic compounds like glutathione.[34] Also, the enzyme protein disulfide-isomerase seems to be involved in the reduction of the 186-201 disulfide bridge as suggested by the finding that bacitracin, an inhibitor of the enzyme, or antibodies against the enzyme, inhibit DT action.[49]

Once inside the cytoplasm, the C domain is responsible for ADP-ribosylation of EF-2 at diphthamide, a posttranslationally modified histidine residue.[13] This completely shuts down protein synthesis and kills the cell.[12] The C domain catalyzes this ADP-ribosylation reaction with close todiffusion-limited efficiency (K_{cat}/K_M about 10^8 min^{-1} M^{-1}).[50] It has been shown by statistical methods that a single C domain is sufficient to kill a cell[51] rendering DT to be one of the most toxic molecules in the biosphere. The mechanism of ADP-ribosylation will be discussed in more detail in the next section.

THE MECHANISM OF ACTION OF THE C DOMAIN

The last step in the deadly action of DT is the actual cell killing by shutting down protein synthesis. This is performed by the C domain which catalyzes the transfer of an ADP-ribose group from NAD$^+$ to a diphthamide residue of EF-2 as described above.[13]

All crystal structures of DT (mDT, dDT and the isolated C domain)[6,14,15,17] have revealed the location of the endogenous inhibitor of NAD binding, ApUp. Although ApUp makes significant contacts with the R and C domains of DT, we will describe only those interactions between ApUp and the C domain, since the isolated C domain is the biologically relevant moiety.

ApUp is bound to a prominent cleft on the front face of the C domain (Fig. 3.2d). This cleft, referred to as the "active site" cleft, is located between two subdomains of the C domain. The adenine moiety of ApUp binds to a hydrophobic pocket on one side of this cleft, while the uracil ring binds to a hydrophobic pocket on the other side. Specific interactions between the C domain and ApUp are shown in Figure 3.4. Residues of the C domain forming H-bonds to ApUp include His 21 and Gly 22 of the second β-strand, and Gly 34, Thr 42, Gly 44, Asn 45, and Tyr 54 of a long loop extending over the active site which we have termed the "active-site" loop. In addition,

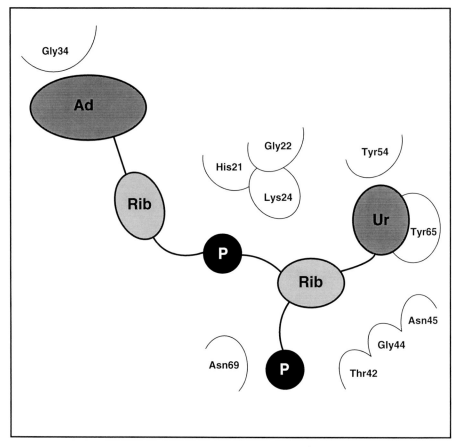

Fig. 3.4. C domain with ApUp. Schematic diagram showing the interactions between residues in the C domain of DT and the inhibitor ApUp. All residues that have at least one atom closer than 3.5 Å to one of the inhibitor atoms are indicated.

the uracil ring of ApUp lies directly above the ring of Tyr 65 making a tight ring stacking interaction.

Similarities in the structures of NAD$^+$, which is the actual substrate in vivo, and ApUp, which is an inhibitor in vitro (Fig. 3.5), suggest that the approximate NAD$^+$-binding site may be inferred from the binding site of ApUp. ApUp and NAD$^+$ are both adenosine-containing dinucleotides which differ only in the connectivities of the phosphates and in the functional groups on the nicotinamide and uracil rings. Almost certainly, the adenine rings of NAD$^+$ and ApUp bind to the same hydrophobic pocket within the C domain of DT. Modeling studies have shown that after alignment of the adenine rings, the uracil and nicotinamide rings can be superimposed. Therefore, results from the ApUp-containing structures can to some extent be applied to NAD$^+$-binding and action. In addition, chemical modification and mu-

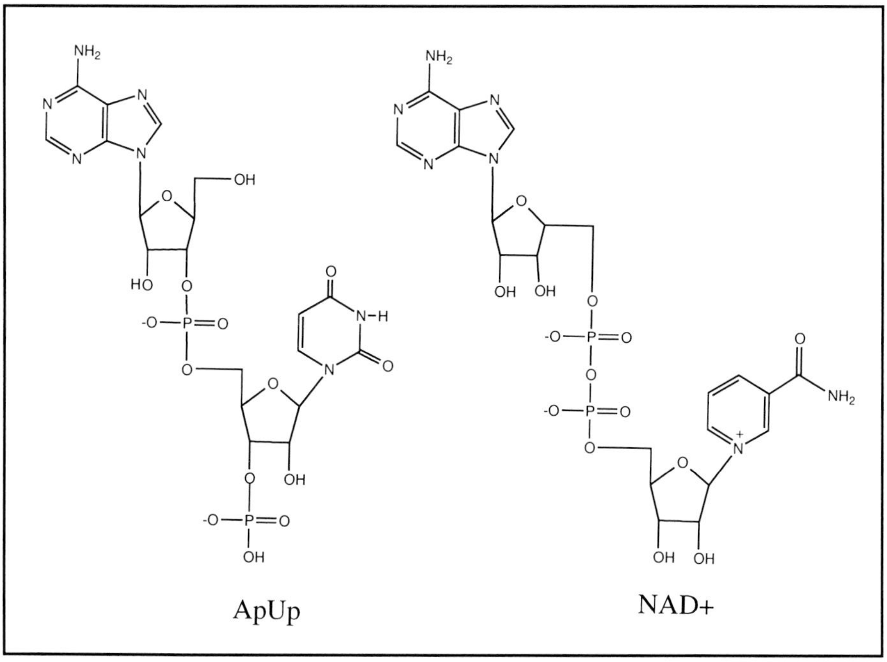

Fig. 3.5. Schematic diagram showing the similarities and differences of the inhibitor ApUp and the substrate NAD⁺.

tagenesis studies support the conclusion that NAD⁺ binds to the same site of the C domain as ApUp does. When His 21 is mutated to residues other than Gln or modified with diethyl pyrocarbonate,[50-52] NAD⁺-binding is severely reduced,[53] perhaps due to a disruption of the stabilizing H-bond pattern. Similarly, when Tyr 65 is mutated to anything other than Phe, NAD⁺-binding is reduced, suggesting that the nicotinamide moiety forms the same ring stacking interaction with Tyr 65 as ApUp does.[54] Perhaps the most convincing evidence that NAD⁺ binds to the C domain in a way similar to ApUp is the position of Glu 148 relative to ApUp. The crystal structures show that the uracil ring of ApUp is approximately 5 Å away from the carboxylate group of Glu 148. A study in which the nicotinamide ring of NAD⁺ was UV-crosslinked to the side chain of Glu 148 revealed that this residue is near the nicotinamide-binding site within the C domain.[55] Site-directed mutagenesis of Glu 148 also revealed that while this residue is not important for NAD⁺-binding, the carboxylate group of Glu 148 is essential for catalysis of the ADP-ribosylation reaction.[56]

Based on the stereochemistry of the ADP-ribose-EF-2 linkage[57] and kinetic studies,[54,56,58] the ADP-ribosylation reaction has been proposed to proceed by an S_N2-type mechanism where the nicotinamide ring of NAD⁺ is displaced by the incoming diphthamide nucleophile. In this

Table 3.2. Summary of mutations within the C-domain that affect NAD-binding and catalysis of ADP-ribosylation (Data from refs. 50, 53, 55 and 62)

		ADP–ribosylation	
Mutation	**K_d(NAD$^+$)**	**rel K_m (NAD$^+$)**	**rel K_{cat}/K_m**
wild type[50]	15.1mM	1 (52 mM)	1 (7.4×10^7 min^{-1} M^{-1})
H21N	163	1.5	0.20
H21R	*	nd	nd
H21A	*	nd	nd
H21L	*	nd	nd
H21Q	*	9.2	0.014
wild type[62]	15.6mM	1 (11.5 mM)	1 (1.2×10^8 min^{-1} M^{-1})
W50F	27.5	1.15	0.77
W50A	nd	10.3	0.00002
W153F	49.4	3.3	0.20
W153A	> 500	9.5	0.030
wild type[53]	15 mM	1 (40 mM)	1 (3.0×10^7 min^{-1} M^{-1})
Y65F	26	0.24	0.83
Y65A	> 800	5.5	0.0015
wild type[55]	9.3 mM	1 (9.0 mM)	1 (6.5×10^6 min^{-1} M^{-1})
E148D	8.7	1.2	0.0083
E148Q	12.7	1.4	0.0040
E148S	12.6	0.98	0.0032

* beyond the limits of detection. nd–not determined. Shown are the results from four separate site-directed mutagenesis studies.

reaction, the carboxylate group of Glu 148 has been proposed to activate the π-imidazole nitrogen of diphthamide for a nucleophilic attack on the C1 atom of the ribose to which the nicotinamide is linked.[54] A glutamic acid residue at a location corresponding to that of Glu 148 of DT has been found in other ADP-ribosylating toxins and may be a common if not a constant feature of this class of enzymes.[59-61]

In summary, the crystal structure of the isolated C domain of DT in complex with ApUp has enabled the development of a binding model between the C domain and NAD$^+$. This model is supported by biochemical studies. The crystal structure of DT in complex with NAD$^+$ has just been completed[65] and will further elucidate the catalytic actions of the C domain.

ACKNOWLEDGMENTS

We acknowledge with pleasure the work of the following scientists at different stages of these investigations: Drs. Steven Blanke, Wei-Hai Chen, Seunghyon Choe, Paul Curmi, Bauke Dijkstra, Gary Fujii, Katherine Kantardijeff, Edwin Westbrook, Hangjun Zhang. We also thank NIH (GM-31299), EMBO long-term fellowship (MSW), Swiss National Science Foundation (MPS) and Ciba-Geigy Jubilaeums-Stiftung (MPS) for support for this research.

DEDICATION

This review is dedicated to the memory of Dr. A.M. Pappenheimer, a leader in research on diphtheria toxin, who died in March 1995.

ABBREVIATIONS

DT (diphtheria toxin); mDT (monomeric DT), dDT (dimeric DT); tDT (trimeric DT); C domain (catalytic domain; in the literature also referred to as DTA and Fragment A); T domain (transmembrane domain); R domain (receptor-binding domain); ApUp (adenylyl-3',5'-uridine-monophosphate); EF-2 (elongation factor 2); $\Delta G°_{dim}$ (Gibbs free energy for dimerization); $\Delta G°_{sol}$ (free Gibbs energy for solvation).

NOTE ADDED IN PROOF

In a recent revieiw (ref. 64), domain swapping has been analyzed for a number of proteins. The recently determined structure of dimeric DT in complex with NAD (ref. 65) expands on the statements made above.

REFERENCES

1. Wood WB Jr. From miasmas to molecules. In: Wood WB Jr., ed., From miasmas to molecules. New York and London: Columbia University Press, 1961; 1-3. (Much of the historical material on the study of diphtheria in the opening section of the present review is adapted from this book.)
2. Roux E, Yersin A. Contribution a l'etude de la diphtherie (2e memoire). Annales de l'Institut Pasteur 1888; III:273.
3. Pope CG, Stevens M. Isolation of a crystalline protein from highly purified diphtheria toxin. Lancet 1953; II:1190.
4. Collier RJ, Westbrook EM, McKay DB et al. X-ray grade crystals of diphtheria toxin. J Mol Biol 1982; 257:5283-5.
5. Fujii G, Choe S, Bennett MJ et al. Crystallization of diphtheria toxin. J Mol Biol 1991; 222:861-64.
6. Choe S, Bennett MJ, Fujii G et al. The crystal structure of diphtheria toxin. Nature 1992; 357:216-22.
7. von Behring E. Ueber ein neues Diphtherieschutzmittel. Deutsche medizinische Wochenschrift 1913; XXXIX:873.
8. Bohlen C. Diphtheria sweeps Russia and Ukraine at quickening pace. (World Health Organization says it is reaching epidemic proportions).

New York Times January 29, 1993; 142:A1(N), A8(L), col 1, 27 col in.

9. Maurice J. Russian chaos breeds diphtheria outbreak. Science 1995; 267:1416-17.

10. Greenfield L, Bjorn MJ, Horn G et al. Nucleotide sequence of the structural gene for diphtheria toxin carried by *Corynebacteriophage Beta*. Proc Natl Acad Sci USA 1983; 80:6853-7.

11. Freeman VJ. Studies on the virulence of the bacteriophage-infected strains of *Corynebacterium diphtheriae*. J Bacteriol 1951; 61:675-88.

12. Collier RJ. Diphtheria toxin: mode of action and structure. Bacteriol Rev 1975; 39:54-85.

13. van Ness BG, Howard JB, Bodley JW. ADP-ribosylation of elongation factor 2 by diphtheria toxin. NMR spectra and proposed structures of ribosyl-diphthamide and its hydrolysis products. J Biol Chem 1980; 255:10710-6.

14. Bennett MJ, Eisenberg D. Refined structure of monomeric diphtheria toxin at 2.3 Å resolution. Prot Sci 1994; 3:1464-75.

15. Bennett MJ, Choe S, Eisenberg D. Refined structure of dimeric diphtheria toxin at 2.0 Å resolution. Prot Sci 1994; 3:1444-63.

16. Bennett MJ, Choe S, Eisenberg D. Domain swapping: entangling alliances between proteins. Proc Natl Acad Sci USA 1994;91:3127-31.

17. Weiss MS, Blanke SR, Collier RJ et al. Structure of the isolated catalytic domain of diphtheria toxin. Biochemistry 1995; 34:773-81.

18. Marquardt M, Deisenhofer J, Huber R et al. Crystallographic refinement and atomic models of the intact immunoglobulin molecule Kol and its antigen-binding fragment at 3.0 Å and 1.0 Å resolution. J Mol Biol 1980;141:369-91.

19. Jones EY, Stuart DI, Walker NPC. Structure of tumour necrosis factor. Nature 1989; 338:225-8.

20. Eck MJ, Sprang SR. The structure of tumour necrosis factor-alpha at 2.6 Å resolution. Implications for receptor binding. J Biol Chem 1989; 264:17595-605.

21. Carroll SF, Barbieri JT, Collier RJ. Dimeric forms of diphtheria toxin: purification and characterization. Biochemistry 1986; 25:2425-30.

22. Richmond TJ, Richards FM. Packing of α-helices: Geometrical constraints and contact areas. J Mol Biol 1978; 119:537-55.

23. Eisenberg D, McLachlan AD. Solvation energy in protein folding and binding. Nature 1986; 319:199-203.

24. van den Berg L, Rose D. Effect of freezing on the pH and composition of sodium and potassium phosphate solutions: the reciprocal system KH_2PO_4-Na_2HPO_4-H_2O. Arch Biochem Biophys 1959; 81:319-29.

25. Carroll SF, Barbieri JT, Collier RJ. Diphtheria toxin: purification and properties. Meth Enzym 1986; 165:68-76.

26. Erickson HP. Cooperativity in protein-protein association. The structure and stability of the actin filament. J Mol Biol 1989; 206:465-74.

27. Sandvig K, Olsnes S. Diphtheria toxin entry into cells is facilitated by low pH. J Cell Biol 1980; 87:828-32.

28. Lory S, Collier RJ. Diphtheria toxin:nucleotide binding and toxin hetero-geneity. Proc Natl Acad Sci USA 1980; 77:267-71.

29. Proia RL, Wray SK, Hart DA et al. Characterization and affinity labeling of the cationic phosphate-binding (nucleotide binding) peptide located in the receptor-recognition region of the B-fragment of diphtheria toxin. J Biol Chem 1980; 255:12025-33.

30. Sandvig K, Olsnes S. Rapid entry of nicked diphtheria toxin into cells at low pH. Characterization of the entry process and effects of low pH on the toxin molecule. J Biol Chem 1981; 256:9068-76.

31. Tsuneoka M, Nakayama K, Hatsuzawa K et al. Evidence for involvement of furin in cleavage and activation of diphtheria toxin. J Biol Chem 1993; 268:26461-65.

32. Chiron MF, Fryling CM, Fitzgerald DJ. Cleavage of *Pseudomonas* exo-toxin and diphtheria toxin by a furin-like enzyme prepared from beef liver. J Biol Chem 1994; 269:18167-76.

33. Gordon VM, Klimpel KR, Arora N et al. Proteolytic activation of bacte-rial toxins by eukaryotic cells is performed by furin and by additional cellular proteases. Infect Immun 1995; 63:82-87.

34. Madshus IH, Wiedlocha A, Sandvig K. Intermediates in translocation of diphtheria toxin across the plasma membrane. J Biol Chem 1994; 269:4648-52.

35. Shen WH, Choe S, Eisenberg D et al. Participation of lysine 516 and phenylalanine 530 of diphtheria toxin in receptor recognition. J Biol Chem 1994; 269:29077-84.

36. Naglich JG, Metherall JE, Russell DW et al. Expression cloning of a diph-theria toxin receptor: identity with a heparin-binding EGF-like growth factor precursor. Cell 1992; 69:1051-61.

37. Higashiama S, Lau K, Besner GE et al. Structure of heparin-binding EGF-like growth factor. Multiple forms, primary structure, and glycosylation of the mature protein. J Biol Chem 1992; 267:6205-12.

38. Morris RE, Gerstein AS, Bonventre PF et al. Receptor-mediated entry of diphtheria toxin into monkey kidney (Vero) cells: electron microscopic evaluation. Infect Immun 1985; 50:721-27.

39. Almond BD, Eidels L. The cytoplasmic domain of the diphtheria toxin receptor (HB-EGF precursor) is not required for receptor-mediated en-docytosis. J Biol Chem 1994; 269:26635-41.

40. Blewitt MG, Chung LA, London E. Effect of pH on the conformation of diphtheria toxin and its implications for membrane penetration. Biochem-istry 1985; 24:5458-64.

41. Dumont ME, Richards FM. The pH-dependent conformational change of diphtheria toxin. J Biol Chem 1988; 263:1567-74.

42. Donovan JJ, Simon MI, Draper RK et al. Diphtheria toxin forms trans-membrane channels in planar lipid bilayers. Proc Natl Acad Sci USA. 1981; 78:172-76.

43. Kagan BL, Finkelstein A, Colombini M. Diphtheria toxin fragment forms large pores in phospholipid bilayer membranes. Proc Natl Acad Sci USA. 1981; 78:4950-54.

44. Sandvig K, Olsnes S. Diphtheria toxin-induced channels in Vero cells selective for monovalent cations. J Biol Chem 1988; 263:12352-59.

45. Eriksen S, Olsnes S, Sandvig K et al. Diphtheria toxin at low pH depolarizes the membrane, increases the membrane conductance and induces a new type of ion channel in Vero cells. EMBO J 1994; 13:4433-39.

46. Parker MW, Pattus F, Tucker AD et al. Structure of the membrane-pore-forming fragment of colicin A. Nature 1989; 337:93-96.

47. Silverman JA, Mindell JA, Zhan H et al. Structure-function relationship in diphtheria toxin channel: I. Determining a minimal channel-forming domain. J Membr Biol 1994; 137:17-28.

48. Papini E, Rappuoli R, Murgia M et al. Cell penetration of diphtheria toxin. Reduction of the interchain disulfide bridge is the rate-limiting step of translocation in the cytosol. J Biol Chem 1993; 268:1567-74.

49. Mandel R, Ryser HJ, Ghani F et al. Inhibition of a reductive function of the plasma membrane by bacitracin and antibodies against protein disulfide-isomerase. Proc Natl Acad Sci USA 1993; 90:4112-16.

50. Blanke SR, Huang K, Wilson BA et al. Active-site mutations of the diphtheria toxin catalytic domain: role of histidine-21 in nicotinamide adenine dinucleotide binding and ADP-ribosylation of elongation factor 2. Biochemistry 1994; 33:5155-61.

51. Yamaizumi M, Mekada E, Uchida T et al. One molecule of diphtheria toxin fragment A introduced into a cell can kill the cell. Cell 1978; 15:245-50.

52. Johnson VG, Nicholls PJ. Histidine 21 does not play a major role in diphtheria toxin catalysis. J Biol Chem 1994; 269:4349-54.

53. Papini E, Schiavo G, Sandona D et al. Histidine 21 is at the NAD$^+$ binding site of diphtheria toxin. J Biol Chem 1989; 264:12385-88.

54. Blanke SR, Huang K, Collier RJ. Active-site mutations of diphtheria toxin: role of tyrosine-65 in NAD binding and ADP-ribosylation. Biochemistry 1994; 33:15494-500.

55. Carroll SF, Collier RJ. NAD binding site of diphtheria toxin: Identification of a residue within the nicotinamide subsite by photochemical modification with NAD. Proc Natl Acad Sci USA 1984; 81:3307-11.

56. Wilson BA, Reich KA, Weinstein BR et al. Active-site mutations of diphtheria toxin: effects of replacing glutamic acid-148 with aspartic acid, glutamine, or serine. Biochemistry 1990; 29: 8643-51.

57. Oppenheimer NJ, Bodley JW. Diphtheria toxin: Site and configuration of ADP-ribosylation of diphthamide in elongation factor 2. J Biol Chem 1981; 256:8579-81.

58. Wilson BA, Collier RJ. Diphtheria toxin and *Pseudomonas aeruginosa* exotoxin A: active-site structure and enzymatic mechanism. Curr Top Microbiol Immunol 1992; 175:27-41.

59. Carroll SF, Collier RJ. Active site of *Pseudomonas aeruginosa* exotoxin A. Glutamic acid 553 is photolabeled by NAD and shows functional homology with glutamic acid 148 of diphtheria toxin. J Biol Chem 1987; 262:8707-11.

60. Sixma K, Pronk SE, Kalk KH et al. Crystal structure of a cholera toxin-related heat-labile enterotoxin from *E. coli*. Nature 1991; 351:371-77.

61. Marsischky GT, Wilson BA, Collier RJ. Role of glutamic acid 988 of human poly-ADP-ribose polymerase in polymer formation. Evidence for active site similarities to the ADP-ribosylating toxins. J Biol Chem 1995; 270:3247-54.

62. Wilson BA, Blanke SR, Reich KA et al. Active-site mutation of diphtheria toxin. Tryptophan 50 is a major determinant of NAD affinity. J Biol Chem 1994; 269:23296-301.

63. Kraulis P. MOLSCRIPT: a program to produce both detailed and schematic plots of protein structures. J Appl Cryst 1991; 24:946-50.

64. Bennett MJ, Schlunegger MP, Eisenberg D. 3D Domain swapping: A mechanism for oligomer assembly. Protein Science 1995; 4:2455-2468.

65. Bell CE, Eisenberg D. Crystal structure of diphthenia toxin bound to nicotinamide adenine dinucleotide. Biochem 1996; 35:1137-1149.

═══ CHAPTER 4 ═══

INSECTICIDAL δ-ENDOTOXINS FROM *BACILLUS THURINGIENSIS*

Jade Li

During sporulation the Gram-positive bacterium *Bacillus thuringiensis* (*B.t.*) devote up to 30% of its cellular protein to synthesize the insecticidal proteins called δ-endotoxins, in the protoxin form.[1,2] These crystallize spontaneously in the cell, producing a parasporal inclusion. The inclusions are released upon cell lysis at the end of sporulation, ingested by susceptible insects and dissolve in the midgut under generally alkaline and reducing conditions. Gut proteases convert the protoxins to the active toxins, which bind to specific sites on the brush border epithelial surface[3] and create leakage channels in the cell membrane.[4] Colloid-osmotic swelling and lysis of these cells[4] results in feeding cessation and breach of the gut-haemocoel barrier, killing the insect through starvation or septicaemia.[2]

The *B.t.* δ-endotoxins belong to two multigenic families, *cry* and *cyt.*,[1,5] which differ significantly in polypeptide size and insecticidal target range. Cry δ-endotoxins are synthesized as protoxins of molecular weights around 130 kD or around 70 kD and processed to active toxins of 66-67 kD.[1] In the larger protoxins the active region is located in the N-terminal half of the molecule. Cyt δ-endotoxins are smaller protoxins of 28-29 kD, which are processed to active toxins of 23-25 kD.[6] The toxicities of Cry δ-endotoxins are highly specific to insects in the three orders of Lepidoptera (butterflies and moths), Diptera (mosquitoes and black flies), and Coleoptera (beetles and weevils), owing to their ability to recognize specific receptor determinants that are uniquely exposed on the midgut epithelial surface.[1] Cyt δ-endotoxins are exclusively toxic to dipteran larvae in vivo[6-9] but broadly cytolytic in vitro.[4,6,7,10,11] They insert into pure lipid bilayers without the necessary mediation of any membrane proteins.[12,13] Furthermore pre-incubation with phospholipid liposomes neutralizes their cytolytic activity, provided that these contain an unsaturated fatty acyl chain in the *syn*-2

Protein Toxin Structure, edited by Michael W. Parker. © 1996 R.G. Landes Company.

position.[10,14,15] Despite these differences, Cry and Cyt δ–endotoxins form pores of similar sizes in the insect cell membranes, of 5-10 Å radius,[4] and their insecticidal process involve the same stages of solubilization, proteolytic activation, binding and membrane insertion.[2] In addition, Cyt δ-endotoxins and the Diptera-specific Cry δ-endotoxins show a synergism which enhances their combined potency against the dipteran larvae by several fold.[9,16,17]

More than 50 Cry δ-endotoxins have been sequenced and characterized.[18] These are divided into major classes according to sequence similarity and insecticidal specificity: the CryI class is Lepidoptera specific; CryIII, Coleoptera specific; CryIV, Diptera specific; while the CryII class is Lepidoptera and Diptera dual-specific.[1] Active Cry δ–endotoxins in different specificity classes share up to five strongly conserved blocks in their amino acid sequences. In the first Cry δ-endotoxin structure obtained, of the Coleoptera specific CryIIIA,[19] these five sequence blocks were seen to build up the core of the molecule, encompassing all the domain interfaces between its three domains. Therefore other Cry δ-endotoxins sharing these conserved sequences can be expected to be show a similar domain organization and topological fold.[19] Recently, the Lepidoptera specific CryIA(a) was found to display a similar fold.[20] The structural similarities between these related proteins and their dissimilarities, which are also significant, provide important insights into the basis of specificity and possible mechanism for pore formation. In particular, both structures indicate strongly that pores made by the Cry family of toxins will be constructed from α-helices.[19,20]

Two Cyt δ-endotoxins have been sequenced and characterized. CytA and CytB share 39% sequence identity and should be similar in structure.[21] The structure of CytB has been determined in the protoxin form.[22] It shows a single domain protein with an entirely distinct fold compared to the Cry toxins. The nature of proteolytic activation was revealed by locating the protease processing sites[6] in the protoxin structure.[22] Reinterpretation of the biochemical and mutagenesis data[6,23,24] led to a hypothesis regarding the conformational changes upon membrane binding, and indicated that the Cyt δ-endotoxins will probably use their β–strands to form the pore.[22]

The toxicities of δ-endotoxins in vivo are targeted at insects which include many agronomically and medically important pests, such as the gypsy moth (a lepidopteran), Colorado potato beetle (a coleopteran), and the mosquito (dipteran) vectors of malaria. They have been in field use for more than 30 years as effective and environmentally safe bio-pesticides,[25,26] and recently transgenic plants have been engineered that express δ–endotoxins at insect resistant levels (see for example ref. 27). A deeper understanding of the structure-function relationships is still needed in order to improve and sustain their utility and help counter potential development of insect resistance.[28] Atomic structures of δ-endotoxins[19,20,22,29] with distinct insecticidal specificities provide a powerful tool for achieving this understanding.

CRY δ-ENDOTOXIN STRUCTURES

The example of CryIIIA will be used to outline the domain organization and topological fold of Cry δ-endotoxins. The CryIA(a) structure will be presented next for detailed comparisons. Functional implications of the structural similarities and dissimilarities between these toxins will be related to biochemical and genetic evidence in the next section.

STRUCTURE OF THE COLEOPTERA SPECIFIC CryIIIA

CryIIIA from *B.t.* ssp. *tenebrionis* is specifically toxic to coleopteran insects such as the Colorado potato beetle.[30] The 644-residue protoxin is processed by sporulation-associated proteases or by papain in vitro to remove 57 residues from the N terminus, producing a 67 kD active toxin.[31,32] The structure of activated CryIIIA was determined at 2.5 Å resolution[19] in the original crystal form of the parasporal inclusions.[33]

CryIIIA is a wedge-shaped molecule comprising three distinct domains (see Fig. 4.1). Domain I, from the N terminus of the activated toxin to residue 290, is a seven-helix bundle in which a central helix is completely surrounded by six outer helices, forming a left-handed supercoil (Fig. 4.2). Domain II, from residue 290 to 500, contains three antiparallel β-sheets packed around a hydrophobic core with a triangular cross-section. Domain III, from residue 501 to 644 at the C-terminus, is a β-sandwich of two antiparallel sheets. Domains II and III are stacked one on top of another on the same side of domain I; domains I and III make up the bulky end of the wedge.

Domain I

In domain I, the central helix α5 is oriented with its C-terminus toward the bulky end of the molecule. Viewed from this end, the outer helices are arranged counterclockwise in the order of α1, α2, α3, α4, α6, and α7 (Fig. 4.2), with helices α6 and α7 adjacent to the β-sheet domains. α2 is interrupted by a nonhelical section and only the N-terminal half, α2a, is packed against α5. The helices are long, especially α3 to α7, which contain respectively 8, 7, 6, 9 and 7 complete helical turns and therefore long enough to span the 30 Å thick hydrophobic region of a membrane bilayer. The six outer helices are clearly amphipathic and bear a strip of hydrophobic residues (defined by $\Delta G \geq 0$ for transfer from oil to water) down their entire length on the side facing helix α5. The central helix in CryIIIA has a less amphipathic character, and is relatively hydrophobic. Only two charged residues are found which are near the C-terminus. These are stabilized by hydrogen bonding, from the side chain of Lys209 to a main chain O atom on α5, and from the side chain of Asp210 to Tyr92 and Tyr143 which

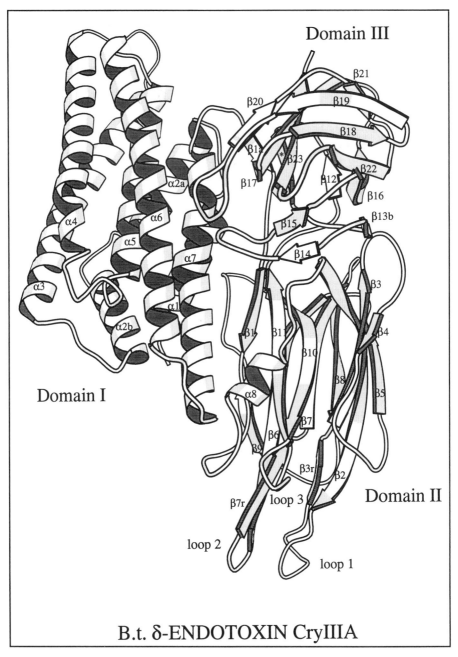

Domain III

Domain I

Domain II

loop 2

loop 3

loop 1

B.t. δ-ENDOTOXIN CryIIIA

Fig. 4.1. Ribbon drawing of the CryIIIA structure, showing the general fold which is also found in CryIA(a) and probably prototypical of the Cry δ-endotoxin family. The most outstanding difference between the two known structures lies in the three hairpin loops, labelled loop 1, loop 2 and loop 3, which extend from the three sheets in domain II to the molecular apex. All α-helices and β-strands are labeled (* indicates b13), and their sequence positions are given in Figure 4.3.

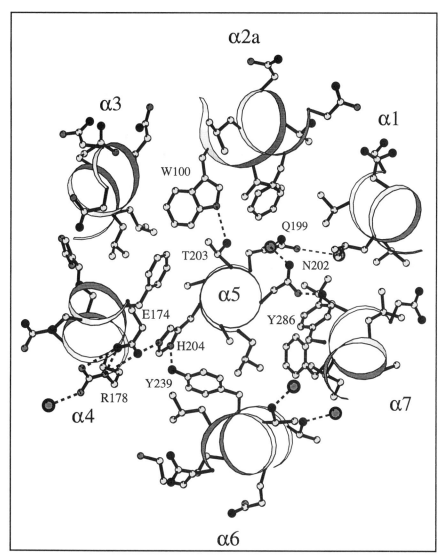

Fig. 4.2. *Cross-sectional view of the domain I helix bundle in CryIIIA. The section passes through the middle of α5 and shows the residues 199QAANTHL205 on this helix, which correspond to the residues 163QAANLHL169 present in α5 of CryIA(a) and CryIA(c) and studied by mutagenesis[88] in CryIA(c). The inter-helical spaces are occupied mostly by nonpolar residues; polar residues present are stabilized by hydrogen bonding to one another and to the buried water molecules shown as large gray spheres of heavy outline. For clarity hydrogen bonds to main chain atoms are omitted.*

are at the N-terminus of α3 and C-terminus of α4, respectively. Uncharged hydrophilic residues do occur in the middle of α5, as in the highly conserved QAAN sequence (residues 199-202). But the polar side chains are stabilized by hydrogen bonding to some of the buried water molecules within the helix bundle, on the side facing α1 and α7 (see Fig. 4.2). The tip of His204 on α5 is exposed to solvent and it participates in a chain of hydrogen bonds and salt bridges involving Glu174 and Arg178 on α4, Arg231 on α6, and Asp590 on a symmetry related molecule.

The helices are close packed. Between α5 and the outer helices the distances of closely approach are 7-13 Å, and the inter-helical angles are 14-29°. The inter-helical space contains 27 aromatic residues packed in an edge-to-face fashion, and all polar interactions are satisfied as described for α5. The concentric arrangement in this helix bundle resembles the pore-forming domain of colicin A,[34,35] where two hydrophobic helices are shielded from the solvent by eight amphiphilic helices, though the colicin helices are generally shorter. This resemblance, in addition to the length and amphiphilicity of the domain I helices, is strongly suggestive of a pore forming function.

Domain II

An approximate structural duplication exists between two of the three sheets, sheet 1 comprising strands β2 to β5 and sheet 2 comprising β6 to β9. Both have the "Greek-key" topology, and from both sheets the inner strands, β2 and β3, and β6 and β7, extend as a β–ribbon or hairpin by about 20 Å to the apex of the molecule. The connecting loop between edge strands crosses each sheet on the solvent face. Sheet 3 is made up by two separate segments. The C-terminal segment of domain II contributes strands β10 and β11 which also form a hairpin; the N-terminal segment of domain II contributes β1, which is hydrogen bonded to β11 to make this a three-stranded sheet, and a small helix α8, which takes the topological place of the missing fourth strand but lies to the solvent side of β10. β1 and α8 of sheet 3 are in contact with helix α7 of domain I.

The three sheets of domain II are packed as three sides of a prism, with their 11 strands oriented axially relative to the prism and with the hydrophobic side chains from their inner faces filling the interior. This was a novel fold, later named a β–prism[36] and observed in the structure of VMO-I with more exact 3-fold symmetry.[37] In domain II of the Cry δ–endotoxin structures,[19,20] such packing brings the tips of the β–hairpins from all three sheets into a cluster at the molecular apex, a region of about 12 Å radius in CryIIIA. The clustering of loops bears a resemblance to the complementarity determining regions of the immunoglobins. Indeed genetic evidence (see below) correlates insect specificity determination with this domain.

Domain III

The β–sandwich consists of an outer sheet which faces the solvent and an inner sheet which faces the other two domains. Four antiparallel strands in both sheets are arranged according to a jelly-roll topology: β19 and β18, β22 and β16 in the outer sheet, and the β20 and β21 half strands, β17, β23, and β13 and β15 half-strands in the inner sheet. A fifth strand is added to the outer sheet as two half strands, β13b running antiparallel to β16 and β12 parallel to it. A fifth strand in the inner sheet is formed by β1a running parallel to β13 and β14 running antiparallel to β15.

Location of conserved sequence blocks

The core of the CryIIIA molecule is built up from the five sequence blocks that are conserved among Cry δ–endotoxins of all specificity classes.[1] Block 1 (see Fig. 4.3), which is the most conserved segment in all Cry sequences, corresponds to the buried helix α5 in domain I and its N- and C-terminal extensions. Block 2 overlaps with the C-terminal half of α6, followed by α7, β1a and β1, which form the interface between the helical domain I and the sheet domains. Block 3 overlaps with the last strand, β11 in domain II and the N-terminal strands in domain III, up to the loop before β15, covering the interface between the sheet domains. The latter part of block 3 together with block 4 and block 5 form three contiguous strands, β13, β17 and β23 respectively, in the inner sheet of domain III. Block 5 also represents the conserved C-terminus of the minimal toxic region in Cry δ–endotoxins.[1,38] Its position as the middle strand of the five-stranded inner sheet in domain III prevents further processing from the C-terminus of the 130 kD protoxins. Conservation of the core residues implies that the structure of related toxins sharing the conserved sequences will be built around a similar core. Therefore the CryIIIA structure was proposed to represent the general fold of the Cry family.

STRUCTURE OF THE LEPIDOPTERA SPECIFIC CRYIA(A) AND COMPARISON WITH CRYIIIA

The 133 kD protoxin of CryIA(a) is present in parasporal inclusions of *B.t.* ssp. *kurstaki* HD-1, which are tetragonal bipyramids with $P4_1$ symmetry.[39] The proteolytic activated CryIA(a) contains residues 29 to 618.[40] This structure has been solved to 2.25 Å resolution independently of CryIIIA and compared with the latter in detail.[20]

Overall similarity

The structure of CryIA(a) is substantially similar to CryIIIA,[20] as expected from the 36% amino acid sequence identity in their active regions. CryIA(a) also consists of three domains, each domain showing the same topological fold as its CryIIIA counterpart, and the domains

```
CryIA(a)        α1===========          α2a=======          α2b==========
        29 IETGYTPIDISLSLTQFLLSE---FVP--GAGFVLGLVDI---IWGIFGPSQWDAFPVQIEQL
               *   :*   ::          *   *:::*       :    **      *:** *:* *
        58 DSSTTKDVIQKGISVVGDLLGVVGFPF-GGALVSFYTNFLNTIWPS--EDPWKAFMEQVEAL
CryIIIA         α1============         α2a===========      α2b==========


CryIA(a)    =      α3============================        α4================
        84  INQRIEEFARNQAISRLEGLSNLYQIYAESFREWEADPT---NPALREEMRIQFNDMNSA
            :*:*  ::*:* *:  *:** *    :  *      : *: :*      **     *  * : :*
       117  MDQKIADYAKNKALAELQGLQNNVEDYVSALSSWQKNPVSSRNPHSQGRIRELFSQAESH
CryIIIA     =      α3===========================          α4===============


CryIA(a)    =========     ........α5==========================....    α6============
       141  LTTAIPLLAVQNYQVPLLSVYVQAANLHLSVLRDVSVFGQRWGFDAATINSRYNDLTRLI
              *   :*    *:* :*: * **** **  :*:*    ::*:  **::   *   *   **
       177  FRNSMPSFAISGYEVLFLTTYAQAANTHLFLLKDAQIYGEEWGYEKEDIAEFYKRQLKLT
CryIIIA     ========          α5====================          α6============


CryIA(a)    ===================    α7====================================    β1a=
       201  GNYTDYAVRWYNTGLERVWGPDSRDWVRYNQFRRELTLTVLDIVALFSNYDSRRYPIRTV
             ::***  *:***:**::: *    ** :* :*** ****** :*** ** * **
       237  QEYTDHCVKWYNVGLDKLRGSSYESWVNFNRYRREMTLTVLDLIALFPLYDVRLYPKEVK
CryIIIA     ================          α7===========================      β1a


CryIA(a)    ===....β1== α8a==           α8====        β2===========
       261  SQLTREIYTNPVLEN-----FDGSFRGMAQRIEQNIRQPHLMDILNSITIYTDVHRG---
            :***:::*:*::        :     :        **: ** *** *  *   *    *
       297  TELTRDVLTDPIVGVNNLRGYGTT----FSNIENYIRKPHLFDYLHRIQFHTRFQPGYYG
CryIIIA        β1=                 α8===        β2=========== loop 1


CryIA(a)      β3r= β3=====       β4==       β5===     β6========
       313  ---FNYWSGHQITASPVGFSGPEFAFPLFGNAGNAAPPVLVSLTGLGIFRTLSSPLYRRI
            ****** :: *    *   *:: **    *    *  ::* :
       353  NDSFNYWSGNYVSTRPSIGSNDIITSPFYGNKSSEPVQNLEF-NGEKVYRAVANTNLAVW
CryIIIA      β3r= β3=====       β4=       β5===      β6============


CryIA(a)          β7r== β7=====         β8===      β9==
       370  ILGSGPNNQELFVL-DGTEFS-FASLTTNLP---STIYRQRG--TVDSLDVIPPQDNSVP
               *   ::   *** :  *    *: * :**:* :**:       *
       412  -----PS--AVYSGVTKVEFSQYNDQTDEASTQTYDSKRNVGAVSWDSIDQLPPETTDEP
CryIIIA      loop 2    β7r=== β7======    β8======    β9===


CryIA(a)       β10===       β11=======   β12=  β13==  ....β13b
       423  PRAGFSHRLSHVTMLS-QAA-GAVYTLRAPTFSWQHRSAEFNNIIPSSQITQIPLTKSTN
            *:**  *   *    * *   : :*  *:* :*  * *** * : **:** *
       465  LEKGYSHQLNYVMCFLMQGSRG-----TIPVLTWTHKSVDFFNMIDSKKITQLPLVKAYK
CryIIIA     β10=======      loop 3  β11======     β12=  β13==     β13b


CryIA(a)    ==...β14=..............β15=    β16=====    ..β17======    β18======
       481  LGSGTSVVKGPGFTGGDILRRTSPGQISTLRVNITAPLSQRYRVRIRYASTTNLQFHTSI
            *  ** *** ** ******:  *     *: *     **:** ** ****::: *  *:
       520  LQSGASVVAGPRFTGGDIIQCTENGSAATIYVTPDVSYSQKYRARIHYASTSQITFTLSL
CryIIIA     =      β14           β15=  β16=======    β17=======    β18====


CryIA(a)       β19===          β20=    β21=    β22=====      β23==
       541  DGRPINQGNFSATMSSGSNLQSGSFRTVGFTTPFNFSNGSSVFTLSAHVFNSGNEVYIDR
            **  * **  *   *      **   *:***::*   : :      *: ****:
       580  DGAPFNQYYFDKTINKGDTLTYNSFNLASFSTPFELSGN--NLQIGVTGLSAGDKVYIDK
CryIIIA     β19=====         β20=    β21    β22====       β23==


CryIA(a)    ====
       601  IEFVPAEVTFEAEYDLER
            ***:* :
       638  IEFIPVN
CryIIIA     ======
```

are associated in a similar manner. Upon the best superposition of all three domains, the two molecules show an r.m.s. deviation of less than 1.5 Å for approximately 400 equivalent Cα-atoms out of the 580 well-defined Cα-atoms. The degree of structural similarity between corresponding domains parallels their level of sequence identity (Fig. 4.3). Domains III with 41% identity show the greatest similarity, followed by domains I with 35% identity, while domains II with only 23% identity show the greatest structural divergence. When individual pairs of corresponding domains are superposed, the relative orientation of the domains within CryIA(a) is found to differ slightly from that in CryIIIA, by 2.9° for domains II relative to the superposed domains III, 3.7° for domains I relative to domains III, and 6.5° for domains II relative to domains I. These small rotations do not prevent many of the domain contacts, especially between domains I and II, from being conserved.[20]

Comparison of domains I

The backbone of domains I superposes well except two regions that show backbone differences. Helix α2a is shifted by 1.5 Å towards its N-terminus and shorter by one turn in CryIA(a) compared to that in CryIIIA, and the loop between α3 and α4 is shorter by three residues in CryIA(a). The outer helices in CryIA(a) also show a pronounced amphipathic character, but this is more mild in the central α5. The C-terminal part of α4, and N-terminal part of α5 and the connecting loop of this helix hairpin form one of the least polar regions of this domain in CryIA(a). Three of the buried water molecules in the interhelical space occupy similar positions in the two proteins.[20]

Comparison of domains II

Except for β2 and β5 in sheet 1, the central parts of the β–strands superpose well.[20] Two salt bridges are found in equivalent positions: one across sheet 2 (R359-D409 in CryIA(a), R401-D451 in CryIIIA), and another (R292-D298 in CryIA(a), R329-D335 in CryIIIA) between the α8-β2 loop and strand β2. However, the most striking differences between the two proteins are found in this domain, as the β ribbons emerge from the three sheets with different lengths and different shapes of their apical loops. Loop 1 at the tip of sheet 1, is smaller in CryIA(a) (residues 310-313) than in CryIIIA (residues 349-354); loop 2 is much larger in CryIA(a) (residues 367-379) than in CryIIIA (residues 412-413), and loop 3 is also larger (residues 438-446) than in CryIIIA (residues 481-486). Finding the greatest structural

Fig. 4.3. (shown left) Sequence alignment of CryIA(a) and CryIIIA with secondary structure positions. The five sequence blocks that are conserved among Cry δ-endotoxins are indicated by dotted lines above and below.

divergence between two toxins from different specificity classes in this part of domain II supports a role of this region for receptor recognition.[19] The longest loop of CryIA(a), loop 2, makes few interactions with the rest of the molecule and contains six disordered residues (370-375) in the middle; its flexibility may be important for receptor recognition.[20] The longest loop of CryIIIA, loop 1, is however in packing contacts[19] and therefore ordered in the crystal structure.

Comparison of domains III

Domains III of CryIA(a) and CryIIIA are most similar and can be superposed with a r.m.s. deviation of only 1.18 Å for 153 out of the 157 Cα atoms.[20] On the surface of the inner sheet, a large number of hydrogen bonds and salt bridges involving the charged residues on β17 and β23, and other residues in their vicinity, are conserved between these proteins.[20] This is so even though four of the Arg residues in β17 of CryIA(a) (hence the nickname "arginine face," see ref. 41) are substituted by Lys and His in CryIIIA. The inner Arg residues in β17 of CryIA(a) form side chain-to-main chain hydrogen bonds with the carbonyl oxygen of Arg254 in the loop between α7 and β1a, and with those of Ser571 and Ala606 at the ends of adjacent strands β20 and β23. The equivalent hydrogen bonds occur in CryIIIA with the main chain partner replaced by other amino acids. This network of salt bridges and hydrogen bonds must be important to the stability of this region. Furthermore, the solvent structure around domains III is also conserved, with at least 14 bound water molecules in common.[20]

Surface electrostatic potential

Most of the solvent accessible surface around CryIA(a) has a positive potential, except the side of its domain I, where helices α2a, α3 and the N-terminal part of α4 are exposed, which has a significant negative potential.[20] In CryIIIA most of the accessible surface has a negative potential, with patches of positive potential occurring over exposed basic residues, such as over the "arginine face" of domain III, the α8-β2 loop and loop 3 of domain II (Li, unpublished). The qualitative nature of the differences indicates that the surface potential probably does not play a major role in orienting the δ-endotoxins during the initial binding with a receptor.

RELATION OF STRUCTURE TO CRY δ-ENDOTOXIN FUNCTION

The substantial structural similarity of the Lepidoptera-specific CryIA(a) to the Coleoptera-specific CryIIIA confirmed that Cry δ-endotoxins of different specificities will share a common fold. More distantly related Cry toxins, with sequence identity levels falling below 25% compared to these examples, will show progressively greater structural divergence in the peripheral elements, but the prototypical fold should still be recognisable.[42] Therefore the two prototype structures provide one framework for interpreting the biochemical and genetic data from different Cry δ-endotoxins. Using the CryIIIA structure as a model, a common structural basis for the insecticidal function of different Cry δ-endotoxins was proposed.[19] Domain I with the long and amphiphilic helices was clearly equipped for membrane insertion and pore formation. Domain II with the cluster of apical loops was proposed for receptor binding. Domain III was thought to be important for the structural integrity or stability of the proteins. That the greatest structural similarity between CryIA(a) and CryIIIA occurs in domain III and the greatest divergence in domain II[20] is consistent with these proposals. Recently Grochulski et al[20] proposed that the high level of structural conservation in domain III could reflect an important although as yet undefined functional role for this domain, such as in oligomerization.

Receptor binding

Receptor recognition is the primary source of insecticidal specificity,[41] although incomplete processing[44] or protein instability[45,46] can also inhibit toxicity. Receptor binding is assayed in vitro as the specific and saturable binding to brush border membrane vesicles (BBMV),[43,47-50] which are right-side-out vesicles derived from the lumenal surface of midgut epithelial cells. Each Cry δ-endotoxin can bind to more than one population of BBMV sites with different affinities (apparent K_d in the order of 0.1 to 10 nM) and site concentrations (B_{max} in the order of 1 to 10^2 pmoles per mg vesicle protein), and different toxins can compete for the same high affinity site or recognize distinct sites in the same insect.[47-49] However total binding to BBMV includes two other components: specific irreversible binding and nonspecific binding.[51,52] Nonspecific binding is due to direct partition of toxin molecules into the hydrophobic phase,[53] which can be accounted for as the binding is observed in excess competitor. Specific, irreversible binding reflects membrane insertion which occurs within 1 sec[55] from initial binding to the receptor. It can cause over-estimation of receptor affinity by one order of magnitude,[56] but this can be corrected by applying steady-state kinetics to the sequential process of reversible followed by irreversible binding,[52] or by measuring binding on immobilized receptor protein separated from the membrane.[55] The irreversible binding has been estimated by pulse-chase measurements of toxin dissociation time-course and extent,[52,56,57] and this component was shown to correlate with the potency of the toxin.[52,57]

The specificity-determining regions defined by studying reciprocal hybrids between closely related toxins of different specificities[46,58-64] map mainly into domain II. In one study the toxic hybrids arising from exchanges within domain II were shown to compete with the parent for the same BBMV site, and their relative toxicities are inversely correlated with the K_d.[63] Although domain III can influence the toxicity of hybrids towards some test insects,[46,61,64] this influence is unrelated to receptor binding. Hybrids between CryIC and CryIE were shown to recognize different sites on *Spodoptera exigua* BBMV in accordance with the origin of their domain II, although they showed toxicity to this insect only if their domain III originated from CryIC.[46] It was suggested that domain III of CryIC is required to stabilize the activated toxin in the gut of this insect.[46]

Recently site-directed mutants have been created in the apical loops (loop 1: β2-β3r; loop 2: β6-β7r; loop 3: β10-β11) of domain II as identified or predicted by the crystal structures, to locate the residues critical to receptor binding and toxicity.[56,57,65,66] In CryIIIA, loops 1 and 3 were shown to be involved in receptor binding.[57] Alanine substitutions in loop 1 (N353A/D354A, or R345A/Y350A/Y351A) or loop 3 (Q482A/S484A/R485A, or M481-G486->A) reduced the binding affinity to BBMV by at least four-fold, while similar substitutions for the two residues of loop 2 had no effect. The loop 1 mutants showed lower toxicities in direct correlation with the decreased affinity, however the block mutant of loop 3 became simultaneously less toxic to one coleopteran insect but more toxic to another compared with wild-type CryIIIA. Therefore binding in loop 3 can discriminate between different receptors in two insects. The greater toxicity of the loop 3 block mutant to one insect was correlated with increased irreversible binding, indicating that membrane insertion is the second rate limiting step after initial binding.[57]

In the Lepidoptera and Diptera dual-specific CryIC, more than 50 mutants were created in the tips of predicted loop 1 (Q317-F320) and loop 2 (Q374-P377).[65] A single amino acid substitution in these two loops can profoundly alter toxicity and differential specificity to the dipteran *Aedes aegypti* larvae and the lepidopteran Sf9 cell line. In loop 1, the differential specificity was not restricted to any single residue, but in loop 2 the F377A mutation alone abolished dipteran toxicity. Dual specificity was only retained by the most conserved substitutions, and in both loops aromatic amino acids appear to play an important role in specificity. In view of the great sensitivity to single changes in these loops, it was remarked that perhaps the structural duplication of sheet 1 and 2 allows an expanded target specificity to be evolved based upon a common structural motif.[65]

In two Lepidoptera-specific toxins, loop 2 mutations were also shown to affect binding. In CryIA(a), which has a long and highly mobile loop 2,[20] mutations in the N-terminal half of loop 2 [Δ(365LYRRIIL371) or (365LYRRIIL371)->A] caused greater than 10-fold reduction in competitive binding affinity and a greater than 1000-fold loss in toxicity, whereas in the C-terminal half of the same loop [(385PNNQK389)->A] they had no effect.[66] In the closely related CryIA(b) deletion [Δ(370PFNIGI375)] or mutations [F371A, G374A] in the tip of loop 2, reduced toxicity by 400-fold by reducing only the irreversible component of the binding.[55] Taken together, the mutagenesis results demonstrate that residues in the domain II apex are directly responsible for recognizing specificity determinants on the cell surface, and that membrane insertion begins with some of the loops still bound to these determinants.

A 120 kD glycoprotein receptor for the Lepidoptera-specific CryIA(c) was recently purified from the susceptible insect *Manduca sexta* via toxin-affinity chromatography, cloned and identified as the insect aminopeptidase N with a signal sequence for glycosylphosphatidylinositol anchor to the membrane.[67,68] Incorporation of the partially purified receptor fraction in phospholipid vesicles reduced the threshold CryIA(c) concentrations required to induce [86]Rb+ release[69] by about 1000-fold, bringing it to the nM range observed in insecticidal activities in vivo. The sugar N-acetyl galactosamine (NAcGal) is part of the specificity determinants on *M. sexta* aminopeptidase N that is uniquely recognized by CryIA(c),[55,67,70,71] so that NAcGal selectively abolishes CryIA(c) cytotoxicity[70,71] and specifically inhibits the CryIA(c) binding to the BBMV[71] and to this glycoprotein immobilized on ligand blots[71] or optical biosensors.[55] In two other insects the receptors for CryIA(c) were also shown to be aminopeptidase N and glycoproteins containing NAcGal.[72,73] The importance of aromatic amino acids for receptor binding, shown by chemical modification of CryIA(c)[74] and mutagenesis of other toxins,[57,65] may reflect their role in sugar binding. However the binding surface on CryIA(c) for the sugar determinants remains to be mapped. Binding proteins for several other Cry δ-endotoxins have been described or cloned,[57,75-79] but their receptor function and specificity determinants are not known.

Pore formation

δ-endotoxins of different specificities kill insects by forming pores in the cell membrane that are permeable to small ions and solutes[4,54,80,81] but exclude macromolecules, thus causing colloid-osmotic lysis.[4] In vitro CryIA(a),[20] CryIA(c),[82,83] CryIC,[84] CryIIIA,[82] and CryIIIB2[85] form ion channels in planar bilayers. In planar bilayers incorporating BBMV, CryIA(c),[86] CryIC and CryID[87] form channels at nM toxin concentrations (similar to concentrations required for insecticidal activities in vivo), and these channels are distinguishable by their permeability properties from the endogenous channels of the insect. Pore formation have also been assayed by the collapse of short-circuit current across the lepidopteran midgut,[41,56] by inactivation of ion-dependent amino acid uptake into BBMV,[80,81,88] and by a light scattering assay of BBMV volume change in response to permeability increases to sucrose and salts.[54]

The long and amphiphilic helices in domain I[19,20] appear capable of membrane penetration and pore formation. It was proposed that receptor binding and the consequent interaction of toxin with the membrane trigger a major conformational change, whereby helix hairpins from helix bundle insert into the membrane and form the pore.[19] Such a role is supported by structural resemblance of domain I to the pore forming domains of colicin A[34,35] and diphtheria toxin,[89] which are also helix bundles and contain a relatively hydrophobic helix hairpin that were proposed or have been demonstrated to insert into the membrane.[90,91] In CryIIIA and CryIA(a) α4-α5 is a relatively hydrophobic hairpin linked at the end which will face the insect cell during the initial binding by the apical loops of domain II. Isolated domain I fragments from CryIA(c)[83] and CryIIIB2[85], and the central α5-peptide from CryIA(c)[92] and CryIIIA[93] have been found to partition into model membranes and form channels. The α5-peptide was shown to insert with its N-terminus toward the hydrophobic interior,[93] in agreement with the proposed hairpin orientation.

Limited mutagenesis data are available which focus on the pore forming potential of a few regions within domain I of CryIA(a), CryIA(b) and CryIA(c). Their amino acid sequences differ at only one or two positions in this domain. In CryIA(c), mutagenesis of the QAANLHL sequence (residues 163-169) in the middle of helix α5 resulted in mutants with very low or no toxicity to three test insects.[88] Many of the nontoxic mutants produced unstable proteins which are readily proteolysed and this could account for the inactivity (Geng, unpublished, as cited in ref. 97). The mutant proteins, sufficiently stable to undergo in vitro assays, retained the ability to bind BBMV but lost the ability to inhibit K+-dependent amino acid uptake into BBMV,[88] therefore demonstrating that helix α5 functions in pore formation. The integrity of α5 is important, and disruption by Pro in the middle (Q163P, A164P) severely inhibited the toxicities of CryIA(c)[88] and CryIA(a)[94], as well as membrane insertion by the α5-peptide.[93]

The middle of α5 is mildly amphiphilic, and mutations on the hydrophobic side to polar and charged residues (L167S/L169S, A164D) resulted in low toxicity.[88] However retaining the amphiphilic character appeared insufficient for pore forming activity.[88] The L167F and L167K mutant proteins both lost the ability to inhibit K+-dependent leucine uptake, while the H168R mutant on the hydrophobic side retained this ability. Grochulski et al [20] noted that Leu167 (cf. Thr203 of CryIIIA, Fig. 4.2) is in a very hydrophobic environment packed between two aromatic rings and three bulky aliphatic side chains from α2a and α4, so substitution by a charged residue Lys or a larger residue Phe or Met can disrupt the tertiary structure of the helix bundle.[88,94] Conversely, His168 has the tip of its side chain exposed to solvent (cf. His204 of CryIIIA, Fig. 4.2), so mutation to the positive Arg with a long side chain is tolerated, but mutation to Asp places a negative charge close to the helix axis and abolishes toxicity.[88] At the C-terminus of α4 replacement by Glu or Arg of the buried Phe148, which faces α5, prevented the expression of the CryIA(b) protoxin in *E. coli* .[95]

In the C-terminal part of the long and markedly amphiphilic α6 (see Figs. 4.1 and 4.2), 27 mutations in residues 207-215 of CryIA(a), including substitutions by Pro (R209P, T213P) and substitutions of hydrophobic residues by charged ones (W210R, Y211R, Y211D), failed to affect the toxicity significantly.[96] This cast doubts on α6 having a direct role in pore formation.[96]

Several Cry δ–endotoxins are susceptible to proteolysis in an inter-helical loop which falls without exception on the "domain III end" of the helix bundle (Fig. 4.1): the α3-α4 loop in CryIIA (97) and CryIIIA (32), and the a5-a6 loop in CryIVA and CryIVB.[98] These cleavages would leave intact the helix hairpins linked on the side facing the cell membrane. However, whether they serve to facilitate the hairpin re-lease from the helix bundle,[19] or are consequences of the conforma-tional change, is unclear. In CryIVB an arginine at the cleavage site was replaced by mutagenesis.[98] The mutant protein was no longer proteolysed by trypsin nor cytolytic in vitro; however it could be cleaved by the insect proteases at an adjacent residue and was three times as toxic to the *Aedes egypti* larvae in vivo.[98]

Two loops at the "domain II end" of the helix bundle were inves-tigated. In the α4-α5 loop, mutations of Tyr153 to neutral or posi-tively charged residues were tolerated, but introduction of a negative charge (Y153D), which can impede the hairpin insertion in face of an intracellular-negative membrane potential, caused a 20-fold reduction in toxicity and a same order of reduction in pore formation measured by voltage clamping, with no effect on receptor binding.[95] In the α2a-α3 loop, mutation of Arg93 at the N-terminus of α3 by residues other than Lys caused great loss of toxicity.[88] However comparison with ef-fects of mutations in adjacent residues indicates that the charge stabi-lization afforded by Arg93 rather than net positive charge in the re-gion was the important factor in toxicity.[20] At the N-terminus of α3, mutations of Ala92 are well tolerated except to a negatively charged residue (A92D, A92E) which abolished toxicity to three insects.[88,95] The side chain of this Ala points toward the α4-α5 loop,[20] so a nega-tive charge here can affect interaction of the hairpin with the mem-brane. The A92E protein showed greater inhibition of pore formation than Y153D in the voltage clamping assay and less irreversible bind-ing to BBMV.[95] The A92D protein was reported to be uncompetitive or less competitive against wild type CryIA(c) for binding to the BBMV of two insects,[88] which could be due to its inability to convert revers-ible to irreversible binding. The α2a-α3 loop is distant from the ligand binding loops of domain II, but it may function in membrane dock-ing of the toxin following receptor binding and this step can also limit the toxicity.[88]

In summary, mutagenesis results in domain I support insertion by a helix hairpin α4-α5 which is linked on the side facing the cell mem-brane during initial receptor binding, and an adjacent α2b-α3 loop may be involved in the post-binding interaction with the cell membrane prior to insertion. Some mutations on α5 that can destabilize the helix packing also inhibited pore formation. This is a paradox, since conformational changes from the helix bundle as initial state must be required for pore formation. Whether these mutations might adversely affect steps leading to membrane insertion, or destabilize helix packing in the pore structure, is unknown.

Possible roles of domain III

Four contiguous strands in the inner β-sheet of domain III, β17, β23, the β13 and β15 half-strands, and the β1b and β14 half-strands, are formed by sequences from the conserved blocks 2 to 5.[1,19] The greatest structural similarity between CryIIIA and CryIA(a) occurs in this domain,[20] including a conserved network of salt bridges and hydrogen bonds and the solvent structure. This domain is clearly important to the protein stability. Mutations in the β13-β14 loop (in conserved block 3) of CryIA(b) resulted in poor expression in *E. coli*, while those expressed at reasonable levels had no effect on toxicity.[99] In CryIA(b) and CryIVA, replacing single charged residues in β23 (block 5) by neutral or positively charged amino acids, which would eliminate salt bridges, caused poor expression and thermal instability,[100,101] but had no direct effect on toxicity in vivo. In CryIA(a), β17 (block 4) contains four Arg residues at positions 521, 523, 525 and 527. All except Arg521 which points into the solvent, are involved in a network of salt bridges and hydrogen bonds, especially Arg525.[20] Replacing all four by Lys or Glu, replacing either Arg523 or Arg525 by Glu or Gly, or replacing only Arg525 by Lys, caused instability and loss of expression in *E. coli*, however mutations of Arg521 alone did not affect expression.[41] This confirms the importance of charge stabilization.

Replacing the outer Arg residues on β17 by Lys was found to inhibit toxicity in vivo and permeability increase across the midgut, without destabilizing the structure.[41] Some of these mutations also inhibited the permeability increases in isolated BBMV to KCl.[102] It was proposed that, while the pore structure is made from elements from domain I, the Arg residues on β17 regulate the ion conductance through the pore by a gating mechanism.[41,102] Since crystal structures of CryIA(a) and CryIIIA in the soluble state showed no direct interactions between β17 and domain I (Fig. 4.1) except one hydrogen bond to the α7-β1 loop at the boundary between domains I and II, a gating function involving β17 could only arise after a significant structural rearrangement.

STRUCTURE OF THE MOSQUITOCIDAL AND CYTOLYTIC TOXIN CYTB

CytB from *B.t.* ssp. *kyushuensis*[103] is the sole protein component responsible for the mosquitocidal activity of the parasporal inclusions. CytA is present in the inclusions of *B.t.* ssp. *israelensis* as one of the crystallites in a cluster with those of three Cry δ-endotoxins having

Diptera specificity.[8] Although the Cyt δ-endotoxins are broadly cytolytic in vitro, and their membrane insertion appears to depend on direct interactions with the hydrophobic zone of the phospholipid bilayer,[10,14,15] the phospholipid binding and pore forming activities are latent in the 29 kD protoxin of CytB and require activation by proteolytic processing occurs from both N- and C-terminus, using larval gut enzymes or proteinase K in vitro.[13,104] Analogously, the detectable cytolytic activity of CytA is attributable to the presence of a processed fraction.[9] The processing sites using proteinase K are known for both CytA and CytB, from N-terminal sequencing of the fragments combined with mass spectrometry.[6]

CytB Monomer Structure

Crystal structure of CytB was determined in the protoxin form,[22,105] using recombinant protein expressed in *E. coli.*.[21] The CytB monomer is a single domain protein of α/β architecture, but it has a novel connectivity whereby two outer layers of α-helix hairpins are wrapped around a five-stranded, mixed β-sheet. As shown in Figure 4.4, the N-terminal strand β1 is separated by a gap from the half-strand β2 forming the first edge of the sheet. The chain then forms, in succession, helix hairpin A–B on one face of the sheet, strand β3 at the other edge of the sheet, helix hairpin C-D on the back face of the sheet, another half-strand β4 to complete the first edge of the sheet, then the three inner strands in the sheet in a β–meander, and finally a small helix F on the molecular surface and an extended C-terminal tail which crosses the hairpin A–B on the outside. Another small helix E is inserted in the loop connecting strands β6 and β7.

Helices of CytB are amphiphilic but rather short. The three inner strands of the β-sheet, however, are respectively 36, 43 and 53 Å in length. They, rather than the helix hairpins, are sufficiently long to span the hydrophobic zone of a membrane bilayer. Because the short helices only half covered the sheet face, these long strands also have a amphiphilic character. The residues on their "helix A–B face" are predominantly hydrophilic, while those on the "helix C–D face" are predominantly hydrophobic. There is a buried salt bridge between Glu192 on β6 and Lys218 on β7 which allows the amphiphilic character of the sheet to extend underneath the tip of helix hairpin A–B. Asn74 in the tip of hairpin A–B forms a hydrogen bond from its Oδ1 atom to Nζ of Lys218, which strengthens the buried salt-bridge and stabilizes the packing of those helices against the sheet. The amphiphilic pattern is broken toward the other end of the long strands, where mostly hydrophobic residues are found on both faces of the sheet.

Protoxin Dimer

The CytB protoxin forms a crystallographic dimer[22] by intertwining their N-terminal arms. In the gap between the N-terminal strand

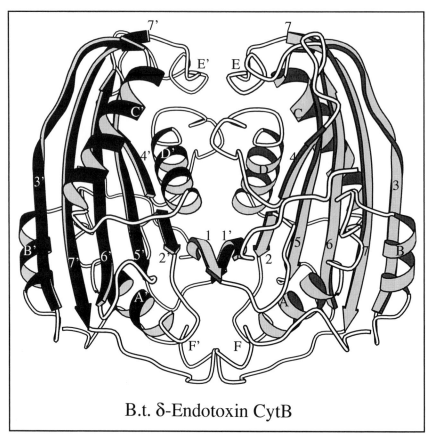

B.t. δ-Endotoxin CytB

Fig. 4.4. Ribbon drawing of the CytB protoxin dimer. The two monomers are shaded light and dark gray, respectively, over their helices and strands. In each monomer a five-stranded β-sheet is sandwiched between two helix hairpins A-B and C-D. In addition, N-terminal arms of two monomers are intertwined, forming the central pair of strands in a twelve-stranded β-sheet of the dimer. In front of this highly twisted sheet, helix hairpin A-B of one monomer and hairpin C'-D' of the other monomer are visible. Proteolytic activation excises the intertwined N-terminal strands to release the active toxin as a monomer, and it removes the C-terminal tails to uncover the three-layered core of the monomer. Thus freed from dimer contacts are the αD-β4 and β6-β7 the loops, which are shown by mutagenesis and by protease probes in liposomes to act as hinge regions in the conformational changes associated with membrane binding.

β1 and the edge of the sheet β2, strand β1' from a symmetry-related monomer is inserted. Hydrogen bonding between β1 and β1' enlarges the five-stranded sheets of the two monomers into one twelve-stranded β-sheet for the dimer (Fig. 4.4). On the helix A–B face, a hydrophobic cavity of 3.2 Å radius is formed by the pairing of the β1 strands and F helices. On the helix C-D face, dimer formations brings into contact the pair of loops connecting helix D to b4, and those connecting β6 and β7, with formation of intermolecular salt bridges. Altogether 3,132 Å², or just over 25% of the monomer surface is buried by the contacts, making the protoxin dimer rather stable. Indeed, the CytB protoxin remains as a dimer after solubilization from the parasporal inclusions at alkaline pH, as shown by mass spectrometry.[6]

EXPECTED STRUCTURAL SIMILARITY BETWEEN CytA AND CytB

Comparison of the sequence alignment between CytA and CytB,[21] with the secondary structure positions in CytB,[22] reveals that the buried parts of the monomer structure as well as the dimer interfaces are formed by segments of high sequence identity. Therefore CytA must have a similar structure to CytB and its protoxin probably also exists as a dimer. Morphological similarity between the CytB crystal, space group P6₁22, and the parasporal inclusion of CytA when that is expressed alone in a *Bacillus* cell,[17] supports this notion.

FUNCTIONAL IMPLICATIONS OF THE CytB STRUCTURE

Nature of proteolytic activation

The proteolytic activation of Cyt δ-endotoxins can be explained by dimer dissociation following cleavage of the intertwined β1 strands. Proteinase K which activates CytB has cleavage sites[6] on the C-terminal side of β1 which would excise the central pair of strands; trypsin which does not activate it[13] has cleavage sites on the N-terminal side of β1 which would leave the dimer intact. The dissociation releases the αD-β4 and β6-β7 loops from constraints imposed by dimer contacts. The proteolytic processing also removes the C-terminal tail[6] to uncover the three-layered core. Both these effects would facilitate the conformational change required for membrane insertion.

Mutagenesis in CytA

In the homologous toxin CytA, every charged residue has been individually replaced by Ala,[23] and selected Ala residues have been replaced by Pro.[24] A minority of mutations that inhibited expression[23,24] were mapped to critical intramolecular contacts, such as between the tip of helix hairpin A–B and the underlying salt-bridge on the sheet, that would destabilize the folding, or intermolecular contacts that would interfere with dimer formation or crystal packing.[22] The mutant proteins expressed have been characterized for larvicidal activity in vivo, cytolytic activity and phospholipid binding in vitro,[23,24] and in one case for channel formation in planar lipid bilayers (Armstrong, unpublished, as cited in ref. 5).

Postulated conformational change

All the mutations that map to the helix surfaces had no effect.[22,23] This and their relative shortness argue against a direct role for the helices in pore formation. The charged-to-Ala mutations that inhibited toxicity without inhibiting expression map to the αD-β4 and β6-β7 loops along the edge of the β–sheet adjoining the helix pair C-D, except one which maps to the C-terminal end of β7 on the sheet face. Since the charged residues in these loops are not conserved between CytA and CytB, they probably don't play a specific role but add a hydrophilic character to the loops. That reducing hydrophilicity here can reduce toxicity suggests that these loops need to remain on the membrane surface as other parts of the toxin molecule are inserted into the membrane.[22] Replacing two Ala in the same loops by Pro were also found to reduce toxicity, suggesting that a degree of main chain flexibility here is needed for pore formation.[24] Therefore it is postulated[22] that these loops at the edge between the sheet and helix hairpin C-D act as hinges in a conformational change, by which helix hairpin C-D peels away from the sheet to lie on the membrane surface. This would expose the hydrophobic face of the sheet underlying these helices and forcing parts of the sheet to partition into the membrane. Consistent with its having a defective hinge, the Glu204 to Ala mutant in the loop made fewer channels in planar lipid bilayers than the wild-type toxin, and the mutant channels had lower conductance, and opened and closed slowly rather than flickering rapidly like the wild-type channels (Armstrong, unpublished, as cited in ref. 5). Further support for the proposed conformational change is obtained when binding to phospholipid liposomes was found to generate new cleavage sites in CytA and CytB to several proteases, and these sites, inaccessible in the soluble toxins, map to the same loops along this edge of the sheet (Du and Knowles, unpublished, as cited in ref. 22).

The Lys225 to Ala mutation which alone maps to the sheet face showed the most severe inhibition of all membrane-directed activities and appears to be blocked at an early step.[22,23] Therefore the region at the C-terminal end of β7 on the sheet is suggested as possibly the site of initial contact with the membrane.[22]

Membrane pore based on the β-sheet

The nature of proteolytic activation implies that the activated CytB approaches and binds to the insect cell membrane as a monomer. Kinetic analysis of pore formation by CytA in the Malpighian tubules of the fly indicates that the membrane bound Cyt δ-endotoxins oligomerize to form the pore.[11] CytA recovered from erythrocytes by detergent extraction was primarily monomeric at low toxin concentrations, with the aggregate fraction rising as the toxin concentration increased.[106] This is consistent with activated toxin binding as monomers and oligomerization taking place in the membrane. Since only the β-strands are long enough to span the hydrophobic zone of the membrane, and only mutations around or in the sheet, but not those on the short helices, affected pore formation, it is most likely that the Cyt δ-endotoxins formed pores by rearranging parts of their β-sheet.

A plausible form of the pore based on the β-sheet is a closed β-barrel, such as observed in the porin structures (see for example ref. 107), or proposed for the pore forming toxin aerolysin.[108,109] Using the known geometric rules for β-barrel parameters,[110] a hypothetical β-barrel with a pore size of 5-10 Å radius estimated[4] for these toxins would comprise 14 to 18 strands, which would be tilted to the membrane normal by about 27-45°. If all three of the long strands β5, β6 and β7 are used to form the pore, or if they and β3 are used with helices A–B packed against them on the lumen face, then the hypothetical pore would be formed from between 4 and 6 monomers.[22]

CONCLUDING REMARKS

On the basis of the crystal structures of *B.t.* δ-endotoxins and re-interpretation of the biochemical and genetic data, the Cry δ-endotoxins are expected to form transmembrane pores using the amphiphilic helices from domain I, including at least the α4-α5 helix pair, while the Cyt δ-endotoxin pores are expected to comprise β-strands. The orientation in which the toxin molecules will penetrate the insect cell membrane have also been deduced, for Cry δ-endotoxins by identifying the apex of domain II as the receptor combining region, and for Cyt δ-endotoxins by locating the hinge regions for the gross conformational change to the edge of the β-sheet adjoining helix pair αC-αD.

Further work is needed to understand the interactions of δ-endotoxin with the membrane. Firstly, while the Cry δ-endotoxins recognize specific determinants on protein receptors,[3,67-71] the Cyt δ-endotoxins are believed to bind to unsaturated phospholipids,[10,14,15] which are present in greater abundance in dipteran cell membranes than in other cells.[111] The structure of a δ-endotoxin in complex with its specificity determinant from the membrane receptor is needed to visualize the initial binding. Secondly, the mechanisms by which initial binding is coupled to membrane insertion is unknown, and require investigations combining site-directed mutagenesis and biophysical techniques. Thirdly, the δ-endotoxin pores probably involve oligomerization, from several indications. The experimentally estimated pore radius of 5-10 Å[4] requires lining by more helices than present in a single Cry toxin molecule; the kinetics of pore formation by Cyt toxins shows high-order concentration dependence;[11] and channel activities of the toxins often show multiple subconductance states which suggest protein insertion in a multimeric manner.[12,20] The oligomeric state and uniformity of pore size are not established. However if conditions are found that favor a particular oligomeric assembly, then crystallographic methods may be applicable to reveal a pore structure directly.

ACKNOWLEDGMENTS

I would like to thank D.J. Ellar, J. Carroll and P.A. Koni for collaborations in solving the CryIIIA and CytB structures, and B.H. Knowles, R. Henderson, A. G.W. Leslie, P.R. Evans, J. Fermi, A. Lesk, C.J. Oubridge, K. Nagai and E. Bone for discussions.

REFERENCES

1. Höfte H, Whitely HR. Insecticidal crystal proteins of *Bacillus thuringiensis*. Microbiol Rev 1989; 53:242-55.

2. Knowles BH. Mechanism of action of *Bacillus thuringiensis* insecticidal d-endotoxins. Adv Insect Physiol 1990; 24:275-308.

3. Hofmann C, van der Bruggen H, Höfte H et al. Specificity of *Bacillus thuringiensis* δ-endotoxins is correlated with the presence of high-affinity binding sites in the brush border membrane of target insect midguts. Proc Natl Acad Sci USA 1988; 85:7844-48.

4. Knowles BH, Ellar DJ. Colloid–osmotic lysis is a general feature of the mechanism of action of *Bacillus thuringiensis* δ-endotoxins with different insect specificity. Biochim Biophys Acta 1987; 924:509-18.

5. Knowles BH, Nicholls CN, Armstrong G et al. Broad spectrum cytolytic toxins made by *Bacillus thuringiensis*. Abstract, Fifth International Colloquium on Invertebrate Pathology and Microbial Control, Society for Invertebrate Pathology, Adelaide, Australia, 1990; 283-87.

6. Koni PA, Ellar DJ. Biochemical characterization of *Bacillus thuringiensis* cytolytic δ-endotoxins. Microbiol 1994; 140:1869-80.

7. Thomas WE, Ellar DJ. *Bacillus thuringiensis* var *israelensis* crystal δ-endotoxin: Effects on insect and mammalian cells in vitro and in vivo. J Cell Sci 1983; 60:181-97.

8. Armstrong JL, Rohrmann GF, Beaudreau GS. Delta-endotoxin of *Bacillus thuringiensis* subsp. *israelensis*. J Bacteriol 1985; 161:39-46.

9. Chilcott CN, Ellar DJ. Comparative study of *Bacillus thuringiensis* var. *israelensis* crystal proteins in vivo and in vitro. J Gen Microbiol 1988; 134:2551-58.

10. Gill SS, Singh GJP, Hornung JM. Cell membrane interaction of *Bacillus thuringiensis* subsp. *israelensis* cytolytic toxins. Infect Immun 1987; 55:1300-8.

11. Maddrell SHP, Lane NJ, Harrison JB et al. The initial stages in the action of an insecticidal δ-endotoxin of *Bacillus thuringiensis* var. *israelensis* on the epithelial cells of the Malpighian tubules of the insect, *Rhodenius prolixus*. J Cell Sci 1988; 90:131-44.

12. Knowles BH, Blatt MR, Tester M et al. A cytolytic δ-endotoxin from *Bacillus thuringiensis* var. *israelensis* forms cation-selective channels in planar lipid bilayers. FEBS Lett 1989; 244:259-262.

13. Knowles BH, White PJ, Nicholls CN et al. A broad-spectrum cytolytic toxin from *Bacillus thuringiensis* var. *kyushuensis*. Proc Roy Soc B 1992; 248:1-7.

14. Thomas WE, Ellar DJ. Mechanism of action of *Bacillus thuringiensis* var. *israelensis* insecticidal d-endotoxin. FEBS Lett 1983; 154:362-68.

15. Drobniewski FA, Ellar DJ. Toxin-membrane interactions of *Bacillus thuringiensis* δ-endotoxin. Biochem Soc Trans 1988; 16:39-40.

16. Tabashnik B. Evaluation of synergism among *Bacillus thuringiensis* toxins. Appl Envir Microbiol 1992; 58:3343-46.

17. Wu D, Johnson JJ, Federici BA. Synergism of mosquitocidal toxicity between CytA and CryIVD proteins using inclusions produced from cloned genes of *Bacillus thuringiensis*. Molec Microbiol 1994; 13:965-92.

18. Schnepf HE. *Bacillus thuringiensis*. toxins: regulation, activities and structural diversity. Curr Opin Biotechnol 1995; 6:305-12.

19. Li J, Carroll J, Ellar DJ. Crystal structure of insecticidal δ-endotoxin from *Bacillus thuringiensis* at 2.5 Å resolution. Nature 1991; 353:815-21.

20. Grochulski P, Masson L, Borisova S et al. *Bacillus thuringiensis* CryIA(a) insecticidal toxin: crystal structure and channel formation. J Mol Biol 1995; 254:447-64.

21. Koni PA, Ellar DJ. Cloning and characterization of a novel *Bacillus thuringiensis* cytolytic delta-endotoxin. J Mol Biol 1993; 229:319-27.

22. Li J, Koni PA, Ellar DJ. Structure of the mosquitocidal δ-endotoxin CytB from *Bacillus thuringiensis* ssp. *kyushuensis* and implications for membrane pore formation. J Mol Biol 1996; 257:129-52.

23. Ward ES, Ellar DJ, Chilcott CN. Single amino acid changes in the *Bacillus thuringiensis* var. *israelensis* δ-endotoxin affect the toxicity and expression of the protein. J Mol Biol 1988; 202:527-35.

24. Armstrong G. Directed mutagenesis of the CytA δ-endotoxin of *Bacillus thuringiensis* ssp. *israelensis*. PhD. thesis, Cambridge University, 1991.

25. Milner R J. History of *Bacillus thuringiensis*. Agric Ecosys Environ 1994; 49:9-13.

26. Margalit J. Biological control by *Bacillus thuringiensis* subsp. *israelensis* (*B.t.i.*): history and present status. Israel J Entomol 1989; 23:3-8.

27. Wunn J, Kloti A, Burkhardt PK, Biswas G C G et al. Transgenic Indica rice breeding line IR58 expressing a synthetic CryIA(b) gene from *Bacillus thuringiensis* provides effective insect pest control. Bio/Technology 1996; 14:171-76.

28. Tabashnik B E. Evolution of resistance to *Bacillus thuringiensis*. Ann Rev Entomol 1994; 39:47-79.

29. Cody V, Luft JR, Jensen E et al. Purification and properties of insecticidal δ-endotoxin CryIIIB2 from *Bacillus thuringiensis*. Proteins: Struct Funct Genet 1992; 14:324.

30. Krieg A, Huber AM. Langenbruch GA et al. Neue Ergebnisse über *Bacillus thuringiensis* var. *tenebrionis* unter besonderer Berucksichtigung seiner Wirkung auf den Kartoffelkafer (*Leptinotarsa decemlineata*). Anz Schdlingskde, Pflanzenschutz, Umweltschutz 1984; 57:145-50.

31. Höfte H, Seurinck J, van Houtven A et al. Nucleotide-sequence of a gene encoding an insecticidal protein of *Bacillus thuringiensis* var. *tenebrionis* toxic against Coleoptera. Nucl Acids Res 1987; 15:7183.

32. Carroll J, Li J, Ellar DJ. Proteolytic processing of a coleopteran-specific delta-endotoxin produced by *Bacillus thuringiensis var.* tenebrionis. Biochem J 1989; 261:99-105.

33. Li J, Henderson R, Carroll J et al. X-ray analysis of the crystalline parasporal inclusion in *Bacillus thuringiensis* var. *tenebrionis*. J Mol Biol 1988; 199:543-45.

34. Parker MW, Pattus F, Tucker AD et al. Structure of the membrane-pore-forming fragment of colicin A. Nature 1989; 337:93-96.

35. Parker MW, Postma JPM, Pattus F et al. Refined structure of pore-forming domain of colicin A at 2.4 Å resolution. J Mol Biol 1992; 224:639-57.

36. Chothia C, Murzin AG. New folds for all-beta proteins. Structure 1993; 1:217-22.

37. Shimizu T, Vassylyev DG, Kido S et al. Crystal structure of vitelline membrane outer layer protein I (VMO-I): a folding motif with homologous Greek key structures related by an internal three-fold symmetry. EMBO J 1994; 13:1003-10.

38. Martens JWM, Visser B, Vlak JM et al. Mapping and characterization of the entomocidal domain of the *Bacillus thuringiensis* CryIA(b) protoxin. Mol Gen Genet 1995; 247:482-87.

39. Holmes KC, Monro RE. Studies on the structure of parasporal inclusions from *Bacillus thuringiensis*. J Mol Biol 1964; 14:572-81.

40. Borisova S, Grochulski P, van Faassen H et al. Crystallization and preliminary X-ray diffraction studies of the lepidopteran-specific insecticidal crystal protein CryIA(a). J Mol Biol 1994; 243:530-32.

41. Chen XJ, Lee MK, Dean DH. Site-directed mutations in a highly conserved region of *Bacillus thuringiensis* δ-endotoxin affect inhibition of short circuit current across *Bombyx mori* midguts. Proc Natl Acad Sci USA 1993; 90:9041-45.

42. Chothia C, Lesk AM. The relation between the divergence of sequence and structure in proteins. EMBO J 1986; 5:823-26.

43. Hofmann C, Lüthy P, Hütter R et al. Binding of the delta endotoxin from *Bacillus thuringiensis* to brush-border membrane vesicles of the cabbage butterfly (*Pieris brassicae*). Eur J Biochem 1988; 173:85-91.

44. Oppert B, Kramer KJ, Johnson DE et al. Altered protoxin activation by midgut enzymes from a *Bacillus thuringiensis* resistant strain of *Plodia interpunctella*. Biochem Biophys Res Commun 1994; 198:940-47.

45. Almond BD, Dean DH. Suppression of protein structure destabilizing mutations in *Bacillus thuringiensis* δ-endotoxins by second site mutations. Biochem 1993; 32:1040-46.

46. Bosch D, Schipper B, van der Kleij J et al. Recombinant *Bacillus thuringiensis* crystal proteins with new properties: possibilities for resistance management. Bio/Technology 1994; 12:915-18.

47. Hofmann C, Lüthy P, Hütter R et al. Binding of the delta endotoxin from *Bacillus thuringiensis* to brush-border membrane vesicles of the cabbage butterfly (*Pieris brassicae*). Eur J Biochem 1988b; 173:85-91.

48. Van Rie J, Jansens S, Höfte H et al. Specificity of *Bacillus thuringiensis* δ-endotoxins. Importance of specific receptors on the brush border membrane of the mid-gut of target insects. Eur J Biochem 1989; 186:239-47.

49. Van Rie J, Jansens S, Höfte H et al. Receptors on the brush border membrane of the insect midgut as determinants of the specificity of *Bacillus thuringiensis* delta-endotoxins. Appl Environ Microbiol 1990; 56:1378-85.

50. Slaney AC, Robbins HL, English L. Mode of action of *Bacillus thuringiensis* toxin CryIIIA: an analysis of toxicity in *Leptinotarsa decemlineata* (Say) and *Diaprotica undecimpunctata howardi* Barber. Insect Biochem Molec Biol 1992; 22:9-18.

51. Sanchis V, Chaufaux J, Pauron D. A comparison and analysis of the toxicity and receptor binding properties of *Bacillus thuringiensis* CryIC δ-endotoxin on *Spodoptera littoralis* and *Bombyx mori*. FEBS Lett 1994; 353:259-63.

52. Liang Y, Patel SS, Dean DH. Irreversible binding kinetics of *Bacillus thuringiensis* CryIA δ-endotoxins to gypsy moth brush border membrane vesicles is directly correlated to toxicity. J Biol Chem 1995; 270:24719-24.

53. Yunovitz H, Yawetz A. Interaction between the δ-endotoxin produced by *Bacillus thuringiensis* ssp. *entomocidus* and liposomes. FEBS Lett 1988; 230:105-8.

54. Carroll J, Ellar DJ. An analysis of *Bacillus thuringiensis* delta-endotoxin action on insect-midgut-mmembrane permeability using a light-scattering assay. Eur J Biochem 1993; 214:771-78.

55. Masson L, Lu Y-J, Mazza A et al. The CryIA(c) receptor purified from *Manduca sexta* displays multiple specificities. J Biol Chem 1995; 270:20309-315.

56. Rajamohan F, Alcantara E, Lee MK et al. Single amino acid changes in domain II of *Bacillus thuringiensis* CryIAb delta-endotoxin affect irreversible binding to *Manduca sexta* midgut membrane vesicles. J Bacteriol 1995; 177:2276-82.

57. Wu S-J, Dean DH. Functional significance of loops in the receptor binding domain of *Bacillus thuringiensis* CryIIIA δ-endotoxin. J Mol Biol 1996; 255:628-40.

58. Ge AZ, Shivarova NI, Dean DH. Location of the *Bombyx mori* specificity domain on a *Bacillus thuringiensis* delta-endotoxin protein. Proc Natl Acad Sci USA 1989; 86:4037-41.

59. Ge AZ, Rivers D, Milne R et al. Functional domains of *Bacillus thuringiensis* insecticidal crystal proteins: refinement of the *Heliothis virescens* and *Trichoplusia ni* specificity domains on CryIAc. J Biol Chem 1991; 266:17954-58.

60. Widner RW, Whiteley HR. Location of the dipteran specificity region in a lepidopteran-dipteran crystal protein from *Bacillus thuringiensis*. J Bacteriol 1990; 172:2826-32.

61. Schnepf HE, Tomczak K, Ortega JP et al. Specificity-determining regions of a lepidopteran-specific insecticidal protein produced by *Bacillus thuringiensis*. J Biol Chem 1990; 265:20923-20930.

62. Nakamura K, Oshie K, Shimizu M et al. Construction of chimeric insecticidal proteins between the 130-kDa and 135-kDa proteins of *Bacillus thuringiensis* subsp. *aizawai* for analysis of structure-function relationship. Agric Biol Chem 1990; 54:715-24.

63. Lee MK, Milnes RE, Ge AZ et al. Location of a *Bombyx mori* receptor binding region on a *Bacillus thuringiensis* δ-endotoxin. J Biol Chem 1992; 267:3115-21.

64. Masson L, Mazza A, Gringorten L et al. Specificity domain localization of *Bacillus thuringiensis* insecticidal toxins is highly dependent on the bioassay system. Mol Microbiol 1994; 14:851-60.

65. Smith GP, Ellar DJ. Mutagenesis of two surface-exposed loops of the *Bacillus thuringiensis* CryIC delta-endotoxin affects insecticidal specificity. Biochem J 1994; 302:611-16.

66. Lu H, Rajamohan F, Dean DH. Identification of amino acid residues of *Bacillus thuringiensis* δ-endotoxin CryIAa associated with membrane binding and toxicity to *Bombyx mori*. J Bacteriol 1994; 176:5554-59.

67. Knight PJK, Crickmore CN, Ellar DJ. The receptor for *Bacillus thuringiensis* CryIA(c) delta-endotoxin in the brush border membrane of the lepidopteran *Manduca sexta* is aminopeptidase N Mol Microbiol 1994; 11:429-36.

68. Knight PJK, Knowles BH, Ellar DJ. Molecular cloning of an insect aminopeptidase N that serves as a receptor for *Bacillus thuringiensis* CryIA(c) toxin. J Biol Chem 1995; 270:17765-70.

69. Sangadala S, Walters FS, English LH et al. A mixture of *Manduca sexta* aminopeptidase and phosphatase enhances *Bacillus thuringiensis* insecticidal CryIA(c) toxin binding and $^{86}Rb^+$-K^+ efflux in vitro. J Biol Chem 1994; 269:10088-92.

70. Knowles BH, Thomas WE, Ellar DJ. Lectin-like binding of *Bacillus thuringiensis* var. *kurstaki* lepidopteran-specific toxin is an initial step in insecticidal action. FEBS Lett 1984; 168:197-202.

71. Knowles BH, Knight PJK, Ellar DJ. N-acetyl galactosamine is part of the receptor in insect gut epithelia that recognizes an insecticidal protein from *Bacillus thuringiensis*. Proc Roy Soc London B 1991; 245:31-35.

72. Gill SS, Cowles EA, Francis V. Identification, isolation, and cloning of a *Bacillus thuringiensis* CryIA(c) toxin-binding protein from the midgut of the lepidopteran insect *Heliothis virescens*. J Biol Chem 1995; 270:27277-82.

73. Valaitis AP, Lee MK, Rajamohan F et al. Brush border membrane aminopeptidase-N in the midgut of the gypsy moth serves as the receptor for the CryIA(c) δ-endotoxin of *Bacillus thuringiensis* . Insect Biochem Molec Biol 1995; 25:1143-51.

74. Cummings CE, Ellar DJ. Chemical modification of *Bacillus thuringiensis* activated delta–endotoxin and its effect on toxicity and binding to *Manduca sexta* midgut membranes. Microbiol 1994; 140:2737-47.

75. Oddou P, Hartmann H, Geiser M. Identification and characterization of *Heliothis virescens* midgut membrane proteins binding *Bacillus thuringiensis* δ–endotoxins. Eur J Biochem 1991; 202:673-80.

76. Vadlamudi RK, Weber E, Ji I, Ji TH et al. Cloning and expression of a receptor for an insecticidal toxin of *Bacillus thuringiensis*. J Biol Chem 1995; 270:5490-95.

77. Sanchis V, Ellar DJ. Identification and partial purification of a *Bacillus thuringiensis* CryIC δ-endotoxin binding protein from *Spodoptera littoralis* gut membranes. FEBS Lett 1993; 316:264-68.

78. Belfiore CJ, Vadlamudi RK, Osman YA et al. A specific binding protein from *Tenebrio molitor* for the insecticidal toxin of *Bacillus thuringiensis* subsp. *tenebrionis*. Biochem Biophys Res Commun 1994; 200:359-64.

79. Feldmann F, Dullemans A, Waalwijk C. Binding of the CryIVD toxin of *Bacillus thuringiensis* subsp. *israelensis* to larval dipteran midgut proteins. Appl Environ Microbiol 1995; 61:2601-605.

80. Wolfersberger MG. Neither barium nor calcium prevents the inhibition by *Bacillus thuringiensis* δ-endotoxin of sodium- or potassium gradient-dependent amino acid accumulation by tobacco hornworm midgut brush border membrane vesicles. Arch Insect Biochem Physiol 1989; 12:267-77.

81. Hendrickx K, de Loof A, van Mellaert H. Effects of *Bacillus thuringiensis* delta–endotoxin on the permeability of brush border membrane vesicles from tobacco hornworm (*Manduca sexta*) midgut. Comp Biochem Physiol C Comp Pharmacol 1990; 95:241-45.

82. Slatin SL, Abrams CK, English, L. δ-Endotoxins form cation-selective channels in planar lipid bilayers. Biochem Biophys Res Common 1990;169:765-72.

83. Walters FS, Slatin, SL, Kulesza CA et al. Ion-channel activity of N-terminal fragments from CryIA(c) delta-endotoxin. Biochem Biophys Res Commun 1993; 196:921-26.

84. Schwartz JL, Garneau L, Savaria D et al. Lepidopteran-specific crystal toxins from *Bacillus thuringiensis* form cation-selective and anion-selective channels in planar lipid bilayers. J Membr Biol 1993; 132:53-62.

85. Von Tersch MA, Slatin SL, Kulesza CA et al. Membrane-permeabilizing activities of *Bacillus thuringiensis* coleopteran-active toxin CryIIIB2 and CryIIIB2 domain I peptide. Appl Environ Microbiol 1994; 60:3711-17.

86. Martin FG, Wolfersberger MG. *Bacillus thuringiensis* δ-endotoxin and larval *Manduca sexta* midgut brush border membrane-vesicles act synergistically to cause very large increases in the conductance of planar lipid bilayers. J Exper Biol 1995; 198:91-96.

87. Lorence A, Darszon A, Diaz C et al. δ-Endotoxins induce cation channels in *Spodoptera frugiperda* brush border membranes in suspension and in planar lipid bilayers. FEBS Lett 995; 360:217-20.

88. Wu D, Aronson AI. Localized mutagenesis defines regions of the *Bacillus thuringiensis* δ-endotoxin involved in toxicity and specificity. J Biol Chem 1992; 267:2311-17.

89. Choe S, Bennett MJ, Fujii G et al. The crystal structure of diphtheria toxin. Nature 1992; 357:216-22.

90. Duché D, Parker MW, González-Mañas J-M et al. Uncoupled steps of the colicin A pore formation demonstrated by disulfide bond engineering. J Biol Chem 1994; 269:6332-39.

91. Zhan H, Choe S, Huynh PD et al. Dynamic transitions of the transmembrane domain of diphtheria toxin: disulfide tapping and fluorescence proximity studies. Biochemistry 1994; 33:11254-63.

92. Cummings CE, Armstrong G, Hodgman TC et al. Structural and functional studies of a synthetic peptide mimicking a proposed membrane inserting region of a *Bacillus thuringiensis* δ-endotoxin. Mol Membr Biol 1994; 11:87-92.

93. Gazit E, Bach D, Kerr ID et al. The alpha–5 segment of *Bacillus thuringiensis* δ-endotoxin—in vitro activity, ion-channel formation and molecular modeling. Biochem J 1994; 304:895-902.

94. Ahmad W, Ellar DJ. Directed mutagenesis of selected regions of a *Bacillus thuringiensis* entomocidal protein. FEMS Microbiol Lett 1990; 68:97-104.

95. Chen X-J, Curtiss A, Alcantara E et al. Mutations in domain I of *Bacillus thuringiensis* delta-endotoxin CryIA(b) reduce the irreversible binding of toxin to *Manduca sexta* brush border membrane vesicles. J Biol Chem 1995; 270:6412-19.

96. Aronson AI, Wu D, Zhang C. Mutagenesis of specificity and toxicity regions of a *Bacillus thuringiensis* protoxin gene. J Bact 1995; 177:4059-65.

97. Nicolls CN, Ahmad W, Ellar DJ. Evidence for two different types of insecticidal P2 toxins with dual specificity in *Bacillus thuringiensis* subspecies. J Bacteriol 1989; 171:5141-47.

98. Angsuthanasombat C, Crickmore N, Ellar DJ. Effects on toxicity of eliminating a cleavage site in a predicted interhelical loop in *Bacillus thuringiensis* CryIVB delta-endotoxin. FEMS Microbiol Lett 1993; 111:255-62.

99. Yamamoto T, Powell GK. *Bacillus thuringiensis* crystal proteins: recent advances in understanding its insecticidal activity. In: Kim L, ed. Advanced Engineered Pesticides. New York: Marcel Dekker, Inc. 1993; 3-42.

100. Nakamura K, Nyrau-Nishioka RM, Shimizu M et al. Insecticidal activity and processing in larval gut juices of genetically engineered 130-kDa proteins of *Bacillus thuringiensis* subsp. *aizawai*. Biosci Biotech Biochem 1992; 56:1-7.

101. Nishimoto T, Yoshisue Y, Ihara K et al. Functional analysis of block 5, one of the highly conserved amino acid sequences in the 130-kDa CryIVA protein produced by *Bacillus thuringiensis* subsp. *israelensis*. FEBS Lett 1994; 348:249-54.

102. Wolfersberger MG, Chen XJ, Dean DH. Site-directed mutations in the third domain of *Bacillus thuringiensis* δ-endotoxin CryIAa affects its ability to increase the permeability of *Bombyx mori* midgut brush border membrane vesicles. Appl Environ Microbiol 1996; 62:279-82.

103. Ishii T, Ohba M. The 23-kilodalton CytB protein is solely responsible for mosquito larvicidal activity of *Bacillus thuringiensis* serovar *kyushuensis*. Current Microbiol 1994;.29:91-94.

104. Drobniewski FA, Ellar DJ. Purification and properties of a 28-kilodalton haemolytic and mosquitocidal protein toxin of *Bacillus thuringiensis* subsp. *darmstadiensis* 73-E10-2. J Bact 1989; 171:3060-67.

105. Li J, Koni PA, Ellar DJ. Crystallization of a membrane pore-forming protein with mosquitocidal activity from *Bacillus thuringiensis* subsp. *kyushuensis*. Proteins: Struct Funct Genet 1995; 23:290-93.

106. Chow E, Singh FJP, Gill SS. Binding and aggregation of the 25-kilodalton toxin of *Bacillus thuringiensis* subsp. *israelensis* to cell membranes and alteration by monoclonal antibodies and amino acid modifiers. Appl Envir Microbiol 1989; 55:2779-88.

107. Cowan SW, Schirmer T, Rummel G et al. Crystal-structures explain functional properties of two *Escherichia coli* porins. Nature 1992; 358: 727-33.

108. Parker MW, Buckley JT, Postma JPM et al. Structure of the *Aeromonas* toxin proaerolysin in its water-soluble and membrane-channel states. Nature 1994; 367:292-95.

109. Wilmsen H-U, Leonard KR, Tichelaar W et al. The aerolysin membrane channel is formed by heptamerization of the monomer. EMBO J 1993; 11:2457-63.

110. McLachlan AD. Gene duplication in the structural evolution of chymotrypsin. J Mol Biol 1979; 128:49-79.

111. Fast PG. A comparative study of the phospholipids and fatty acids of some insects. Lipids 1966; 1:209-15.

CHAPTER 5

STRUCTURE AND ASSEMBLY OF THE CHANNEL-FORMING *AEROMONAS* TOXIN AEROLYSIN

Michael W. Parker, J. Thomas Buckley,
F. Gisou van der Goot and Demetrius Tsernoglou

*A*eromonas hydrophila is a water-borne Gram-negative bacterium associated with gastroenteritis and opportunistic infections.[1] The organism secretes a protein toxin called aerolysin that appears to be a major virulence factor of the bacterium.[2] The toxin is synthesized as a preproprotein with a typical signal sequence that is removed during transit across the inner membrane of the *Aeromonas* bacterium.[3] The protoxin then appears to fold and dimerize in the periplasm and leaves the cell in a separate step that requires a group of more than 14 genes of the General Secretory Pathway.[4,5]

Once outside the cell, proaerolysin is activated by proteolytic cleavage of a C-terminal peptide. The protein binds to a receptor protein on the cell surface of the target cell which facilitates oligomerization of the toxin.[6,7] These oligomers insert into the membrane, producing voltage-gated channels that lead to osmotic lysis of the cell and eventual cell death. There is nothing in the amino acid sequence of the water-soluble toxin to explain how it can cross the outer membrane of the bacteria, or how it can insert into eukaryotic membranes to form voltage-gated ion channels. The recently determined crystal structure of proaerolysin has shed light on many aspects of the toxin's mechanism of action.[8]

PRIMARY STRUCTURES

The aerolysin structural genes from three *Aeromonas* species (*A. hydrophila*, *A. trota* and *A. salmonicida*) have been sequenced.[9-11] The polypeptide chain of each is approximately 470 amino acids long and there is greater than 80% pairwise sequence identity. Aligned

Protein Toxin Structure, edited by Michael W. Parker. © 1996 R.G. Landes Company.

sequences of the aerolysins are shown in Figure 5.1. Mass spectrometry of aerolysin has demonstrated the absence of any post-translational modifications.[12] There are two features of the amino acid composition worth remarking on: the large number of glycine residues (9.7%) and the large number of aromatic residues (12.4%). The glycine residues

Fig. 5.1. Sequence alignment of three Aeromonas pro-aerolysins and a related toxin (alpha toxin) from C. septicum. Invariant residues are shown in blocked type and domain classification numbers are shown directly above the sequences. The location of helices (denoted H) and β-sheet (denoted B) are shown above the domain classifications. Sequence data are from the following sources: A. hydrophila,[9] A. salmonicida,[10] A. trota[11] and C. septicum. [54]

```
                                                                   50
                    HHHHHB  BB          BBBBBHHHH HHH HHHHHH H   BBBBB
                    1111111111 1111111111 1111111111 1111111111 1111111111
  A. hydrophila     AEPVYPDQLR LFSLGQGVCG DKYRPVNREE AQSVKSNIVG MMGQWQISGL
  A. salmonicida    HEPVYPDQVK WAGLGTGVCA SGYRPLTRDE AMSIKGNLVS RMGQWQITGL
  A. trota          AEPIYPDQLR LFSLGEDVCG TDYRPINREE AQSVRNNIVA MMGQWQISGL
  C. septicum       .......... .......... .......... .......... ..........

                                                                   100
                     BBBBBHH HH BBB        BBBBBB          BBB   HHHH
                    1111111111 1111111111 1111111111 112222222 2222222222
  A. hydrophila     ANGWVIMGPG YNGEIKPGTA SNTWCYPTNP VTGEIPTLSA LDIPDGDEVD
  A. salmonicida    ADRWVIMGPG YNGEIKQGTA GETWCYPNSP VSGEIPTLSD WNIPAGDEVD
  A. trota          ANNWVILGPG YNGEIKPGKA STTWCYPTRP ATAEIPVLPA FNIPDGDAVD
  C. septicum       .......... .......... .......... .......... ...... DN

                                                                   150
                    HHHHHHH    HHHHHHHHH HHH                  BB BBBBB    B   BB
                    2222222222 2222222222 2222222222 2222222222 2222222222222222
  A. hydrophila     VQWRLVHDSA NFIKPTSYLA HYLGYAWVGG NHSQYVGEDM DVTRD-GDG----WV
  A. salmonicida    VQWRLVHDND YFIKPVSYLA HYLGYAWVGG NHSPYVGEDM DVTRV-GDG----WL
  A. trota          VQWRMVHDSA NFIKPVSYLA HYLGYAWVGG DHSQFVGDDM DVIQE-GDD----WV
  C. septicum       LKAKIIQDP- EFIRNWANVA HSLGFGWCGG TANPNVGQGF EFKREVGAGGKVSYL

                                                                   200
                    BBBB                  BBB BBBBBBBBB   BBB B    BBBBBBBBBB
                    2222222222 2222222222 2222222233 3333333333 3333344444
  A. hydrophila     IRGN---NDGGCD GYRCGDKTAI KVSNFAYNLD PDSFKH-GDVT QSDRQLVKTV
  A. salmonicida    IKGN---NDGGCS GYRCGEKSSI KVSNFSYTLE PDSFSH-GQVT ESGKQLVKTI
  A. trota          LRGN---DGGKCD GYRCNEKSSI RVSNFAYTLD PGSFSH-GDVT QSERTLVHTV
  C. septicum       LSARYNPNDPYAS GYRAKDRLSM KISNVRFVID NDSIKL-GTPK VKKLAPLNSA

                                                                   250
                    BBBBBB     BB BBBBB  BBBBBBBBBB  H HHHHHH   BBB
                    4444444444 4444444444 4443333333 3333333333 3333333333
  A. hydrophila     VGWAVNDSDT PQSGY-DVTLR YDTATNWSKT NTYG-LSEKVT TKNKFKWPL--V
  A. salmonicida    TANATNYTDL PQQVV--VTLK YDKATNWSKT DTYS-LSEKVT TKNKFQWPL--V
  A. trota          VGWATNISDT PQSGY-DVTLN YTTMSNWSKT NTYG-LSEKVS TKNKFKWPL--V
  C. septicum       -SFDLINESK TESKL-SKTFN YTTSKTVSKT DNFK-FGEKIG VKTSFKVGLEAI

                                                                   300
                         BBB      HHHHHHB BBBBBBBBBB B      B BBB-BBBBBBB
                    3333333333 3333333333 3333444444 4444444444 4444444443
  A. hydrophila     GETEL--SIEIA ANQSWASQNG GSTTTSLSQS VRPTVPARSK IPV-KIELYKA
  A. salmonicida    GETEL--AIEIA ASQSWASQKG GSTTETVSVE ARPTVPPHSS LPV-RVALYKS
  A. trota          GETEV--SIEIA ANQSWASQNG GAVTTALSQS VRPVVPARSR VPV-KIELYKA
  C. septicum       ADSKVETSFEFN AEQGWSNTNS TTETKQESTT YTATVSPQTK KRL-FLDVLGS

                                                                   350
                    BBBBBBBBBB BBBBBBBBBB BB           BBBB BBBBB    H
                    3333333333 2222222222 2222222222 2222222222 2222222222
  A. hydrophila     DISYPYEFKA DVSYDLTLSG FLRWGGNAWY THPDNRPNWN HTFVIGPYKD
  A. salmonicida    NISYPYEFKA EVNYDLTMKG FLRWGGNAWY THPDNRPTWE HTFRLGPFRG
  A. trota          NISYPYEFKA DMSYDLTFNG FLRWGGNAWH THPEDRPTLS HTFAIGPFKD
  C. septicum       QIDIPYEGKI YMEYDIELMG FLRYTGNARE DHTEDRPTVK LKFGKNG-MS

                                                                   400
                    HHHHHHHHHH H  HHHHH    HHHHHHHHH  HHHHHHHH HHH BBBBBB
                    2222222222 2222222222 2222222222 2222222222 2222222223
  A. hydrophila     KASSIRYQWD KRYIPGEVKW WDWNWWTIQQN GLSTMQNNLA RVLRPVRAGI
  A. salmonicida    QGEQHPLPVD KRYIPGEVKW WDWNWTISEY GLSTMQNNLG RVLRPIRSAV
  A. trota          KASSIRYQWD KRYLPGEMKW WDWNWAIQQN GLATMQDSLA RVLRPVRASI
  C. septicum       AEEHLKDLYS HKNINGYSEW -DWKWVDEKF GYLFKNSYDA LTSRKLGGII

                                                                   450
                    BBBBBBBBBB BBBBBBB  B B               BBB  HHH
                    3333333334 4444444444 4444444444 4444444444444 4444444444
  A. hydrophila     TGDFSAESQF AGNIEIGAPV PLAADSKVRR ARSVDG-----AGQG LRLEIPLDAQ
  A. salmonicida    TGDFYAESQF AGDIEIGQP- -QTRSAKAAQ LRSASA-----EEVA LT-SVDLDSE
  A. trota          TGDFRAESQF AGNIEIGTPV PLGSDSKVRR TRSVDG-----ANTG LKLDIPLDAQ
  C. septicum       KGSFTNINGT KIVIREGKEI PLP-DKKRRG KRSVDSLDARLQNEG IRIE-NIETQ

                    HHHHH   BBB BBBBBBB
                    4444444444 4444444444
  A. hydrophila     ELSGLGFNNV SLSVTPAANQ
  A. salmonicida    ALANEGFGNV SLTIVPVQ-
  A. trota          ELAELGFENV TLSVTPA-RN
  C. septicum       DVP--GFRLN SITYNDKKLI
```

may provide needed flexibility of the toxin when it converts from a water-soluble to a transmembrane state. The aromatic residues may play a role in various protein-protein, protein-carbohydrate and protein-lipid interactions required of the toxin. The sequences are quite hydrophilic as judged by plots of the sum of the hydrophobic moment and mean hydrophobicity profiles and there are no obvious hydrophobic stretches. One of the least homologous regions is that of the C-terminal propeptide (residues 428 to 470), perhaps because it does not play any role in the insertion process after cleavage occurs[13] and thus can tolerate random mutations more readily.

THE 3-D STRUCTURE

A ribbon representation of the crystal structure of the toxin monomer is shown in Figure 5.2. The monomer has a distinct bilobal shape with one large, elongated lobe about 100 Å long and a much smaller

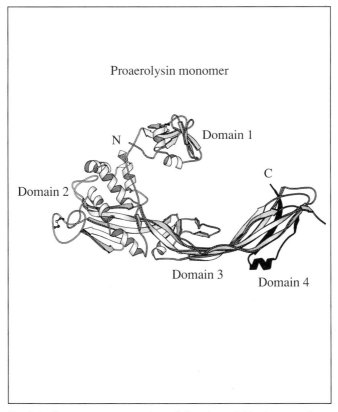

Fig. 5.2. Cartoon representation of the chain fold of proaerolysin showing the location of the domains. The activation peptide, located in Domain 4, is in dark shade. This figure was produced by the computer program MOLSCRIPT.[57]

lobe protruding out of it. We have described the polypeptide fold in terms of four domains, each of which may play different roles in the activity of the toxin (Fig. 5.2). Domain 1 (residues 1 to 82) is a small domain consisting of two beta-sheets packed against some alpha-helix. It contains a disulphide bridge between cysteine residues 19 and 75. The domain is connected to the larger lobe of the molecule by an extended stretch of chain. The major lobe of proaerolysin consists of three distinct domains which share some common beta-strands although they are discontinuous in sequence. The linear packing of these domains generates the distinctive elongated shape of the molecule. Domain 2 (residues 83 to 178, 311 to 398) is the largest domain and consists of a five-stranded sheet packed against a framework of helices. There is a disulphide bridge between cysteine residues 159 and 164. Domain 3 (residues 179 to 195, 224 to 274, 299 to 310, 399 to 409) is shaped like a box, with the front formed by a five-stranded antiparallel beta-sheet and the back by a long loop that has a small amount of two-stranded sheet. Domain 4 (196 to 223, 275 to 298, 410 to 470) is made up of a beta sandwich, with a small helix near the C-terminus. We have been unable to find structures similar to proaerolysin or its domains in the Brookhaven Protein Database,[14] leading us to conclude that proaerolysin adopts a novel fold.

The location of the secondary structure with respect to the sequence is shown in Figure 5.1. The toxin is rich in beta-sheet (42%), with 23 identifiable strands making up eight antiparallel sheets. One strand (residues 289 to 322) straddles the major lobe of the molecule almost from tip to tip and measures about 94Å from N- to C-terminus. This is the longest strand in the Brookhaven Database.[14] Three continuous regions of sequence straddle the large lobe from tip to tip. Two of these regions (162 to 209 and 288 to 324) include long strands and the third region (378 to 424) starts off as a helix in Domain 2 and continues into Domains 3 and 4 as a 55Å long strand. The strands adopt a helical twist so that the hydrophilic face of the sheet in Domain 2 undergoes a 180 degree rotation as it reaches Domain 4. The toxin also contains a significant amount of helical structure (21%) with 13 helices varying between 3 and 12 residues in length.

Analysis of the distribution of side chains in the structure indicates that there is significant clustering of aromatic residues.[8] An unusual feature of the aromatic residues is that approximately 70% of them are at least partially accessible to solvent. This includes all the histidine residues and the majority of tryptophan and tyrosine residues.

AEROLYSIN IS A DIMER

Aerolysin is both a dimer in the crystal and in solution.[15,16] A ribbon diagram of the dimer is shown in Figure 5.3. The monomer-monomer interface is fairly extensive, with an area of approximately 2,500 Å^2 buried per monomer which represents about 10% of acces-

Fig. 5.3. Ribbon diagram of the proaerolysin dimer with each monomer denoted by different shading. This figure was produced by the computer program MOLSCRIPT.[57]

sible surface area lost on subunit association. There are a total of six salt bridges and 23 potential hydrogen-bonding interactions between the monomers. There is a nonuniform distribution of polar and non-polar residues in the interfacial regions, unlike the distribution normally found in protein subunit interfaces,[18] which leads to distinct regions of nonpolar residues in the interface.[8] The unusual clasping interaction of Domain 1s shown in Figure 5.3 requires the loop from the large lobe (consisting of Domains 2, 3 and 4) be connected to the more distant Domain 1 of the dimer rather than the closest Domain 1.

The criss-crossing of Domain 1s across the molecular diad is reminiscent of the situation in diphtheria toxin where the carboxyl-terminal R domain is also crossed over and which led to a "domain-swapping" hypothesis of oligomer evolution.[19] The exchange of identical domains between monomers preserves the various interdomain interactions, but also provides important intermolecular interactions between monomers of the dimer. In the case of aerolysin, the same domain-swapping principle could also be of great significance in the formation of the oligomeric structure that precedes membrane insertion. It has been proposed that Domain 1 undergoes significant conformational changes in order to attach itself to neighboring monomers in the formation of the oligomeric state.[8]

The stability and water solubility of the toxin are due in part to its ability to dimerize. When the dimer is dissociated, rigid tertiary interactions are lost and the protein becomes vulnerable to complete destruction by proteases.[17] Interactions between extensive patches of hydrophobic residues observed in the monomer interface are probably a significant driving force in the association of the monomers. Recent work has established that the dimer can be dissociated by very low concentrations of ionic detergents pointing to the importance of hydrophobic interactions between monomers in solution.[16]

MOBILITY

Other workers have shown a correlation between conformational flexibility and the ability of a protein to cross a membrane.[20] We have already pointed out that there is a high glycine content in the molecule. This residue is often associated with flexibility. We have also located the presence of ten cavities within the protein core of the proaerolysin dimer, ranging in volume between 21 Å^3 and 114 Å^3 (unpublished results). The presence of cavities may be important to early unfolding events which unmask regions involved in oligomerization and membrane penetration.

INHIBITION BY ZINC IONS

Zinc ions can inhibit lysis of erythrocytes by a range of membrane-damaging proteins including streptolysin, *Staphylococcal* toxin, *C. perfringens* toxin and complement C9.[22,23] In the case of aerolysin, zinc has been shown to inhibit oligomerization, and in planar lipid bilayers it induces reversible closure of preformed channels.[24] Two to three zinc ions appear to bind cooperatively to close the channel and experiments with site-directed mutants suggest one binds to His 121.[25]

We have soaked proaerolysin crystals in the presence of zinc salts and used X-ray diffraction to observe four zinc ions binding to the dimer. One binds to His 121 of Domain 2, supporting the lipid bilayer results, and another two bind in the monomer-monomer interface linking His 186 in Domain 3 of one monomer and to Asp 92 in

Domain 2 of the other monomer (unpublished results). Zinc may stabilize the dimer by binding to the interface, preventing dissociation and thereby inhibit oligomerization. Since aerolysin with His 186 replaced by asparagine behaves just like wild type in planar lipid bilayers, zinc binding to this residue cannot be responsible for closure of preformed channels.[25] Other potential zinc-binding sites cannot be excluded since they may either be hidden in the crystal state, or only become exposed on channel formation.

RECEPTOR BINDING

The first stage in toxic activity is the binding of the aerolysin molecule to a receptor on the surface of the host cell's membrane.[6,7] The receptor is not essential for channel formation in vitro, as aerolysin will form channels in planar lipid bilayers.[24] The role of the receptor is probably to concentrate the protein on the cell surface and hence greatly increase the likelihood of oligomerization. The high affinity receptor for aerolysin on the rat erythrocytes has recently been identified as a 47 kDa glycosylphosphatidylinositol-anchored glycoprotein (S. Cowell and J.T. Buckley, unpublished results).[7] Earlier work showed that both proaerolysin and aerolysin bound with equal but lower affinity towards glycophorin.[6] A mutant protein in which His 332 is replaced by aspargine has reduced affinity for rat erythrocytes.[27] His 332 is located on a surface loop above the beta-sheet in Domain 2, close to a cluster of aromatic residues centered around Trp 373 (Fig. 5.4). Such aromatic clusters are often involved in protein-protein[18] and protein-carbohydrate interactions.[29] Several other amino acid changes in this region have since been shown to reduce binding to rat erythrocytes (S. Cowell & J.T. Buckley, unpublished observations).

ACTIVATION

Proaerolysin is completely unable to oligomerize or form channels.[24,30,31] It can be activated by a variety of proteases.[32] Trypsin has been shown by mass spectrometry to cut the protein after Lys 427.[12,33] *Aeromonas* proteases likely act in the same region.[32] The cleavage site region is not visible in the final electron density maps, presumably because of the high mobility of the long loop in Domain 4 in which it resides. The region C-terminal to the site of cleavage, that can be seen in the crystal structure (residues 440 to 470; see Fig. 5.2), covers a large area of approximately 1,100 Å^2 in Domain 4. Dissociation of this region would expose a large hydrophobic patch.[8] The distal peptide is involved in extensive hydrophobic contacts, and it forms 2 salt bridges and 26 potential hydrogen bonds with the rest of Domain 4.[8] Despite this, it has now been shown that the cleaved peptide does separate from the activated toxin and plays no further role in the process of channel formation.[13]

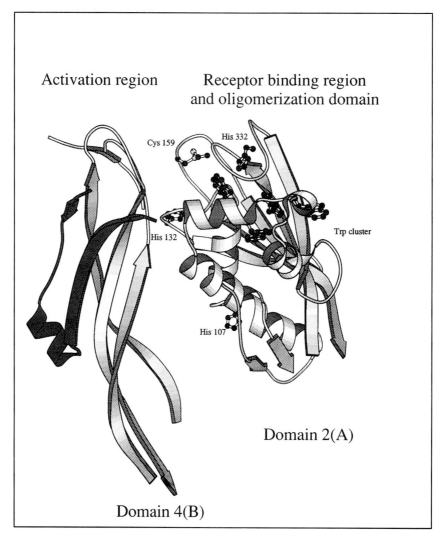

Fig. 5.4. Ribbon picture of the activation and receptor binding regions of the proaerolysin dimer. Key residues implicated in receptor binding and oligomerization are indicated. This figure was produced by the computer program MOLSCRIPT.[57]

The mature toxin appears to have a similar structure to the protoxin, based on the observation that activation by proteases does not lead to major changes in secondary or tertiary structure, as judged by circular dichroism,[12] and from crosslinking and ultracentrifugation studies that indicate that the dimer does not dissociate on activation.[16]

OLIGOMERIZATION

Oligomerization is a critical step for channel formation.[24,27,32,34] Mutagenesis has been used to identify regions involved in oligomerization. Replacing His 107 reduces the ability of the protein to oligomerize and replacing His 132 with asparagine or aspartic acid prevents oligomerization almost completely[25,27] in spite of the fact that both mutant toxins bind to erythrocytes with wild-type affinity and both are as stable as wild-type toxin.[27] In the case of His 132, it has recently been shown that the residue must be protonated for oligomerization to occur.[26] Both these histidines are on the surface of the protein and thus it is unlikely that the mutations disturb the structure in any way. The crystal structure of the His 132 mutant has confirmed this idea (unpublished results). Changing Cys 159 to serine lowers the oligomerization rate (unpublished results). The cysteine and the two histidines lie on the same face, adjacent to the beta-sheet in Domain 2 (see Fig. 5.4).

A tryptophan-rich region has been implicated in oligomerization (Fig. 5.4). Changing Trp 371 or Trp 373 to leucine lowers the protein concentration at which oligomerization takes place by 5- to 10-fold.[34] The mutations reduce the stability of the protoxin dimers so that they are more sensitive to dissociation by SDS.[16] Reference to the structure shows that these two tryptophans form part of an aromatic bowl of residues adjacent to the beta-sheet in Domain 2, but opposite the face described above. The bowl consists of residues Tyr 125, Trp 329, Tyr 330, Trp 339, His 341, Phe 343, Trp 370, Trp 371 and Trp 373. Both Trp 371 and Trp 373 are totally buried within the aerolysin hydrophobic core. The creation of cavities from the mutations may lower the energy barrier for conversion of the dimer into the oligomer. The near U.V. spectra of the two Trp mutants are only modestly different from the spectrum of wild-type, and there is no measurable difference in the rate of secretion of either of the mutant toxins, indicating that any changes in the structure of the protein as a result of replacing these residues have been minor. The crystal structures of the Trp 371 and 373 mutants indicate no significant conformational differences in the proaerolysin structure are caused by these mutations (unpublished results).

The far U.V. circular dichroism spectra of the dimeric and oligomeric forms of aerolysin are very similar, suggesting there is little difference in secondary structure.[17] The near U.V. CD spectrum provides evidence of a change in the position or environment of some aromatic side-chains. Images of aerolysin oligomers derived from electron microscopy (see below) are consistent with a heptameric model in which the monomers are oriented in a parallel fashion as distinct to the antiparallel packing observed in the dimer.[8,35] This model implies that the dimers must dissociate to form oligomers. In support of this, we have recently shown that the mutant M41C, which results in the formation of a disulphide bridge between the two monomers in the aerolysin dimer, is completely inactive in the oxidized state.[28]

CHANNEL FORMATION

Aerolysin forms voltage-gated channels in planar lipid bilayers that are slightly anion selective.[24,25] Voltage-clamp experiments indicate that aerolysin channels appear to have a well-defined conformation and uniform size and remain open between -70 and 70 mV.[24] Image analysis from electron micrographs of two-dimensional crystals of aerolysin in lipid membranes has yielded a low resolution model of the aerolysin channel (Fig. 5.5).[35] The model consists of a central cylindrical-shaped density of outer diameter about 46 Å encircling a water-

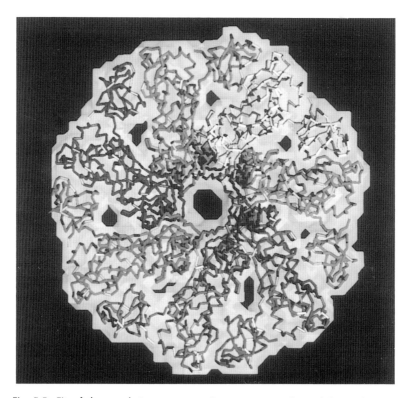

Fig. 5.5: Fit of the aerolysin monomers into an image derived from electron microscopy of the aerolysin ion channel.[35] Each monomer is shown as an alpha-carbon backbone and in different colors. The view is looking directly down the channel towards the membrane. This figure was produced using the program O.[59] See color figure in insert.

filled channel 17 Å in diameter. The cylinder is surrounded by seven arms of density, each made up of two unequal-sized domains. The total diameter of the disk-like complex is about 140 Å and is between 30 and 40 Å thick. The disk lies above the membrane surface at a distance of about 20 Å. We have convincingly fitted seven aerolysin monomers into the e.m. image so that Domains 1 and 2 fit into the disk and Domains 3 and 4 connect the disk to the lipid bilayer as shown in Figure 5.5.[8] The modeling predicts that Domain 4 would span the lipid bilayer so as to form a beta-barrel topology reminiscent of porins.[36,37]

MECHANISM OF TOXIC ACTION

We have recently proposed a detailed mechanism for membrane insertion based on the crystal structure and various biochemical data.[8] The toxin is secreted as a water-soluble dimer which subsequently is concentrated on the surface of the target cell by virtue of its binding to its receptor. Dissociation of the dimer and removal of the activation peptide by extracellular proteases would unmask three-dimensional hydrophobic patches for use in oligomerization and membrane insertion. It is likely that these steps may be promoted by both the electrostatic and detergent properties of the nearby lipid bilayer. Dimer dissociation and oligomerization of aerolysin are probably concerted events as the protease-sensitive regions of aerolysin never seem to be exposed.[16] Our results show that the Domain 4 end of the mature toxin would be highly hydrophobic and thus would be a good candidate for initial insertion into the membrane bilayer. Our model of the channel form of the protein supports this notion.

RELATIONSHIP TO OTHER TOXINS

There are a number of functional similarities between the mechanism of cell killing by aerolysin and by other toxins. The requirement of proteolytic activation for toxin activity is not unique to aerolysin. Activation is also necessary for *Clostridium septicum* alpha-toxin,[39] *Pseudomonas aeruginosa* cytotoxin,[40] *Clostridium botulinum* C2 toxin,[41] the protective antigen component of anthrax toxin[42] and yeast killer toxins.[43] In addition, several intracellularly-acting toxins such as diphtheria toxin, *P. aeruginosa* exotoxin A and shiga toxin depend on proteolytic cleavage at specific sites within the toxin to become active.[44] Nor is the formation of heptameric channels unique to aerolysin; both *S. aureus* alpha toxin[50] and the anthrax protective antigen form such channels.[51] Although the oligomeric state of the pneumolysin channel is much larger than that of aerolysin and the other two toxins, a similar four domain structure is observed in images obtained from electron micrographs of pneumolysin oligomers.[52]

In addition to the common feature of heptameric oligomers, aerolysin shares a surprising number of other features with *Staphylococcus aureus* alpha-toxin. Each is secreted as a water-soluble protein and each heptamerizes to form pores in target cells. Other similarities include the essentiality of histidine residues in oligomerization and pore formation, pores of similar dimensions that are slightly anion-selective and can be inhibited by two to three zinc ions per monomer.[45-48] As well, each contains no hydrophobic stretches of sequence capable of spanning a bilayer, and circular dichroism studies predict a high proportion of beta-sheet.[49]

The similarities go even further beyond the functional ones discussed above. There is a short stretch of sequence in aerolysin that has been reported to be very similar to alpha-toxin,[9,53] *Pseudomonas aeruginosa* cytotoxin,[40] *C. septicum* alpha-toxin,[54] *C. perfringens* epsilon toxin[55] and the family of thiol-activated toxins (for example, perfringolysin O[56]) (Fig. 5.6). This sequence is located in Domain 3 of proaerolysin. Any possible functional significance is unclear at this stage. The alpha-toxin from *C. septicum* possesses significant sequence similarity to the region of aerolysin spanning Domains 2 to 4. Since *C. septicum* is a Gram-positive bacterium, whereas *Aeromonas* sp are Gram-negative, the evolutionary relationship between the two toxins presents a fascinating puzzle.

CONCLUSIONS AND FUTURE STUDIES

The passage from a water-soluble to a membrane-competent state is an essential step in the activity of many toxins.[38] It seems very likely that the exposure of hydrophobic regions is a key event in the transition. Previous structural studies of other toxins have demonstrated that linear regions of hydrophobic sequence, in the form of buried helical hairpins, can become exposed as a result of a pH-triggered conformational change at the membrane surface (see chapters 2 to 4). The structure of proaerolysin points to a very different mechanism of exposure. In this case dissociation of the dimer and peptide cleavage appear to play a major role in exposing hydrophobic surfaces. It will be fascinating to see whether other toxins will be discovered that use similar mechanisms.

At this stage, we have a model of proaerolysin and a hypothetical, albeit based on electron microscopy images, model of the aerolysin channel. Future work is being directed towards structural studies of the activated form of aerolysin and the channel itself. In the meantime, the present models are providing a good basis for the rational design of various experiments such as site-directed mutagenesis to dissect the pathway from water-soluble to membrane channel.

ACKNOWLEDGMENTS

We thank Dr. Franc Pattus for many fruitful discussions and Dr. Kevin Leonard for providing us with a computerized version of the aerolysin channel image. Michael W. Parker is a Wellcome Australian

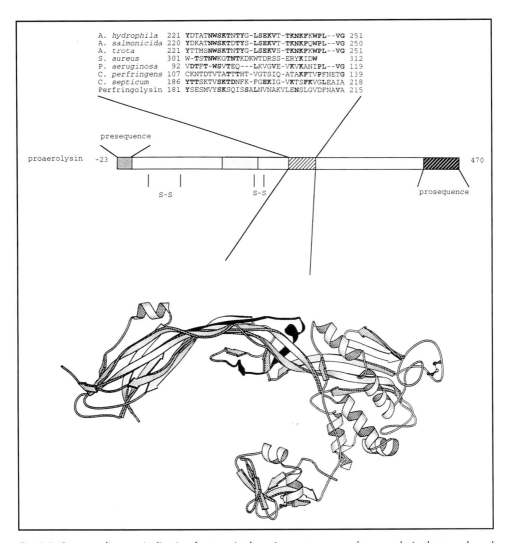

Fig. 5.6. Cartoon diagram indicating features in the primary structure of proaerolysin that are shared with other proteins. Sequences were chosen and extracted from the SWISSPROT database on the basis of significant hits found with the program SCRUTINEER.[58] Invariant residues amongst proaerolysins are shown in bold type. The alignment includes A. hydrophila proaerolysin, A. salmonicida proaerolysin, A. trota proaerolysin, S. aureus alpha toxin, P. aeruginosa cytotoxin, C. perfringens epsilon toxin, C. septicum alpha toxin and perfringolysin O. This region maps onto the darkly-shaded region indicated in the ribbon picture of the proaerolysin monomer.

Senior Research Fellow and acknowledges support from the National Health and Medical Research Council. J. Thomas Buckley acknowledges support from the National Science and Engineering Research Council of Canada. F. Gisou van der Goot acknowledges support from the Swiss National Science Foundation.

REFERENCES

1. Altwegg M, Geiss HK. *Aeromonas* as a human pathogen. CRC Crit Rev Microbiol 1989; 16:253-86.

2. Chakraborty T, Huhle B, Hof H et al. Marker exchange mutagenesis of the aerolysin determinant in *Aeromonas hydrophila* demonstrates the role of aerolysin in *A. hydrophila*-associated infections. Infect Immun 1987; 55:2274-80.

3. Howard SP, Buckley JT. Protein export by a Gram-negative bacterium: production of aerolysin by *Aeromonas hydrophila*. J Bacteriol 1985; 161:1118-24.

4. Pugsley AP. The complete general secretory pathway in Gram-negative bacteria. Microbiol Rev 1993; 57:50-108.

5. Jiang B, Howard SP. The *Aeromonas hydrophila* exeE gene, required both for protein secretion and normal outer membrane biogenesis, is a member of a general secretion pathway. Molec Microbiol 1992; 6:1351-61.

6. Howard SP, Buckley JT. Membrane glycoprotein receptor and hole-forming properties of a cytolytic protein toxin. Biochemistry 1982; 21:1662-67.

7. Gruber HJ, Wilmsen HU, Cowell S et al. Isolation and reconstition of the receptor for the hemolytic toxin aerolysin from rat red blood cell membranes. Molec Microbiol 1994; 14:1093-11.

8. Parker MW, Buckley JT, Postma JPM et al. Structure of the *Aeromonas* toxin proaerolysin in its water-soluble and membrane-channel states. Nature 1994; 367:292-95.

9. Howard SP, Garland WJ, Green MJ et al. Nucleotide sequence of the gene for the hole-forming toxin aerolysin of *Aeromonas hydrophila*. J Bacteriol 1987; 169:2869-71.

10. Husslein V, Huhle B, Jarchau T et al. Nucleotide sequence and transcriptional analysis of the *aerCaerA* region of *Aeromonas sobria* encoding aerolysin and its regulatory region. Molec Microbiol 1988; 2:507-17.

11. Hirono I, Aoki T. Cloning and characterization of three hemolysin genes from *Aeromonas salmonicida*. Microb Pathog 1993; 15:269-82.

12. van der Goot FG, Lakey J, Pattus F et al. Spectroscopic study of the activation and oligomerization of the channel-forming toxin aerolysin: identification of the site of proteolytic activation. Biochemistry 1992; 31:8566-70.

13. van der Goot FG, Hardie KR, Parker MW et al. The C-terminal peptide produced upon proteolytic activation of the cytolytic toxin aerolysin is not involved in channel formation. J Biol Chem 1994; 269:30496-501.

14. Bernstein FC, Koetzle TF, Williams GJB et al. The Protein Data Bank: a computer based archival file for macromolecular structures. J Mol Biol 1977; 112:535-42.

15. Tucker AD, Parker MW, Tsernoglou D et al. Crystallization of a proform of aerolysin, a hole-forming toxin from *Aeromonas hydrophila*. J Mol Biol 1990; 212:561-62.

16. van der Goot FG, Ausio J, Wong KR et al. Dimerization stabilizes the pore-forming toxin aerolysin in solution. J Biol Chem 1993; 268:18272-79.

17. van der Goot FG, Pattus F, Parker MW et al. The cytolytic toxin aerolysin, from the soluble form to the transmembrane channel. Toxicology 1994; 87:19-28.

18. Argos P. An investigation of protein subunit and domain interfaces. Protein Engng 1988; 2:101-13.

19. Bennett MJ, Choe S, Eisenberg D. Domain swapping: entangling alliances between proteins. Proc Natl Acad Sci USA 1994; 91:3127-31.

20. Verner K, Schatz, G. Protein translocation across membranes. Science 1988; 241:1307-13.

22. Montgomery DW, Don LK, Zukoski CF et al. The effect of zinc and other metals on complement hemolysis of sheep red blood cells in vitro. Proc Soc Exp Biol Med 1974; 145:263-67.

23. Avigad LS, Bernheimer AW. Inhibition by zinc of hemolysis induced by bacterial and other cytolytic agents. Infect Immun 1976; 13:1378-81.

24. Wilmsen H-U, Pattus F, Buckley JT. Aerolysin, a hemolysin from *Aeromonas hydrophila*, forms voltage-gated channels in planar lipid bilayers. J Membr Biol 1990; 115:71-81.

25. Wilmsen H-U, Buckley JT, Pattus F. Site-directed mutagenesis at histidines of aerolysin from *Aeromonas hydrophila*: a lipid planar bilayer study. Molec Microbiol 1991; 5:2745-51.

26. Buckley JT, Wilmsen H-U, Lesieur C et al. Protonation of His-132 promotes oligomerization of the channel-forming toxin aerolysin. Biochemistry 1995; in press.

27. Green MJ, Buckley JT. Site-directed mutagenesis of the hole-forming toxin aerolysin: studies on the role of histidines in receptor binding and oligomerization of the monomer. Biochemistry 1990; 29:2177-80.

28. Hardie KR, Schulze A, Parker MW et al. *Aeromonas sp.* secrete proaerolysin as a folded dimer. Molec Microbiol 1995; in press.

29. Vyas NK. Atomic features of protein-carbohydrate interactions. Curr Opin Struct Biol 1991; 1:732-40.

30. Kozaki S, Kato K, Asao T et al. Activities of *Aeromonas hydrophila* hemolysins and their interaction with erythrocyte membranes. Infect Immun 1987; 55:1594-99.

31. Chakraborty T, Schmid A, Notermans S et al. Aerolysin of *Aeromonas sobria*: evidence for formation of ion-permeable channels and comparison with alpha-toxin of *Staphylococcus aureus*. Infect Immun 1990; 58:2127-32.

32. Garland WJ, Buckley JT. The cytolytic toxin aerolysin must aggregate to disrupt erythrocytes, and aggregation is stimulated by human glycophorin. Infect Immun 1988; 56:1249-53.

33. Howard SP, Buckley JT. Activation of the hole-forming toxin aerolysin by extracellular processing. J Bacteriol 1985; 163:336-40.

34. van der Goot FG, Pattus F, Wong KR et al. Oligomerization of the channel-forming toxin aerolysin precedes insertion into lipid bilayers. Biochemistry 1993; 32:2636-42.

35. Wilmsen H-U, Leonard K, Tichelaar W et al. The aerolysin membrane channel is formed by heptamerization of the monomer. EMBO J 1992; 11:2457-63.

36. Weiss MS, Abele U, Weckesser J et al. Molecular architecture and electrostatic properties of a bacterial porin. Science 1991; 254:1627-30.

37. Cowan SW, Schirmer T, Rummel G et al. Crystal structures explain functional properties of two *E. coli* porins. Nature 1992; 358:727-33.

38. Parker MW, Tucker AD, Tsernoglou D et al. Insights into membrane insertion based on studies of colicins. Trends Biochem Soc 1990; 15:126-29.

39. Ballard J, Sokolov Y, Yuan W-L et al. Activation and mechanism of *Clostridium septicum* alpha toxin. Molec Microbiol 1993; 10:627-34.

40. Hayashi T, Kamio Y, Hishinuma F et al. *Pseudomonas aeruginosa* cytotoxin: the nucleotide sequence of the gene and the mechanism of activation of the protoxin. Molec Microbiol 1989; 3:861-68.

41. Schmid A, Benz R, Just I et al. Interaction of *Clostridium botulinum* C2 toxin with lipid bilayer membranes. J Biol Chem 1994; 269:16706-11.

42. Milne JC, Collier RJ. pH-dependent permeabilization of the plasma membrane of mammalian cells by anthrax protective antigen. Molec Microbiol 1993; 10:647-53.

43. Suzuki C, Nikkuni S. The primary and subunit structure of a novel type of killer toxin produced by a halotolerant yeast, *Pichia farinosa.* J Biol Chem 1994; 269:3041-46.

44. Montecucco C, Papini E, Schiavo G. Bacterial protein toxins penetrate cells via a four-step mechanism. FEBS Lett 1994; 346:92-98.

45. Arbuthnott JP, Freer JH, Bernheimer AW. Interaction of *staphylococcal* alpha toxin with artificial and natural membranes. J Bacteriol 1968; 95:1153-68.

46. Füssle R, Bhakdi S, Sziegoleit A et al. On the mechanism of membrane damage by *Staphylococcus aureus* α-toxin. J Cell Biol 1981; 91:83-94.

47. Menestrina G. Ionic channels formed by *Staphylococcus aureus* alpha toxin: voltage dependent inhibition by divalent and trivalent cations. J Membr Biol 1986; 90:177-90.

48. Pederzolli C, Cescatti L, Menestrina GJ. Chemical modification of *S. aureus* α-toxin by diethylpyrocarbonate: role of histidines in its membrane damaging properties. J Membr Biol 1991; 119:41-52.

49. Tobkes N, Wallace BA, Bayley H. Secondary structure and assembly mechanism of an oligomeric channel protein. Biochemistry 1985; 24:1915-20.

50. Gouaux JE, Braha O, Hobaugh MR et al. Subunit stoichiometry of staphyloccal a-hemolysin in crystals and on membranes: A heptameric transmembrane pore. Proc Natl Acad Sci USA 1994; 91:12828-31.

51. Milne JC, Furlong D, Hanna PC et al. Anthrax protective antigen forms oligomers during intoxication of mammalian cells. J Biol Chem 1994; 269:20607-12.

52. Morgan PJ, Hyman SC, Byron O et al. Modeling the bacterial protein toxin, pneumolysin, in its monomeric and oligomeric form. J Biol Chem 1994; 269: 5315-20.

53. Gray GS, Kehoe M. Primary sequence for the α-toxin gene from *Staphylococcus aureus* Wood 46. Infect Immun 1984; 46:615-18.

54. Imagawa T, Dohi Y, Higashi Y. Cloning, nucleotide sequence and expression of a hemolysin gene of *Clostridium septicum*. FEMS Microbiol Letters 1994; 117:287-92.

55. Hunter SEC, Clarke IN, Kelly CD et al. Cloning and nucleotide sequencing of the *Clostridium perfringens* epsilon-toxin gene and its expression in *Escherichia coli*. Infect Immun 1992; 60:102-10.

56. Tweten RK. Nucleotide sequence of the gene for perfringolysin O (theta-toxin) from *C. perfringens*: significant homology with the genes for streptolysin O and perfringolysin. Infect Immun 1988; 56:3235-40.

57. Kraulis PJ. MOLSCRIPT: a program to produce both detailed and schematic plots of protein structures. J Appl Cryst 1991; 24:946-50.

58. Sibbald PR, Argos P. Scrutineer: a computer program that flexibly seeks and describes motifs and profiles in protein sequence databases. CABIOS 1990; 6:279-88.

59. Jones TA, Zou JY, Cowan SW et al. Improved methods for building models in electron density maps and the location of errors in these models. Acta Cryst 1991; A47:110-19.

======= CHAPTER 6 =======

THE ANTHRAX TOXIN

Carlo Petosa and Robert C. Liddington

Anthrax is a disease known since antiquity[1] and one of the first bacterial infections whose etiology was definitively established. The disease is caused by the Gram-positive, aerobic, spore-forming *Bacillus anthracis*, first isolated in 1877 by Robert Koch.[2] The study of anthrax led to the establishment of Koch's postulates, a set of criteria for identifying an organism as the causative agent of a specific infection.[2] Louis Pasteur's use of heat-inactivated anthrax cultures to immunize against the disease is generally credited as the first instance of a bacterial vaccine.[3] Anthrax is primarily a disease of herbivorous animals, particularly sheep and cattle.[4] Humans may acquire the disease from infected animals, typically as a cutaneous infection characterized by black pustules[5] (whence the naming of the disease after the Greek word for "coal"). The pulmonary infection known as wool-sorter's disease results from the inhalation of anthrax spores, often as a result of handling contaminated raw wool, hides or animal hair, and can lead to death within days.[6-8] Though increasingly rare in human populations, anthrax remains of interest for several reasons, including its continuing incidence in animal populations,[9] interest in improving the efficacy of the human vaccine,[4,10] the threat of its use as a weapon of biological warfare (see ref. 11), its potential applications in the development of new therapeutic strategies such as targeted toxins,[12] and as an experimental system for studying molecular pathogenesis.[13]

In this chapter, we first review current knowledge of the biochemistry of intoxication by anthrax. We then describe the components of the anthrax exotoxin, including the "protective antigen" (PA), whose crystal structure we have recently determined. We summarize what has been learned from ion channel studies and electron microscopy and examine the sequence homology of PA with the iota-Ib toxin of *Clostridium perfringens*. Finally we speculate on events that follow the proteolytic activation of PA and propose a model for its membrane-inserting oligomeric form.

Protein Toxin Structure, edited by Michael W. Parker. © 1996 R.G. Landes Company.

INTOXICATION BY ANTHRAX

Two factors account for the virulence of *B. anthracis*: a three-protein exotoxin, and a poly-D-glutamic acid capsule that protects the bacillus against phagocytosis and other bactericidal components of host sera.[14] Production of these factors is controlled by two large extrachromosomal plasmids: pXO1 (184 kb), encoding the tripartite exotoxin,[15] and pXO2 (97 kb), encoding the antiphagocytic capsule.[16,17] Strains lacking either plasmid have reduced virulence.[18]

The three exotoxin proteins are the edema factor (EF), the lethal factor (LF), and the protective antigen (PA), collectively called "anthrax toxin."[19] The most immunogenic is PA[20,21] which derives its name from its use in vaccines. The mature polypeptides have molecular weights in the range 83-90,000 (Table 6.1), and their genes have all been cloned and sequenced.[22-25] Transcription requires the presence of bicarbonate and is regulated by an activator and a repressor encoded by pXO1[13,26-29] The exotoxin proteins lack toxic activity when administered individually,[30] but two binary combinations are toxic: co-injection of PA with EF ("edema toxin") produces edema in the skin of animals and inhibits the function of phagocytic cells,[31] while the combination of PA and LF ("lethal toxin") induces death in experimental animals[32,33] (Table 6.1). In the "A-B" toxin nomenclature*, PA can be considered the B moiety with LF and EF as alternative A moieties. In common with certain clostridial toxins, anthrax toxin is classified as a binary A-B toxin because the two moieties exist as separate gene products, not as domains within a polypeptide.[35]

While the action of edema toxin appears to be responsible for the clinical manifestations of cutaneous anthrax[36,37] the shock and death resulting from systemic anthrax are primarily due to the effects of the lethal toxin: strains unable to produce EF are only ten-fold less virulent than wildtype,[30] whereas strains deficient in LF are nonlethal;[28,30,38] and animals treated with lethal toxin show symptoms closely resembling those characterizing infection with the bacillus.[14,36,39] Mediating the effects of lethal toxin is the macrophage. In vitro, only cells of the macrophage lineage are affected by exposure to lethal toxin:[19] high concentrations lead to cytolysis,[19,40] likely mediated by the overproduction of reactive free radical oxygen intermediates,[41] while lower concentrations can induce cytokine overproduction, proposed to be the cause of the systemic shock and death associated with anthrax.[42] In a decisive study, Hanna and colleagues showed that mice depleted of macrophages are resistant to lethal-toxin challenge, and regain sensitivity upon injection of toxin-sensitive macrophages.[42]

---

* *Many bacterial toxins have been described as "A-B" toxins because they contain one moiety, A, bearing enzymatic activity toxic to the host cell, and another, B, responsible for binding to the cell surface and delivery of the A moiety into the cytosol.[34]*

Table 6.1. The components of the anthrax toxin

Component	Abbr	#res (kDa)	activity	toxicity	immunogenic?
A chains					
edema factor	EF	767 (89)	adenylate cyclase, calmodulin dependent	inactive	–
lethal factor	LF	776 (90)	cytolytic to macrophages Zn-dependent metalloprotease(?)	inactive	±
B chain					
protective antigen	PA	735 (83)	binds receptor, cleaved into 2 fragments: PA$_{20}$ – released from cell surface PA$_{63}$ – binds EF/LF – forms heptamer – forms membrane channels – translocates EF/LF across bilayer	inactive	+++
edema toxin	PA + EF			local edema	+++
lethal toxin	PA + LF			lethal	+++
anthrax toxin	PA + EF + LF			edema, lethal	+++

Abstracted in part from references 20 and 33.

The events which occur during intoxication of the cell are depicted in Figure 6.1. The first step involves the binding of PA to a receptor on the host cell surface. The receptor is found in a variety of mammalian cell lines[14] and appears to be a membrane protein of 85-90 kDa with a high affinity for PA (K_d ~ 1 nM).[43] Once bound to the receptor, PA is cleaved by a cell-surface protease believed to be furin,[44,45] a ubiquitous, subtilisin-like endoprotease.[46] Proteolysis releases an N-terminal 20 kDa fragment, PA_{20}, from the cell surface[47] and exposes a high-affinity binding site (K_d = 10 pM)[14] for LF or EF on the

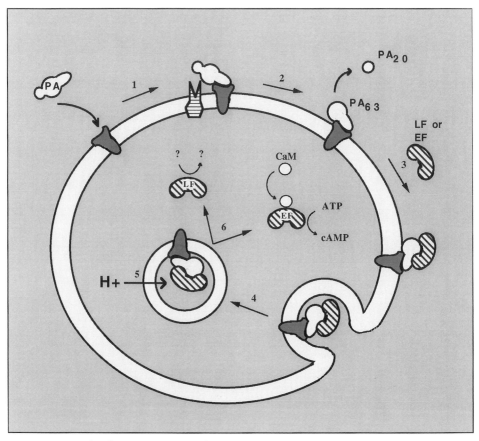

Fig. 6.1. Steps of cell intoxication. (1) The protective antigen binds to a cell surface receptor. (2) Cleavage by a cell surface protease releases an N-terminal 20 kDa fragment, PA_{20}. (3) The lethal or edema factor can then bind to the C-terminal fragment, PA_{63}. (4) The complex undergoes receptor-mediated endocytosis. (5) Acidification of the vesicle leads to insertion of PA_{63} into the membrane, and (6) translocation of EF or LF into the cytosol. EF can then bind calmodulin and catalyze the formation of cAMP. LF is believed to be a zinc-dependent metalloprotease but a substrate has not yet been found. Not shown are the heptameric and membrane-inserted forms of PA_{63}.

63 kDa fragment, PA_{63}. PA_{63} then binds LF or EF, and the entire complex undergoes receptor-mediated endocytosis.[48] Acidification of the vesicle leads to insertion of PA_{63} into the endosomal membrane[13] and translocation of EF or LF bilayer into the cytosol where they exert their toxic effects—adenylate cyclase activity in the case of EF,[37] and (probably) proteolytic activity in the case of LF.[49] It has been recently shown that PA_{63} forms a heptameric, ring-like structure in vitro,[50] which is probably the membrane-inserting form in vivo. It is not yet clear if oligomerization occurs at the cell surface or after acidification of the endosome, and the precise sequence of EF/LF binding, oligomerization and membrane insertion has yet to be established. The activities of the three exotoxin proteins are summarized in Table 6.1.

COMPONENTS OF THE ANTHRAX TOXIN

THE EDEMA FACTOR (EF)

EF contains 767 residues and is a calmodulin-dependent adenylate cyclase.[37,51] The elevation of cAMP levels following the entry of EF into the cytosol accounts for the effects of the edema toxin. The domain organization of EF is shown in Figure 6.2. Catalytic activity resides in the C-terminal two-thirds of the polypeptide,[52] a region which bears homology to the adenylate cyclase of *Bordetella pertussis*.[24,25] The

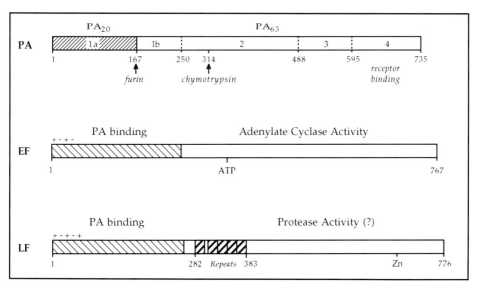

Fig. 6.2. Domain organization of the anthrax toxin components. ATP, the consensus nucleotide binding site (GxxGKS/T) in residues 314-321 of EF. Zn, the consensus zinc binding site (HExxH) in residues 686-690 of LF. The N-termini of EF and LF are characterized by a large number of charged residues (17 of the first 28 residues of EF and 38 of the first 65 in LF are charged).

ATP-binding site is believed to lie within residues 314-321, which forms a consensus nucleotide binding site (GxxxxGKS/T).[53,54] Less well defined is the calmodulin binding site: a synthetic peptide corresponding to residues 499-532 and predicted to form an amphipathic helix binds calmodulin in vitro,[55] while cross-linking experiments place calmodulin in close proximity to residues 613-767.[52] The N-terminal 300 residues bear significant homology to the N-terminus of LF[22] and are responsible for binding PA,[56] accounting for the ability of EF and LF to bind PA competitively.[14] In both EF and LF, this region is characterized by a large number of charged residues, although their significance is not known.

THE LETHAL FACTOR (LF)

Current knowledge of the domain organization of LF is summarized in Figure 6.2. Chimeric proteins consisting of the first 254 residues of LF fused to the catalytic domains of heterologous toxins, such as exotoxin A, are toxic to cells in the presence of PA,[57-60] demonstrating that this domain is sufficient for binding to PA_{63} and translocation to the cytoplasm. Insertion mutations within the C-terminal one-third of the protein eliminate toxicity without affecting binding,[61] suggesting that toxic activity occurs within this region. An intriguing sequence from residues 282-383 contains five imperfect repeats of 19 residues sharing 42-68% identity,[22] but its significance is unknown. LF is able to bind zinc in solution[49,62] and there is strong evidence that it is a Zn-dependent metalloprotease: residues 686-690 match the consensus sequence, HExxH, found in several Zn-binding metalloproteases;[49,63,64] and the toxic effects of LF can be significantly reduced by incubation with certain protease inhibitors, or completely eliminated by point mutation of the consensus Glu or His residues.[49] The HExxH motif is also present in the botulinum and tetanus toxins,[65,66] which possess proteolytic activities specific for proteins involved in the docking/fusion process of vesicle exocytosis.[67] A substrate for LF remains to be identified.

THE PROTECTIVE ANTIGEN (PA)

The crystal structure of PA has been determined at high resolution (manuscript in preparation), and is illustrated in Figure 6.3. The molecule is about 100 Å tall by 50-70 Å wide and 30-40 Å deep. The 735 residues of PA comprise four domains organized predominantly into β-sheets with only a few short helices (≤ 4 turns long). The secondary structure assignment of the PA sequence is shown in Figure 6.4. Domains 1 and 2 are approximately equal in size, with ~250 residues each. Domain 1 contains PA_{20}, the fragment removed by the cell surface protease. The function of domain 2 is not known, but it may play a role in membrane insertion and/or oligomerization. Domain 3 is the smallest of the domains with about 100 residues; its function is

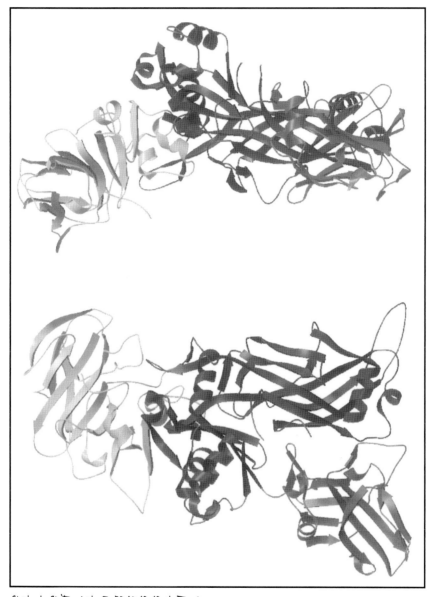

Fig. 6.3. The crystal structure of PA. The domains are colored as follows: 1—yellow, 2—red, 3—blue, 4—green. The molecule has dimensions of roughly 100 Å x 65 Å x 35 Å. Domain 1 contains the residues comprising PA_{20}; domain 2 is the only domain long enough to span a membrane; the function of domain 3 is unknown; and domain 4 is believed to bind the cell-surface receptor. See text and Table 6.2 for further details. See color figure in insert.

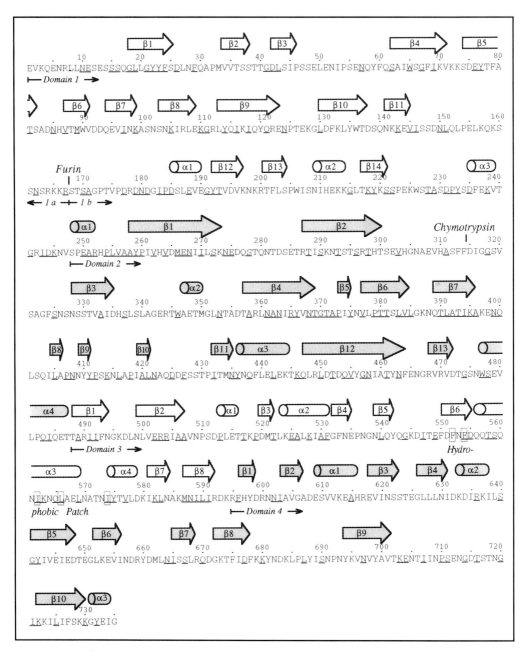

Fig. 6.4. Secondary structure of PA. Strands and helices are shown as arrows and cylinders, respectively. Underlined residues are identical to those in the sequence of iota-Ib. The furin and chymotrypsin cleavage sites are indicated. Residues comprising the hydrophobic patch of domain 3 are boxed.

also currently unknown. Domain 4 contains the C-terminal 140 residues and is probably the receptor-binding domain. The domain structures are discussed in detail below and are summarized in Table 6.2.

DOMAIN 1

Domain 1 consists of residues 1-249. Of these, the first 220 fold into 14 β strands arranged in an antiparallel fashion in two sheets of unequal size (Fig. 6.5). Protease cleavage on the cell surface occurs at the sequence Arg164-Lys-Lys-Arg167[68] which forms an exposed loop (dotted line in Fig. 6.5) between two strands of the larger sheet. Mutation of the cleavage site to prevent proteolysis results in a loss of toxic activity.[69] Residues making up the PA_{20} fragment are indicated in Figure 6.5. These residues form a β-sandwich composed of the small sheet and 6 of the 9 strands of the large sheet. That the cleavage site does not occur between structural domains but within a domain explains why it is possible to "nick" PA with trypsin without causing dissociation of the two resulting polypeptide chains.[14] The remaining three strands of the large sheet (residues 190-220) pack against the last 30 residues of the domain, which seem to have little regular secondary structure. Domain 1 can thus be viewed as having two subdomains: 1a (residues 1-167), which consists of a 5 + 6 stranded β sandwich, and gives rise to the PA_{20} fragment following proteolytic cleavage, and 1b, (residues 168-249), which consists of a 3-stranded sheet, 3 short helices and a random coil. Removal of PA_{20} would not only disrupt hydrogen bonding interactions in subdomain 1b by tearing

Table 6.2. The domains of PA

Domain	Res.	Structure	% identity with iota–Ib	Comments
1	1–249	2 β sheets	29	1–167 = PA_{20} –removed by protease 168–249 – bind EF/LF?
2	250–487	β barrel	40	– loop (304–321) implicated in oligomerization – length suggests role in membrane insertion – pH dependent change (res. 342–55)
3	488–594	α + β	37	– exposed hydrophobic patch – role in oligomerization?
4	595–735	2 β sheets	21	– binds cell surface receptor

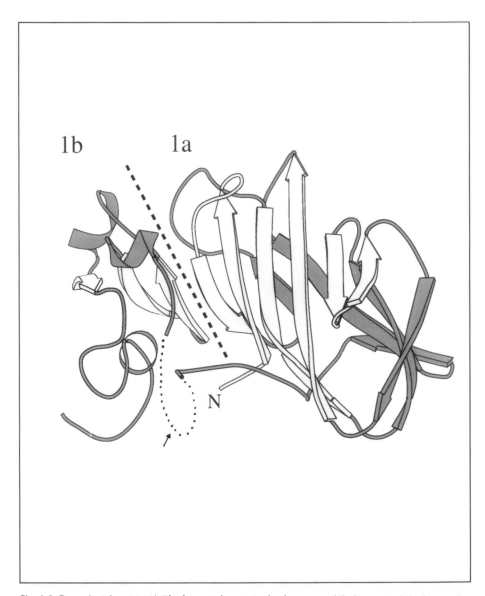

Fig. 6.5. Domain 1 (res 1-249). The large β-sheet is in the foreground (light gray) while the smaller sheet is in the background (dark gray). The N-terminus is marked. The arrow indicates the cleavage site, Arg164-Lys-Lys-Arg167, which is recognized by furin and trypsin and is located in a disordered loop (dotted line). Residues to the right of the dashed line comprise the PA_{20} fragment (subdomain 1a), while residues to the left form subdomain 1b. Removal of PA_{20} disrupts the large β sheet and may induce a conformational change in subdomain 1b. The view shown is similar to that of Figure 6.3a rotated clockwise by 90 degrees.

the large β sheet (as shown by the dashed line in Fig. 6.5), but would also cause several hydrophobic residues in subdomain 1b to become exposed. It thus appears that subdomain 1b is poised to undergo a conformational change once PA_{20} is removed. We do not yet know what functional role the residues in this subdomain play, but since PA cannot bind EF or LF until PA_{20} is removed, they may be involved in EF/LF binding (see also HOMOLOGY WITH IOTA-IB TOXIN, below).

DOMAIN 2

Domain 2 (residues 250-487) is approximately 65 Å tall and is the only domain long enough to span a membrane. It consists of a 9-stranded β-barrel with a helix at either end (Fig. 6.6). The barrel

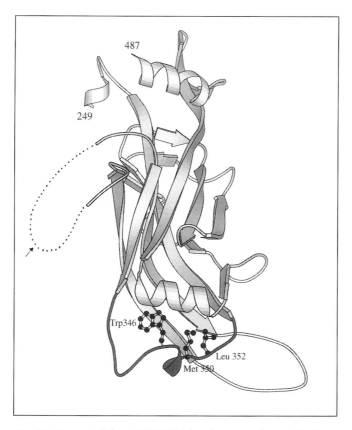

Fig. 6.6. Domain 2 (res. 250-487). The chymotrypsin-sensitive loop (res. 304-21) is shown as a dotted line, and the arrow indicates the Phe residues (313-14) implicated in oligomer formation. The stretch from 342 to 355 is shown in the conformation seen at pH 7.5, with the three hydrophobic residues buried in the barrel interior. At pH 6 this stretch forms a disordered loop. The view shown is similar to that of Figure 6.3a.

contains three very long strands with contour lengths of 39-54 Å, and a large loop composed of residues 304-321 (dotted line). This loop contains two hydrophobic residues at its tip, Phe-313 and Phe-314, and is susceptible to proteolysis by chymotrypsin (which cleaves after aromatic residues).[13,14] As in the case of the trypsin site of domain 1, it is possible to nick PA with chymotrypsin without causing dissociation into two fragments.[68] However, such treatment renders the PA (in combination with LF) nontoxic to cells,[68] despite its ability to bind to LF and to cell surface receptors, and to undergo receptor-mediated endocytosis.[68] A deletion mutant lacking the two Phe residues was found to be defective in its ability to form oligomers.[70] Taken together, these data suggest that the loop has a role in oligomerization and is probably involved in an intermolecular contact within the PA_{63} heptamer. A complementary surface has not yet been identified, but a strong candidate is an exposed hydrophobic patch located on the surface of domain 3, described below.

DOMAIN 3

Domain 3 consists of residues 488-594 and is the smallest of the domains. It consists primarily of a 4-stranded β sheet and 4 α-helices. One helix is part of a triangular strand-helix-loop structure that permits 5 hydrophobic residues to form a flat surface exposed to the solvent, constituting a "hydrophobic patch" (Fig. 6.7). In crystals of PA this patch is occupied by a phenylalanine residue from a neighboring molecule (shown in white). It is this interaction in the crystal that leads us to wonder whether the hydrophobic patch of one PA_{63} monomer might be occupied by Phe-313 or Phe-314 from domain 2 of a neighboring monomer in the heptamer. Further studies are required to determine the significance of this patch and the functional role of the entire domain.

DOMAIN 4

Domain 4 consists of residues 595-735, organized as a sandwich of two 4-stranded β sheets, and connected to domain 3 via a β-hairpin (Fig. 6.8). C-terminal deletion mutants have implicated domain 4 in receptor binding.[71,72] Deletion of 3, 5 or 7 residues from the C-terminus leads to reduced binding of PA to cells, while deletion of 12 or 14 residues leads to complete loss of activity and to sensitivity to proteases. These results can be explained in terms of the structure: deletion of 12 or more residues would remove the outermost strand of a β-sheet (see Fig. 6.8), severely affecting the conformation and rigidity of the preceding loop (residues 703-722) and compromising the ability of the entire domain to fold properly. Deletion of up to seven residues would remove the small helix at the C-terminus and only a portion of the final strand, such that proper folding of the domain could still occur.

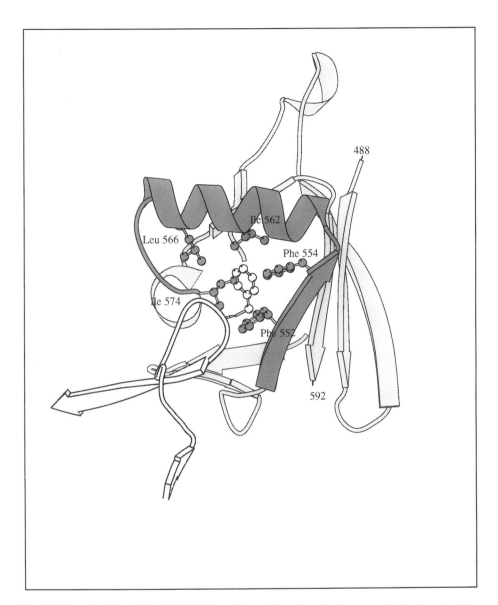

Fig. 6.7. Domain 3 (res. 488–594). This domain consists primarily of a 4-stranded β-sheet (right), a β–hairpin (bottom in background) and four helices, two of which are not shown. The triangular strand-helix-loop structure is highlighted and the residues comprising the hydrophobic patch are labeled. The Phe occupying the hydrophobic patch and shown in the foreground (in white) belongs to a neighboring molecule in the crystal. The view is that of Figure 6.3a seen from the bottom left.

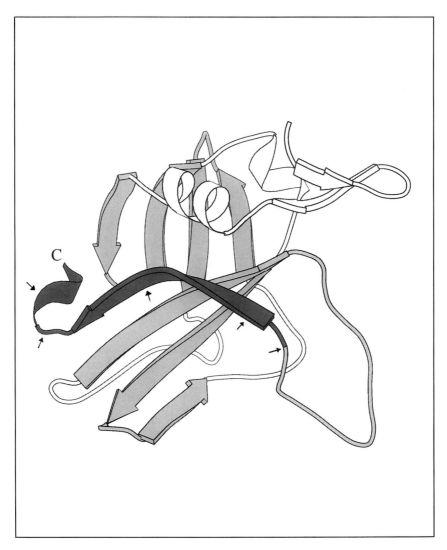

Fig. 6.8. Domain 4 (res. 595-735). The C-terminal 14 residues are highlighted, and the arrows indicate positions 3, 5, 7, 12 and 14 residues from the C terminus (see text for details). The loop connecting the last two strands is in contact with domain 2, while the β hairpin at the top is in contact with domain 3. The view shown is the same as in Figure 6.3a.

While domains 1-3 appear quite intimately associated, domain 4 appears somewhat separated from the rest of the molecule (Fig. 6.3a). The only interdomain contacts are the initial β-hairpin joining domain 4 with domain 3 and a small contact surface with domain 2 (Fig. 6.8). It is therefore conceivable that during membrane insertion (or perhaps at another step) domain 4 separates altogether from domain 2, swinging clockwise in Figure 6.3a, for example.

THE PA_{63} OLIGOMER

Following proteolytic cleavage of PA by trypsin in vitro, PA_{63} associates into an SDS-resistant oligomeric structure which binds EF and LF with high affinity. Such an oligomer has also been shown to form in mammalian cells.[50] An electron micrograph of typical oligomers is shown in Figure 6.9. The oligomer appears as a compact ring made of seven subunits. Two domains can be distinguished, a larger one forming the body of the ring, and a smaller protrusion. The outer diameter of the rings is roughly 140 Å including the protrusions and 105 Å excluding them, while the inner diameter is 20 Å.[50]

In both artificial lipid bilayers[73-75] and cells,[76] PA_{63} inserts into membranes and forms ion-conductive channels. Channel formation in planar lipid bilayers is rapidly accelerated when the pH is lowered from 7.4 to 6.5.[73] In liposomes the formation of channels occurs at pH 6.0, is maximal at pH 4.7, and is only observed in the presence of a pH gradient.[76] Although the precise pH requirements vary with experimental design, channel formation by PA_{63} does not require as low a pH as does that by the B chains of diphtheria, tetanus and botulinum toxins.[77] PA_{63} channels appear to be selective for monovalent cations[73] and can be blocked by the addition of LF.[78] Permeability to tetraheptylammonium ions indicates that the channel diameter is at least 12 Å,[79-81] consistent with the electron microscopy.

HOMOLOGY WITH IOTA-IB TOXIN

PA shares 34% sequence identity with the B-chain of the *Clostridium perfringens* iota toxin, another toxin whose A and B parts exist as separate polypeptides.[82] Like PA, iota-Ib is produced in a proform that is proteolytically activated and internalized by receptor-mediated endocytosis.[83,84] Iota-Ib is believed to coordinate the delivery to the cytoplasm of iota-Ia, which ADP-ribosylates actin.[85-87] Residues in PA identical to those in iota-Ib are underlined in Figure 6.4. We assume that residues performing similar functions in PA and iota-Ib should exhibit a greater degree of similarity than those which interact with the A moieties (EF/LF and iota-Ia) or with the receptors, since these ligands differ (or likely differ**) for the two toxins. The highest degree of similarity occurs in the middle of the sequence (Fig. 6.4 and Table 6.2): domains 2 and 3 have 40 and 37% sequence identity, while domains 1 and 4 have 29 and 21% identity, respectively. Four of the five residues making up the hydrophobic patch of domain 3 have identical corresponding residues in iota-Ib, and the fifth, Phe-552, corresponds to a leucine. Thus, a hydrophobic patch likely also exists in iota-Ib.

** *The cell surface receptors for iota-Ib and PA have not yet been identified.*

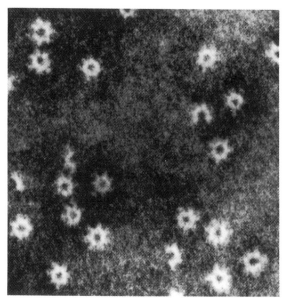

Fig. 6.9. Electron micrograph of PA_{63} oligomers. In most of the rings seven subunits can be distinguished clearly. The inner diameter of the ring is roughly 20 Å, while the outer diameter is about 140 Å. Incomplete rings with less than 7 subunits can also be seen. (Courtesy of Jill C. Milne and coworkers).[50]

In general, the high homology of domains 2 and 3 is consistent with a common role in oligomerization and membrane insertion.

Absence of homology is better illustrated in Figure 6.10, in which each PA residue is assigned a score based on similarity to the corresponding iota-Ib residue. A score of 2 is given for identical residues, 1 for very similar residues (e.g., ile vs. leu, glu vs. asp), -2 for residues missing in the iota-Ib sequence, and 0 for all others. Three low-scoring regions, indicated by arrows, clearly stand out. Two stretches occur in domain 4, consistent with a role in receptor binding. The longest stretch occurs between residues 194-214, which are located in subdomain 1b. The low degree of similarity in this region would be consistent with a role in EF/LF binding, since the A moieties in the two toxins are not homologous.

A HYPOTHETICAL MODEL FOR OLIGOMERIZATION AND MEMBRANE INSERTION

We have studied two crystal forms of PA, one obtained at pH 7.5 and the other at pH 6. The structure shown in Figure 6.3 is that

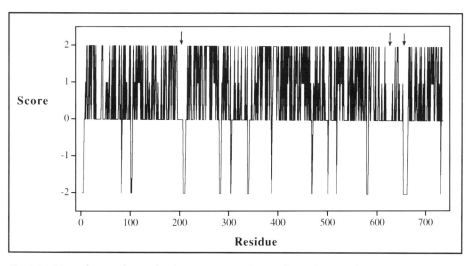

Fig. 6.10. Homology with iota-Ib. The sequences of PA and iota-Ib were aligned as in ref. 82 and residues were scored according to the following scheme: 2 for identity, 1 for high similarity, -2 for "deletions" (residues absent from the iota-Ib sequence) and 0 otherwise. No account was made for "insertions" (residues present in iota-Ib but absent from PA). The gaps indicated by arrows correspond to regions of low sequence similarity. The largest gap occurs between residues 194 and 214, in subdomain 1b. The other two gaps are in domain 4.

observed at pH 7.5 and closely resembles the low pH structure. The largest structural difference occurs in residues 342-355, located at the base of the barrel formed by domain 2 (Fig. 6.6). This region contains three hydrophobic residues, a Trp, a Met and a Leu, in the sequence $x_4WxxxMxLx_3$, where x represents a hydrophilic residue. At pH 7.5 these three residues are buried within the core of the β-barrel but become exposed at pH 6 in a disordered loop. We have not yet established that this change is pH-induced rather than a consequence of the crystal lattice, but it is tempting to hypothesize that the exposure of hydrophobic residues represents a step in pH-induced membrane insertion. Interestingly, the corresponding sequence in iota-Ib $(x_4WxxxLxIx_3)$ displays a similar pattern of amphipathicity.

PA_{63} has recently been crystallized in a high pH water-soluble form (Petosa, unpublished), and the heptameric nature of PA_{63} in the crystal has been established. The results of a crystal packing analysis in combination with electron microscopy measurements[50] have allowed us to build a crude model of the heptamer, which is shown in Figure 6.11, viewed along the 7-fold rotation axis. A side view is shown in cartoon form in Figure 6.12. This model has the long axis of the monomer oriented roughly parallel to the 7-fold axis, with domain 2 facing the channel interior and domains 3 and 4 facing outward. Thus, domains 2 and 3 would comprise the body of the heptameric ring

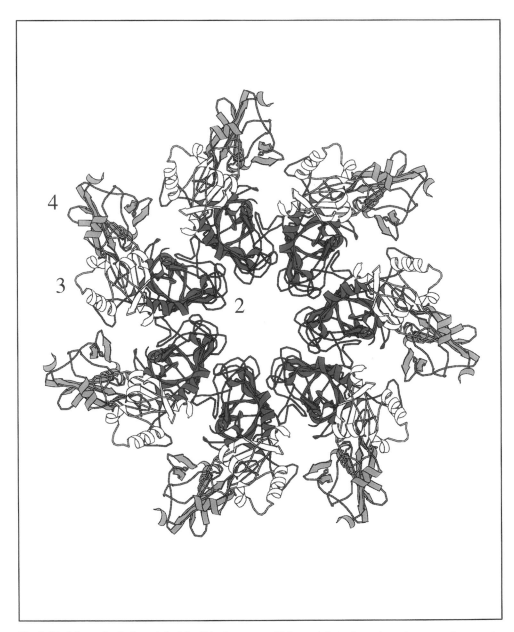

Fig. 6.11. A hypothetical model of the PA$_{63}$ heptamer. This model was based on electron micrographs and on a preliminary analysis of PA$_{63}$ crystals. Domains 2-4 are labeled; subdomain 1b is not shown. The monomers are oriented as they would be seen from the top of Figure 6.3a (minus domain 1). See text for further details.

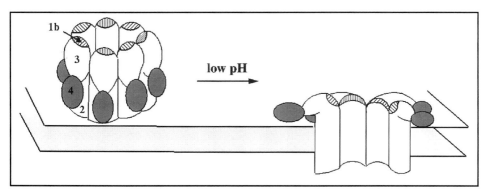

Fig. 6.12. A model of pH-induced membrane insertion. The residues at the bottom of domain 2 probably undergo a significant conformational change during this step. The tenuous interface between domains 2 and 4 suggests that the latter could separate from the main body of the heptamer. This would permit domain 4 to remain on the membrane surface while domain 2 penetrated the bilayer. We envisage LF or EF to bind subdomain 1b at the top of the heptamer.

seen in electron micrographs, and domain 4 would account for the small protrusion (Fig. 6.9). Full-length PA would not be able to form a similar heptamer because a significant overlap would occur among the seven PA_{20} moieties (subdomains 1a), suggesting that perhaps PA_{20} sterically hinders oligomerization. Because of its tenuous interface with domain 2, it would be possible for domain 4 to separate laterally from the body of the heptamer. The putative mobility of domain 4 would be consistent with the variability in the protrusions that is frequently seen in electron micrographs (J. Wall, personal communication).

Our heptamer model has a height of roughly 75 Å (as oriented in Fig. 6.12), which is sufficient to span a membrane. The model has a ring of acidic residues lining the interior, consistent with the cation selectivity of PA_{63} channels. The chymotrypsin-sensitive loop of domain 2 extends outward and could feasibly interact with the hydrophobic patch on domain 3 of a neighboring molecule, explaining this loop's importance in oligomer formation. The receptor-binding domain and the WxxxMxL stretch possibly involved in membrane insertion are located at the bottom of the heptamer, and the residues suspected of binding EF and LF (subdomain 1b) are at the top (Fig. 6.12). We therefore imagine the membrane oriented horizontally below the heptamer, such that membrane insertion would leave the putative EF/LF binding site exposed to the extracellular environment, consistent with the ability of EF/LF to block the conductance of membrane channels formed by PA_{63}.[78] It is not clear exactly how PA_{63} inserts into the membrane, since there are no extensive hydrophobic areas on the surface of our model. We therefore suspect that a significant conformational change occurs during membrane insertion, probably involving the residues at the bottom of domain 2.

We emphasize that the above remarks concerning oligomerization and membrane insertion are speculative, and their verification will require further investigation. We are currently engaged in a crystallographic analysis of PA_{63}, and hope to build an accurate model of the heptamer in the near future.

CONCLUDING REMARKS

Although much progress has been made in anthrax research in recent years, many fundamental questions remain unanswered: what conformational changes occur after PA_{20} is removed? How does PA_{63} interact with LF and EF? How do the monomers of PA_{63} associate to form a heptamer? How does membrane insertion and the formation of ion channels occur? How are LF and EF translocated into the cytoplasm? What is the substrate recognized by LF and how does its processing lead to increased cytokine production? Clearly, the answers to these questions will be of great general interest to the biochemist, the structural biologist and the cell biologist alike.

ACKNOWLEDGMENTS

We thank John Collier, Stephen Leppla, Philip Hanna and João Cabral for critical reading of the manuscript, Kurt Klimpel for sharing experimental results before publication, and Jill Milne for providing us with a copy of her thesis and the electron micrograph shown in Figure 6.9. This work is supported by the U.S. Army under contract number DAMD17-94-C-4047.

REFERENCES

1. Dirckx JH. Virgil on anthrax. Am J Dermatopathol 1981; 3:191-95.
2. Koch R. The aetiology of anthrax based on the ontogeny of the anthrax bacillus. Beitr Biol Pflanz 1877; 2:277.
3. Pasteur L. De l'attenuation des virus et de leur retour a la virulence. CR Acad Sci 1881; 19:429.
4. Hambleton P, Turnbull PCB. Anthrax vaccine development: a continuing story. In: Mizrahi A. Bacterial vaccines. Advances in Biotechnological Processes, Vol 13. New York: Alan R. Liss, 1990; 105-22.
5. Christie AB. Infectious diseases: epidemiology and clinical practice. Edition IV. Edinburgh: Churchill Livingstone, 1987; 983.
6. Bell JH. On anthrax and anthracaemia in wool sorters, heifers and sheep. Br Med J 1880; 2:656.
7. La Force FM. Woolsorters' disease in England. NY Acad Med 1978; 54:956.
8. Albrink WS, Brooks SM, Biron RE et al. Human inhalation anthrax: a report on three cases. Am J Pathol 1960; 36:457.
9. FAO/OIE/Who. Animal Health Year Book. New York: Unipub, 1985.

10. Turnbull PC. Anthrax vaccines: past, present, and future. Vaccine 1991; 9:533-39.

11. Meselson M, Guillemin J, Hugh-Jones M et al. The Sverdlovsk anthrax outbreak of 1979. Science 1994; 266:1202-28.

12. Pastan I, Chaudhary V, Fitzgerald DJ et al. Recombinant toxins as novel therapeutic agents. Ann Rev Biochem 1992; 61:331-54.

13. Leppla SH. Anthrax toxins. In: Moss J, Iglewski B, Vaughan M, Tu AT, eds. Bacterial Toxins and Virulence Factors in Disease. New York: Marcel Dekker, 1995:543-72.

14. Leppla SH. The anthrax toxin complex. In: Alouf JE, Freer JH, eds. Sourcebook of Bacterial Protein Toxins. San Diego: Academic Press, 1991; 277-302.

15. Mikesell P, Ivins BE, Ristroph JD et al. Evidence for plasmid-mediated toxin production in *Bacillus anthracis*. Infect Immun 1983; 39:371-76.

16. Green BD, Battisti L, Koehler TM et al. Demonstration of a capsule plasmid in *Bacillus anthracis*. Infect Immun 1985; 49:291-97.

17. Uchida I, Sekizaki T, Hashimoto K et al. Association of the encapsulation of *Bacillus anthracis* with a 60 megadalton plasmid. J Gen Microbiol 1985; 131:363-67.

18. Uchida I, Hashimoto K, Terakado N. Virulence and immunogenicity in experimental animals of *Bacillus anthracis* strains harbouring or lacking 110 MDa and 60 MDa plasmids. J Gen Microbiol 1986; 132:557-59.

19. Jriedlander AM. Macrophages are sensitive to anthrax lethal toxin through an acid-dependent process. J Biol Chem 1986; 261:7123-26.

20. Hambleton P, Carman JA, Melling J. Anthrax: the disease in relation to vaccines. Vaccine 1984; 2:125-32.

21. Stanley JL, Smith H. The three factors of anthrax toxin: their immunogenicity and lack of demonstrable enzymic activity. J Gen Microbiol 1963; 31:329-37.

22. Bragg TS, Robertson DL. Nucleotide sequence and analysis of the lethal factor gene (lef) from *Bacillus anthracis*. Gene 1989; 81:45-54.

23. Welkos SL, Lowe JR, Eden-McCutchan F et al. Sequence and analysis of the DNA encoding protective antigen of *Bacillus anthracis*. Gene 1988; 69:287-300.

24. Escuyer V, Duflot E, Sezer O et al. Structural homology between virulence-associated bacterial adenylate cyclases. Gene 1988; 71:293-98.

25. Robertson DL, Tippets MT, Leppla SH. Nucleotide sequence of the *Bacillus anthracis* edema factor gene (cya): a calmodulin-dependent adenylate cyclase. Gene 1988; 73:363-71.

26. Bartkus JM, Leppla SH. Transcriptional regulation of the protective antigen gene of *Bacillus anthracis*. Infect Immun 1989; 57:2295-300.

27. Uchida I, Hornung JM, Thorne CB et al. Cloning and characterization of a gene whose product is a transactivator of anthrax toxin synthesis. J Bacteriol 1993; 175:5329-38.

28. Cataldi A, Labruyère E, Mock M. Construction and characterization of antigen-deficient *Bacillus anthracis* strain. Mol Microbiol 1990; 4:1111-17.

29. Hornung JM, Thorne CB. Insertion mutations affecting pXO1-associated toxin production in *Bacillus anthracis*. Abstr 91st Annu Meet Am Soc Microbiol 1991; p.98 Abstr D-121.

30. Pezard C, Berche P, Mock M. Contribution of individual toxin components to virulence of *Bacillus anthracis*. Infect Immun 1991; 59:3472-77.

31. Fish DC, Klein F, Lincoln RE et al. Pathophysiological changes in the rat associated with anthrax toxin. J Infect Dis 118:114-24.

32. Beall FA, Taylor MJ, Thorne CB. Rapid lethal effect in rats of a third component found upon fractionating the toxin of *Bacillus anthracis*. J Bacteriol 1962; 83:1274-80.

33. Smith H, Stoner HB. Anthrax toxic complex. Fed Proc 1967;26:1554-57.

34. Li J. Bacterial toxins. Curr Opin Struct Biol 1992; 2:545-56.

35. Considine R, Simpson L. Cellular and molecular action of binary toxins possessing ADP-ribosyltransferase activity. Toxicon 1991; 29:913-36.

36. Friedlander AM. The anthrax toxins. In: Saelinger C, ed. Trafficking of Bacterial Toxins. CRC Press: Boca Raton, 1990:121-28.

37. Leppla SH. Anthrax toxin edema factor: a bacterial adenylate cyclase that increases cyclic AMP concentrations in eukaryotic cells. Proc Natl Acad Sci USA 1982; 79:3162-66.

38. Pezard C, Duflot, E, Mock M. Construction of *Bacillus anthracis* mutant strains producing a a single toxin component. J Gen Microbiol 1993; 139:2459-63.

39. Stephen J. Anthrax toxin. Pharm Therap 1991; 12:501-13.

40. Hanna PC, Kochi, S, Collier RJ. Biochemical and physiological changes induced by anthrax lethal toxin in J774 macrophage-like cells. Mol Biol Cell 1992; 3:1269-77.

41. Hanna PC, Kruskal BA, Ezekowitz RAB et al. Role of macrophage oxidative burst in the action of anthrax lethal toxin. Molecular Medicine 1994; 1:7-18.

42. Hanna PC, Acosta D, Collier RJ. On the role of macrophages in anthrax. Proc Natl Acad Sci USA 1993;90:10198-201.

43. Escuyer V, Collier RJ. Anthrax protective antigen interacts with a specific receptor on the surface of CHO-K1 cells. Infect Immun 1991; 59:3381-86.

44. Klimpel KR, Molloy SS, Thomas G et al. Anthrax toxin protective antigen is activated by a cell surface protease with the sequence specificity and catalytic properties of furin. Proc Natl Acad Sci USA 1992; 89:10277-81.

45. Molloy SS, Bresnahan PA, Leppla SH et al. Human furin is a calcium-dependent serine endoprotease that recognizes the sequence Arg-X-X-Arg and efficiently cleaves anthrax toxin protective antigen. J Biol Chem 1992; 267:16396-402.

46. Gordon VM, Klimpel KR, Arora N et al. Proteolytic activation of bacterial toxins by eukaryotic cells is performed by furin and by additional cellular proteases. Infect Immun 1995; 63:82-87.

47. Leppla SH, Friedlander AM, Cora EM. Proteolytic activation of anthrax toxin bound to cellular receptors. In: Fehrenbach FJ, Alouf JE, Falmagne

P, Goebel W, Jeljaszewicz J, Jurgen D, Rappuoli, R, ed. Bacterial Protein Toxins. New York: Gustav Fischer, 1988:111-12.

48. Gordon VM, Leppla SH, Hewlett EL. Inhibitors of receptor-mediated endocytosis block the entry of *Bacillus anthracis* adenylate cyclase toxin but not that of *Bordetella pertussis* adenylate cyclase toxin. Infect Immun 1988; 56:1066-69.

49. Klimpel KR, Arora N, Leppla SH. Anthrax toxin lethal factor contains a zinc metalloprotease consensus sequence which is required for lethal toxin activity. Mol Microbiol 1994; 13:1093-100.

50. Milne JC, Furlong D, Hanna PC et al. Anthrax protective antigen forms oligomers during intoxication of mammalian cells. J Biol Chem 1994; 269:20607-12.

51. Leppla SH. *Bacillus anthracis* calmodulin-dependent adenylate cyclase: chemical and enzymatic properties and interactions with eukaryotic cells. Adv Cyclic Nucleotide Protein Phosphorylation Res 1984; 17:189-98.

52. Labruyère E, Mock M, Ladant D et al. Characterization of ATP and calmodulin-binding properties of a truncated form of *Bacillus anthracis* adenylate cyclase. Biochem 1990; 29:4922-28.

53. Goyard S, Orlando C, Sabaier J-M et al. Identification of a common domain in calmodulin-activated eukaryotic and bacterial adenylate cyclases. Biochem 1989; 28:1964-67.

54. Xia Z, Storm DR. A-type ATP binding consensus sequences are critical for the catalytic activity of the calmodulin-sensitive adenylyl cyclase form *Bacillus anthracis*. J Biol Chem 1990; 265:6517-20.

55. Munier H, Blanco FJ, Prêcheur B et al. Characterization of a synthetic calmodulin-binding peptide derived from *Bacillus anthracis* adenylate cyclase. J Biol Chem 1993; 268:1695-701.

56. Little SF, Leppla SH, Burnett JW et al. Structure-function analysis of *Bacillus anthracis* edema factor by using monoclonal antibodies. Biochem Biophys Res Commun 1994; 199:676-82.

57. Arora N, Klimpel KR, Singh Y et al. Fusions of anthrax toxin lethal factor to the ADP-ribosylation domain of *Pseudomonas* Exotoxin A are potent cytotoxins which are translocated to the cytosol of mammalian cells. J Biol Chem 1992; 267:15542-48.

58. Arora N, Leppla SH. Residues 1-254 of anthrax toxin lethal factor are sufficient to cause cellular uptake of fused polypeptides. J Biol Chem 1993; 268:3334-41.

59. Arora N, Leppla SH. Fusions of anthrax toxin lethal factor with shiga toxin and diphtheria toxin enzymatic domains are toxic to mammalian cells. Infect Immun 1994; 62:4955-61.

60. Milne JC, Blanke SR, Hanna PC et al. Protective antigen-binding domain of anthrax lethal factor mediates translocation of a heterologous protein fused to its amino- or carboxy-terminus. Mol Microbiology 1995; 15:661-66.

61. Quinn CP, Singh Y, Klimpel KR et al. Functional mapping of anthrax toxin lethal factor by in-frame insertion mutagenesis. J Biol Chem 1991; 266:20124-30.

62. Kochi SK, Schiavo G, Mock M et al. Zinc content of the *Bacillus anthracis* lethal factor. FEMS Microbiol Lett 1994; 124:343-48.

63. Jongeneel CV, Bouvier J, Bairoch A. A unique signature identifies a family of zinc-dependent metallopeptidases. FEBS Lett 1989; 242:211-14.

64. Vallee BL, Auld DS. Zinc coordination, function, and structure of zinc enzymes and other proteins. Biochemistry 1990; 29:5647-59.

65. Schiavo G, Benfenati F, Poulain B et al. Tetanus and botulinum-B neurotoxins block neurotransmitter release by proteolytic cleavage of synaptobrevin. Nature 1992; 359:832-35.

66. Montecucco C, Schiavo G. Tetanus and botulism neurotoxins: a new group of zinc proteases. Trend Biochem Sci 1993; 18:324-27.

67. Oguma K, Fujinaga Y, Inoue K. Structure and function of *Clostridium botulinum* toxins. Microbiol Immunol 1995; 39:161-68.

68. Novak JM, Stein M-P, Little SF et al. Functional characterization of protease-treated *Bacillus anthracis* protective antigen. J Biol Chem 1992; 267:17186-93.

69. Singh Y, Chaudhary VK, Leppla SH. A deleted variant of *Bacillus anthracis* protective antigen is nontoxic and blocks anthrax toxin in vivo. J Biol Chem 1989; 264:19103-37.

70. Singh Y, Klimpel KR, Arora N et al. The chymotrypsin-sensitivie site, FFD[315], in anthrax toxin protective antigen is required for translocation of lethal factor. J Biol Chem 1994; 269:29039-46.

71. Singh Y, Klimpel KR, Quinn CP et al. The carboxyl-terminal end of protective antigen is required for receptor binding and anthrax toxin activity. J Biol Chem 1991; 266:15493-97.

72. Little SF, Lowe JR. Location of receptor-binding region of protective antigen from *Bacillus anthracis*. Biochem Biophys Res Commun 1991; 180:531-37.

73. Blaustein RO, Koehler TM, Collier RJ et al. Anthrax toxin: channel-forming activity of protective antigen in planar phospholipid bilayers. Proc Natl Acad Sci USA 1989; 86:2209-13.

74. Finkelstein A. The channel formed in planar lipid bilayers by the protective antigen component of anthrax toxin. Toxicology 1994; 87:29-41.

75. Koehler TM, Collier RJ. Anthrax toxin protective antigen: low pH-induced hydrophobicity and channel formation in liposomes. Mol Microbiol 1991; 5:1501-06.

76. Milne JC, Collier RJ. pH-dependent permeabilization of the plasma membrane of mammalian cells by anthrax protective antigen. Mol Microbiol 1993; 10:647-53.

77. Finkelstein A. Channels formed in phospholipid bilayer membranes by diphtheria, tetanus, botulinum, and anthrax toxin. J Physiol (Paris) 1990; 84:188-90.

78. Zhao J, Milne JC, Collier RJ. Effect of anthrax toxin's LF component on ion channels formed by the PA component. J Biol Chem 1995; in press.

79. Blaustein RO, Finkelstein A. Diffusion limitation in the block by symmetric tetraalkylammonium ions of anthrax toxin channels in planar phospholipid bilayer membranes. J Gen Physiol 1990; 96:943-57.

80. Blaustein RO, Finkelstein A. Voltage-dependent block of anthrax toxin channels in planar phospholipid bilayer membranes by symmetric tetraalkylammonium ions. J Gen Physiol 1990; 96:905-19.

81. Blaustein RO, Finkelstein A. Voltage-dependent block of anthrax toxin channels in planar phospholipid bilayer membranes by symmetric tetraalkylammonium ions. Effects on macroscopic conductance. J Gen Physiol 1990; 96:921-42.

82. Perelle S, Gibert M, Boquet P et al. Characterization of *Clostridium perfringens* iota-toxin genes and expression in *Escherichia coli*. Infect Immun 1993; 61:5147-56.

83. Stiles BG, Wilkins TD. *Clostridium perfringens* iota toxin: synergism between two proteins. Toxicon 1986; 24:767-73.

84. Stiles BG, Wilkins TD. Purification and characterization of *Clostridium perfringens* iota toxin: dependence on two nonlinked proteins for biological activity. Infect Immun 1986; 54:683-88.

85. Simpson LL, Stiles BG, Zepeda HH et al. Molecular basis for the pathological actions of *Clostridium perfringens* iota toxin. Infect Immun 1987; 55:118-22.

86 Vandekerckhove J, Schering B, Barmann M et al. *Clostridium perfringens* iota toxin ADP-ribosylates skeletal muscle actin in Arg-177. FEBS Lett 1987; 225:48-52.

87. Popoff MR, Milward FW, Bancillon B et al. Purification of the *Clostridium spiroforme* binary toxin and activity on HEp-2 cells. Infect Immun 1989; 57:2462-69.

======= CHAPTER 7 =======

THE CHOLERA FAMILY OF ENTEROTOXINS

David L. Scott, Rong-Guang Zhang and Edwin M. Westbrook

Cholera continues to plague developing nations which lack the sanitary facilities to prevent bacterial contamination of food and water supplies.[1,2] Seven pandemics have been recorded since 1817 with current major outbreaks in South America and Asia.[3] Untreated, the profuse diarrhea characteristic of the classical form of cholera rapidly leads to dehydration and hypovolemic shock.[4] Antibiotics are of secondary importance to fluid therapy in the short-term management of infected patients. Efforts to prevent the spread of cholera focus on improving sanitation, protecting water supplies, and developing effective vaccines. Current vaccines, using either purified toxin or peptide epitopes, have failed to provide long-term immunity.[5]

Cholera is caused by the secretion of a potent enterotoxin (choleragen) from the gram negative bacteria *Vibrio cholerae*.[6-8] Choleragen belongs to a family of hexameric microbial toxins that are composed of a single A (catalytic) subunit and five B (targeting) subunits.[9,10] This group includes the toxins produced by *Escherichia coli* (heat-labile enterotoxin (LT) and verotoxin (VT)), *Bordetella pertussis* (pertussis (PT)), and *Shigella dysenteria* (Shiga-toxin (ST)). These toxins are either ADP-ribosyltransferases (CT, LT, PT) or N-glycosidases (ST and VT). Cholera toxin, LT, and PT act at the cellular level by ribosylating guanine-nucleotide binding (G) proteins involved in the regulation of adenyl cyclase.[11-13] In contrast, ST and the related verotoxins inhibit protein synthesis by depurinating a specific adenosine base in 28S ribosomal RNA.[14,15]

Intoxication by cholera requires passage of the catalytic subunit across the target cell membrane.[16,17] After choleragen binds cooperatively via its B subunits to G_{M1} [Gal(β1-3)GalNAc(β1-4)(NeuAc(a2-3))Gal(β1-4)Glc(β1-1)-ceramide] gangliosides exposed on the luminal surface of intestinal epithelial cells,[18]

Protein Toxin Structure, edited by Michael W. Parker. © 1996 R.G. Landes Company.

the mature A subunit undergoes reductive dissociation.[19-22] The liberated A1 fragment (residues 1-192 or 1-194), in the presence of NAD[+], ADP, and additional cytosolic factors, ribosylates $G_{s\alpha}$, a GTP-binding regulatory protein associated with adenylate cyclase.[23-29] Increased cyclic-AMP indirectly activates intestinal sodium pumps precipitating a profuse, life-threatening, diarrhea.[30]

Recent reports have described the three-dimensional X-ray structures of choleragen,[31] choleragenoid (the isolated B pentamer of cholera toxin),[32] the highly homologous heat-labile enterotoxin,[33,34] exotoxin A (from *Pseudomonas aeruginosa*),[35,36] verotoxin-1 (B-pentamer only),[37] diphtheria toxin (from *Corynebacterium diphtheriae*),[38] Shiga toxin,[39] and pertussis toxin.[40] The AB_5 toxins (CT, LT, VT, and ST) share a common basic structure despite variable sequence homology. From a biological perspective, cholera toxin remains the most thoroughly studied of the hexameric toxins. The crystalline coordinates of choleragen and choleragenoid are consistent with previous low-resolution models derived from electron microscopy,[41,42] atomic force microscopy,[43] and infrared spectroscopy.[44,45] The X-ray models provide a sensible structural framework for examining the mechanics of ganglioside binding, catalysis, and intoxication at the atomic level. This analysis has been greatly aided by complementary structural work with LT,[33,34,46,47] mutational and chemical experiments, as well as by inference from homologous enzyme systems.[34,48,49]

B SUBUNIT ARCHITECTURE

The 103 amino acids of cholera toxin's B subunit form a robust core module that includes two α–helices and a variety of β structure (Fig. 7.1 and Fig. 7.2). The overall fold consists of six antiparallel β-strands forming a closed β-barrel that is bridged at one end by a long α-helix (residues 58 to 79). This helical cap is gently curved with its hydrophilic face contributing to the boundary wall of the central "pore" of the B_5 pentamer. The B subunit's sole disulfide bridge (Cys9 to Cys86) secures a short solvent exposed amino-terminal helix (residues 4-12) to an interior strand. This cystine is essential for the stabilization of B subunit monomers prior to pentamer formation.[20]

Despite variations in subunit size (ranging from 69 to 103 residues) and sequence homology (< 80%), the overall folds of the B subunits of PT, ST, VT-1, and cholera-like toxins are very similar.[32,34,39,40,48] Deviations from cholera toxin's secondary structure occur primarily near the amino and carboxyl termini. Sixma et al noted that 52 of the main-chain atoms of VT-1 and LT are superposable with a root-mean-square deviation of less than 1.29 Å/atom (corresponding to 75% of the VT-1 B chain and 50% of LT).[48] The toxin fold is also found in other nonrelated proteins including the monomeric *Staphylococcus aureus* nuclease.[50]

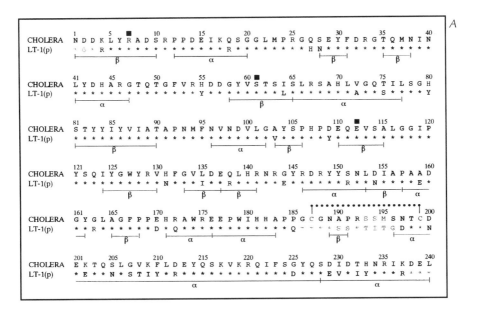

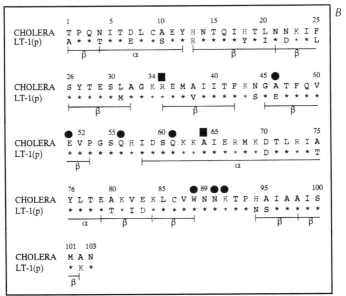

Fig. 7.1. Sequence of the cholera toxin A and B subunits.[98-105] The sequence of the highly homologous heat-labile enterotoxin from E. coli *(porcine variant) is included for comparison (asterisks denote residues identical to that of cholera toxin). (A) Cholera toxin A (catalytic) subunit. Closed circles denote residues implicated in catalysis; the single disulfide bridge is shown by a dotted line.[106-108] (B) Cholera toxin B (binding) subunit. Automated sequencing of polymerase-chain reaction generated amplicons has identified three types of B subunits in CT 01 strains.[109] The sequence shown here is consistent with that determined from the crystal structures.[31,32] Residues important to pentamer stability or to ganglioside binding are indicated with filled squares and ovals, respectively.*

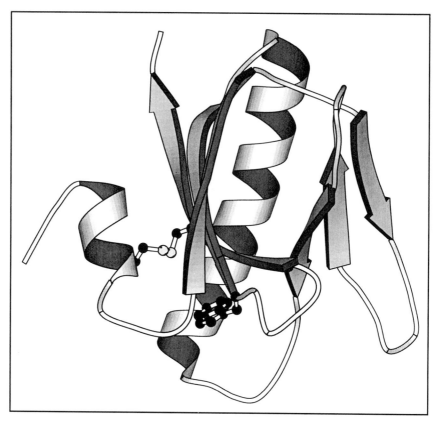

Fig. 7.2. Structure of the B subunit.[110] The secondary structure includes two α-helices, the longer of which "caps" a β-barrel formed from two β-sheets.[111] The single disulfide (Cys9 to Cys86) is shown as well as the sidechain of Trp 88. The basic fold is identical to that described for LT[34] and similar to that of ST,[39] VT[37] and PT.[40]

B-Subunit Pentamer

The B subunits differ among the microbial toxins in their quaternary arrangement and in their receptor-binding specificities.[4,51-54] The homopentamers of choleragenoid, choleragen, LT, and ST are nearly 5-fold symmetric (Fig. 7.3 and Fig. 7.4). In contrast, the pertussis toxin heteropentamer is assembled from four different subunits (single copies of S2, S3, S5 and two copies of S4) and is relatively asymmetric.[40] The amount of surface area buried during oligomer formation differs sharply among the toxins ranging from 1269 Å^2 (PT S2-S4) to 2700 Å^2 (CT or LT). In the case of CT, this represents approximately 40% of the total subunit surface area.[29]

Cholera toxin's doughnut shaped B pentamer has a "vertical" height of 32 Å and a radius of 31 Å. In the absence of the A subunit, a conical central pore, or channel, traverses the entire vertical height

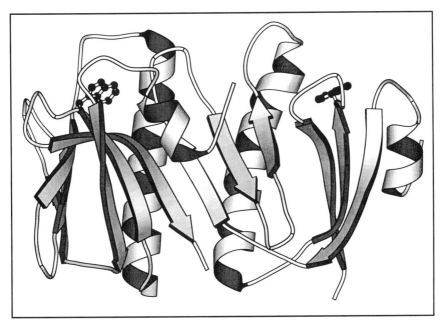

Fig. 7.3. B subunit interactions.[110] Two adjacent B subunits are shown as they appear in the pentamer. Three anti-parallel β-strands from each subunit combine to form a six-stranded sheet. This sheet, along with the central helices, provides a remarkably stable intersubunit interface.[61,63] The sidechains of the respective subunit's Trp88s are also shown indicating the sites of G_{M1} binding. Although ganglioside binding appears to be cooperative,[44,69] the 5-fold symmetric sites are geographically isolated.

parallel to the 5-fold axis (Fig. 7.4).[31,32] The flat 'dorsal' surface perpendicular to the pentamer axis faces the A subunit in the holotoxin and incorporates the free carboxyl termini and the negative end of the central helix dipoles.[55] Whether this potential charge asymmetry has any functional bearing is unknown. The prominent "ventral" flange, created by residues 50 to 64 of each subunit creates a small pocket adjacent to the external wall of the pentamer. The sidechain of Trp88, which lies exposed at the base of this pocket, forms a sandwich with the terminal galactose sugar of the ganglioside receptor (Fig. 7.5).[56,57]

The B subunits of cholera toxin are synthesized as monomers in the bacterial cytoplasm and aggregate in the periplasm.[20,58,59] Choleragenoid is stabilized by approximately 30 inter-subunit hydrogen bonds and 7 salt bridges per monomer with tight interdigitation of hydrophobic groups at the subunit interface (Fig. 7.3). The stable B pentamers of choleragen and cholergenoid are highly resistant to chemical or thermal denaturation.[60-64]

A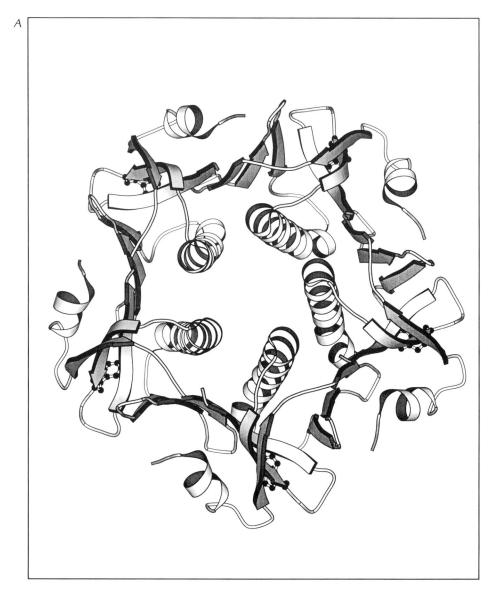

Fig. 7.4. The B pentamer (choleragenoid). The symmetry can be readily appreciated from this view down the 5-fold axis of (A) a backbone tracing of the secondary structure,[110] and (B) a space-filling representation. The orientation in (A) corresponds to that seen from the perspective of the A subunit with the ganglioside-binding domains (Trp88) facing away; it is reversed in (B). The central pore or channel is outlined by the five central helices. In (C) the axis has been tilted 90 degrees.[110] This "side" view emphasizes the similarity of choleragenoid to a "lunar lander" with the ganglioside binding site shown at the base.

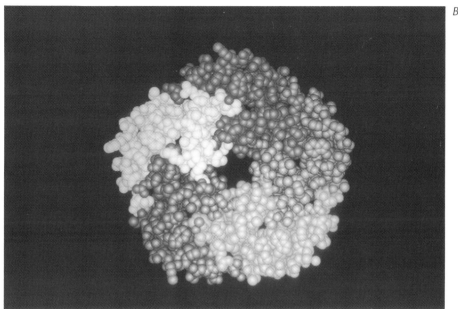

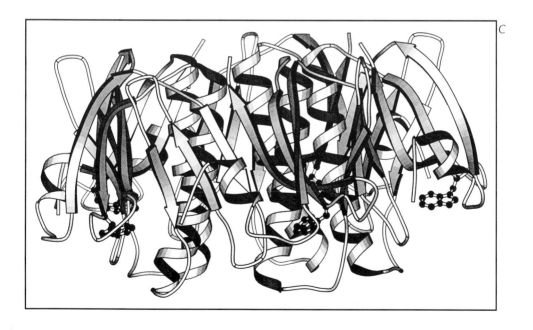

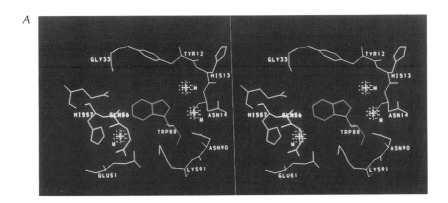

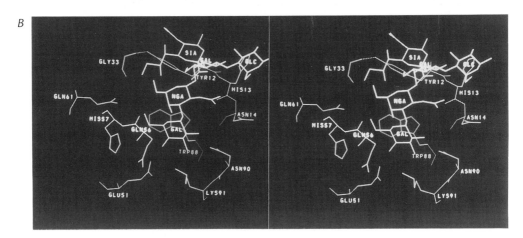

Fig. 7.5. Schematic representation of the ganglioside-binding site of cholera toxin in the absence (A) and presence (B) of G_{M1} pentasaccharide. G_{M1} (1,284 kD) consists of a ceramide linked by a β-glycosidic bond to a polysaccharide. The ganglioside is an integral membrane component with a polar head group (the polysaccharide) extending out of the membrane, and two hydrocarbon tails (the ceramide), extending into the hydrophobic layer of the membrane.[85-87] Physical studies of the interaction of choleragenoid with G_{M1} (fluorescence spectroscopy, photoaffinity labeling, differential calorimetry, differential solubility thermal gel analysis, nuclear magnetic resonance, hydrodynamics) are consistent with the location determined by X-ray crystallography. The 5-fold redundant receptor-binding sites are located on the "ventral" surface of the pentamer opposite the surface that interacts with the A subunit. The binding pocket appears to be "hard-wired" requiring very little adjustment of either side chains or backbone to coordinate G_{M1}. Residues interacting with the pentasaccharide are all derived from a single subunit with the exception of Gly33. The backbone carbonyl oxygen of Gly33 interacts with the hydroxyl goup of Tyr12 stabilizing one wall of the receptor site.

THE CENTRAL PORE (DOUGHNUT HOLE)

The most striking architectural feature of the AB$_5$ toxins is the central pore or channel that runs along the 5-fold axis of the B$_5$ assembly (Fig. 7.4). This is an intriguingly common motif among multimeric proteins (e.g., GroEl,[65] annexin V,[66] etc.). Cholera toxin's central pore (11 Å diameter x 40 Å long) is lined by 5 closely-packed, parallel, amphipathic α-helices that gently bow inward during their course reducing the effective diameter of the channel from 16 Å (amino end) to 11 Å (carboxyl end).

The sidechains of Asp59, Lys62, Lys63, Glu66, Arg67, Asp70, and Arg73 extend into the pore with five-fold redundancy. The sequential alternation of positive and negatively charged residues encourages salt bridge formation between neighboring helices. The amino acids lining the pore are identical in CT and LT with the exception of the substitution of an Asn for Asp70 in some CT strains. The pore walls of Shiga toxin, in contrast, consist entirely of uncharged residues (Asn35, Ser38, Ser42, Ile45, and Thr46).[39] The effective diameter of the pore remains similar in these toxins reflecting its common role in anchoring the A subunit.

The carboxyl-terminus of the A2 chain (residues 212-240) occupies the central pore of the holotoxin. In the absence of the A2 chain, the "empty" pores of cholera-like toxins are highly hydrated.[32,57] The additional solvent encourages favorable sidechain interactions among B subunits and accounts, at least in part, for the similar stabilities of choleragen and choleragenoid.[20,60,61] Holotoxin cannot be reconstituted by mixing A subunits and preformed choleragenoid in vitro.[20] Presumably, this reflects either a physical size and/or electrostatic barrier to passing a charged peptide through the preformed pore. Holotoxin can be recovered, however, after adding isolated A subunits to choleragenoid bound to membrane associated G$_{M1}$ receptors.[64] This raises the intriguing possibility of a receptor-induced conformational change in the pentamer that permits greater access to the central channel.

THE OLIGOSACCHARIDE-BINDING SITE

Choleragenoid contains five equivalent G$_{M1}$ binding sites with measured dissociation constants ranging from 10^{-9} to 10^{-10} M depending upon the analytical technique.[56,67] Derivatives of G$_{M1}$ bearing a fucose residue α-glycosidically linked to the 2-position of the terminal galactose may also serve as receptors.[68] Dissociated B subunits cannot efficiently bind ganglioside.[56]

G$_{M1}$ gangliosides markedly improve choleragenoid's thermal stability and promote cooperative folding of the B subunits (unfolding is shifted from 66 to 87 °C).[44,61,63] A small conformational change accompanying G$_{M1}$ binding has been inferred from circular dichroism spectra.[67] Similar experiments using Fourier-transform infrared spectroscopy, however, have revealed little, if any, change in the secondary structure (< 3-4% change in β-sheet or α-helix content).[44,45]

The G_{M1} pentasaccharide binds within a pocket formed by Glu11, Tyr12, His13, Asn14, Glu51, Gln56, His57, Gln61, Trp88, Asn90, and Lys91 on the ventral flange of the B pentamer (Fig. 7.5). Comparison of the crystal structure of choleragenoid[32] with that of the G_{M1} pentasaccharide complex[57] reveals only modest local conformational changes indicating "hardwiring" of the ganglioside binding to recognize its specific receptor.[32] The coordinates of the respective alpha-carbon traces superpose with a root-mean square deviation of less than 0.5 Å/atom. Direct or water-mediated hydrogen bonds are formed between G_{M1} and the sidechains of Asn14, Glu51, Gln61, Trp88, Asn90, and Lys91. These sidechains are almost ideally oriented in the crystal structure of the uncomplexed choleragenoid to accept the ligand.

Significant backbone movement accompanying pentasaccharide binding is limited to the His13-Asn14 peptide bond and the sidechain of Gln56. In the absence of ganglioside, the carbonyl of His13 points away from Trp88 and the sidechain of Gln56 donates a hydrogen bond to the Oε2 atom of Gln61.[32] Binding of G_{M1} flips the His13-Asn14 peptide bond in order to accommodate a water-mediated hydrogen-bond between the sialic acid and the carbonyl oxygen of His13. The bound pentasaccharide also displaces Gln56 transferring its sidechain hydrogen bond from Gln61 to Glu51. This rearrangement is accompanied by a small shift of residues 54 through 64 towards the oligosaccharide, narrowing the pocket and tightening the pentamer's grip on the receptor.

Although cholera toxin binds to G_{M1} with a Hill coefficient of 1.2, the source of this cooperativity is not obvious from the crystal structures.[44,69] Each receptor-binding site is derived almost entirely from the residues of a single B subunit. The bound pentasaccharide forms a lone solvent-mediated hydrogen bond with the backbone nitrogen of a neighboring subunit's Gly33.[57] Negatively-charged or hydrophobic amino acid substitutions (Glu, Asp, Ile, Val, Leu) at position 33 markedly reduce the affinity for G_{M1}, whereas small or positively-charged substitutions (Ala, Lys, Arg) have little effect.[70] Two spatially adjacent residues (Lys34 and Arg35) are not mutationally sensitive. Curiously, Gly33 is fairly remote from the binding pocket and its contribution to the binding energy presumably weak. The residue's importance may be related to the stabilizing hydrogen-bond formed between its carbonyl oxygen and the hydroxyl group of Tyr12. The precise positioning of Tyr12 is important for two reasons; (1) one edge of the tyrosine ring contacts the pentasaccharide's sialic acid, and (2) hydrogen bonds from the peptide backbone of neighboring residues (Glu11 O, His13 O, and His13 N) anchor the sialic acid. In the absence of ligand, residues 11 through 15 lie on a flexible solvent-exposed loop. Amino-acid substitutions at sequence position 33 may, therefore, affect ganglioside binding by simply compromising the stability afforded by Tyr12.

Despite differences in receptor preferences,[51-54] the binding pockets of the crystalline CT and LT appear to be structurally identical.[32] The contributing amino acids are conserved with the possible exception of residue 13, which is a histidine in CT but may be either a histidine or arginine in LT depending on the strain of *E. coli* from which the toxin is isolated. The side-chain of His13 does not directly participate in the binding of the G_{M1} pentasaccharide. Thus, it is difficult to explain why cholera toxin productively binds only to G_{M1} gangliosides (and fucosylated derivatives) whereas LT is less discriminating. Comparable receptor binding sites have not yet been identified for members of the Shiga family of toxins which differ from CT and LT in preferring the globotriosylceramides G_{b3} and G_{b4}.[3] The S2 and S3 subunits of PT are unique among the AB_5 toxins in attaching to receptors as isolated monomers in vitro.[71] Pertussis toxin targets glycoconjugates with some preference shown for asparagine-coupled oligosaccharides having $\alpha(2\text{-}6)$-linked sialic acid residues.[71,72]

THE A (CATALYTIC) SUBUNIT—OVERVIEW

The A subunit is translated as a single 240 amino acid peptide that is nicked by a bacterial endoprotease to form two chains (A1 and A2) prior to internalization (Fig. 7.1 and Fig. 7.6).[13,21] The two chains of the mature A subunit remain associated by extensive noncovalent forces and a single intrachain disulfide bond (Cys187 to Cys199). The A subunit is modestly hydrophobic but lacks a surface suitable for direct membrane contact. The A subunits of CT, LT, and PT are homologous in the region implicated in catalysis (116 residues near the amino terminus of PT).[34] In contrast to the highly stable B-subunit, choleragen's A subunit is loosely folded with almost complete loss of secondary structure occurring above 46 °C.[45]

Choleragen's A chain catalyzes the reaction (NAD[+] + Acceptor → ADP-ribose-Acceptor + Nicotinamide + H[+]) by acting as both an ADP-ribosyltransferase and a NAD-glycohydrolase.[11,23,73] The ADP-ribose moiety derived from NAD[+] is covalently coupled to the guanidinium group on an arginine residue of Gs_a, a stimulatory G protein of the adenylate cyclase system. Increased production of cyclic AMP precipitates the massive release of ions and fluids into the gut lumen. Ribosylation is a property that is shared with several other microbial toxins including the exotoxins from *Corynebacterium diphtheriae* and *Pseudomonas aeruginosa*. The ADP-ribosylation reaction is stereospecific and strongly influenced by the local environment of the guanidino group.[11,28]

THE A1 CHAIN

The A1 chain of cholera toxin can be divided into three distinct substructures. The first 132 amino acids form a compact globular unit composed of a mixture of α-helices and β-structure (A1$_1$). Catalysis is thought to occur in a well-defined cleft on the free surface of A1$_1$

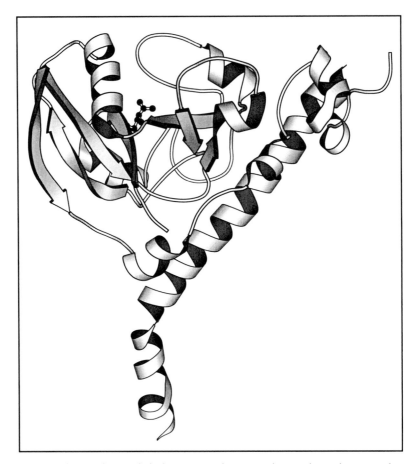

Fig. 7.6. The A subunit of cholera toxin. After proteolytic nicking, the triangular-shaped A1 chain remains covalently connected to the α-helical A2 strand by a single disulfide bond.[13,21] The distal portion of the A2 chain spirals down into the central channel of the B pentamer and provides a strong anchor between the A and B subunits. Catalysis is thought to occur in a crevice located on the surface of the A subunit. Several residues have been shown by mutational experiments to be necessary for ribosylation including Arg7 (shown).[73-76]

(Fig. 7.7).[73-76] This cleft is remote from both the A1/A2 and A/B interfaces and is presumably the binding site for both NAD$^+$ and substrate.[77]

The A1$_2$ substructure (residues 133 to 161) extends 23 Å from the distal free face of A1$_1$ (near the catalytic site) to the A1$_1$/A2 interface. The A1$_2$ "linker" appears to serve as a molecular "tether" between the compact A1$_1$ and A1$_3$ domains. The distal A1$_2$ chain is quite flexible and becomes increasingly disordered as it approaches the protease nick site located along its remote free edge. Nicking does not appear to significantly perturb the local protein structure.[78]

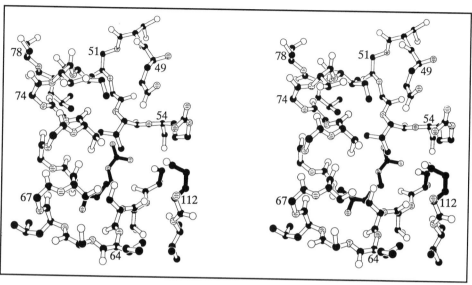

Fig. 7.7. Schematic stereo view of the putative catalytic site. The view presented is similar to that shown in Figure 7.5. The residues implicated in catalyzing the NAD-dependent ADP-ribosylation of substrate line the left wall of a narrow (7Å) cleft at the top of the A1 subunit. The cleft lacks the traditional βαβ NAD-binding fold. Both Glu112 and Ser61, which share a side-chain hydrogen bond, are important for catalysis. The side chain of Arg7, which forms the floor of the cleft, hydrogen bonds to the backbone carbonyl oxygen of Phe52 and the side-chain carboxylate of Asp9. Replacement of Arg7 with a lysine prevents catalysis; the shorter lysine side chain could fail to hydrogen-bond with the normal partners and instead interact with Glu112. The left wall of the crevice is composed of a short stretch of charged α-helix (residues 65 to 77).[73-76]

The carboxyl terminal 31 residues, which surrounds the disulfide bridge linking the A1 and A2 fragments, forms a third globular substructure (A1$_3$). The A1$_3$ substructure contains a high density of hydrophobic residues including a cluster of four prolines (Pro168, Pro169, Pro184, and Pro185) and two tryptophans (Trp174 and Trp179). The A1$_1$ and A1$_2$ interface is almost entirely nonpolar with only a few ionic interactions occurring at the molecular surface. This arrangement tempts speculation that at the lipid-water interface the A1 subunit could 'unravel' to permit the A1$_1$ substructure to make contact with and/or pass through the bilayer.

THE A2 CHAIN

Cholera toxin's A2 chain, which consists of a nearly continuous α-helix broken only by a central 52 degree kink, anchors the enzymatic A1 chain to the B pentamer (Fig. 7.6 and Fig. 7.8). The central kink, stabilized by a hydrogen-bond between the γ-oxygen of Ser228 and the peptide nitrogen of Asp229, redirects the helix prior to its descent into the pentamer pore. Intriguingly, the sequence of the last

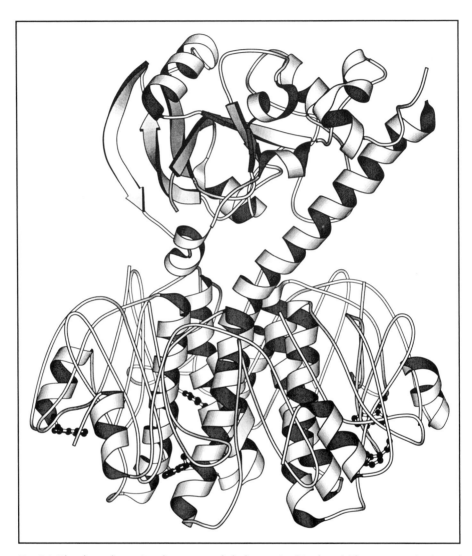

Fig. 7.8. The three-dimensional structure of cholera toxin. (A, above) The structure is viewed from the "side" showing the A subunit as it rests on top of the B pentamer. With the exception of the tethering A2 helix, the interface between the A and B subunits is limited and permits several degrees of rotation around the 5-fold axis.[115] The A2 chain fits snugly into the central pore of choleragenoid where it is surrounded by the five central helices. The carboxyl terminus of the A2 chain emerges from the "ventral" surface of the pentamer where it remains protected by a small protein flange. The sites of ganglioside binding are indicated by the sidechains of Trp88. (B, on facing page) A space-filling representation of cholera toxin on an idealized membrane. Although the manner in which the catalytic A subunit passes into the cell remains controversial, choleragen most likely binds to its target membrane with the A subunit facing away. The G_{M1} gangliosides are also shown emerging from the membrane and binding to the ventral surface of the B pentamer.

4 residues of the A2 chain (KDEL) mimics that of an endoplasmic retention signal (HDEL).[79-81] Cholera's KDEL residues lie outside the ventral opening of the central pore with little or no direct stabilization by the B subunits but are somewhat sheltered by the ventral flange. Presumably, if cholera toxin binds to its G_{M1} receptors with the A1 subunit facing away from the target cell, the carboxyl-terminus of the A2 chain would make contact with the lipid membrane.

Despite the relatively small size of the A2 chain, it shares an extensive interface with the A1 chain. The long amino-terminal A2 helix (residues 196-228) lies in a shallow groove that extends from one corner of the A1 subunit to the A1/B interface. Although after nicking, the disulfide bridge between A1(Cys187) and A2(Cys199) is the sole covalent anchor between the A1 and A2 chains, there are multiple nonpolar interactions throughout the length of the interface.

The A2 chain of Shiga toxin resembles that described for cholera toxin.[39] The A2 chain of LT, in contrast, is divided into three discrete segments; a long amino-terminal helix (residues 197 to 224), a length of extended chain that winds through the pore of the B pentamer (residues 225 to 231), and a small carboxyl-terminal helix (residues 232 to 236+).[34] The less compact structure of the A2 chain places the terminal residues further away from the pore and the sheltering flange. Not surprisingly, the RDEL residues are disordered in the crystal structure of LT. Deletion of these terminal four residues has little effect on B subunit oligomerization but significantly reduces the stability of the holotoxin.[62] The in vivo effect of such a deletion on toxin potency is unknown.

THE A SUBUNIT / B PENTAMER INTERFACE

The arrangement of the B subunits in choleragen differs only modestly from that described for choleragenoid. The presence of the A chain perturbs only those B side chains that extend into the pentamer's central pore or that lie directly at the A-B interface. The A subunit does not appear to have any structural or functional bearing on the ganglioside-binding sites which lie on the remote outer edge of each B subunit.

Few direct stabilizing interactions occur between the A1 chain and the B pentamer. Exceptions include three arginine side chains (Arg33, Arg143, and Arg148) located along the ventral surface of the A1 subunit that form multiple hydrogen bonds with carbonyl and glutamyl oxygens located along the top of the pentamer's central α-helices (Thr78 O, Glu79 O, Glu79 Oϵ1 and Oϵ2). Since these contacts are all located adjacent to the central axis of the pore, substantial rotation of the A subunit with regard to the B pentamer is possible without disruption of the interface (Fig. 7.8).

In contrast to the A1 chain, the A2 chain intimately interacts with all five B subunits. The A2 subunit passes through the central pore of the B pentamer either as a continuous helix (CT) or as an extended chain anchored at its carboxyl terminus by a short turn of helix (LT).[34] The pore diameter is just sufficiently wide to accommodate the A2 chain as a helix, therefore, choleragen's A subunit displaces many of the solvent molecules noted to fill the choleragenoid pore. Stabilizing contacts within the pore between the A2 chain and the B subunits are largely hydrophobic with very few specific hydrogen bonds noted at the resolution of the current structure. The presence of the A subunit has little effect on the stability of the B pentamer which suggests a trade off between the stability of intrapentameric and intersubunit bonds at the A/B interface.[82]

Residues 227 through 236 of the A2 fragment promote the cooperative assembly of choleragen.[62] These residues form part of a long helix that descends into the central pore of the B subunit pentamer. Deletion of the four terminal residues of the A2 chain (residues 236 through 240) weakens the interaction between the enzymatic A subunit and the B pentamer without impairing pentamer formation.[62]

MECHANISM OF ACTION

The mechanism by which the A1 fragment gains access to the target cell's cytoplasmic machinery remains a mystery. Several nonexclusive hypotheses have been advanced including: (1) direct insertion of the

A1 fragment into the membrane/cytoplasm[83] and (2) receptor mediated endocytosis.[17,84] In the absence of ligand, the negatively charged pentasaccharide of G_{M1} lies nearly 21 Å above the lipid/water interface.[18,85-87] Photolabelling and surface pressure experiments indicate that the B subunits do not penetrate deeply into the membrane while bound to their ganglioside receptors.[88-90] Binding of both the holotoxin and choleragenoid is, however, associated with increased membrane permeability.[91-93] Krasilnikov et al reported that at pH 4.5 choleragenoid forms an anionic channel with a mean conductance that varies inversely with pH.[94,95] Using small molecular probes, the effective diameter of the channel was calculated to be 21 ± 2 Å. This value is similar to estimates made from low-resolution electron micrographs of CT bound to lipid films.[41] One possible explanation for the discrepancy between this diameter and that measured from the crystalline toxins (an effective diameter of 11-16 Å depending upon the depth of insertion) is widening of the pore after membrane attachment. This movement would be consistent with the A subunit translocating into the cytoplasm by unfolding and threading its way through the central pore.[83] However, there is no structural evidence to support such a dramatic rearrangement of the stable pentamer interface.

The structure of the choleragenoid-G_{M1} pentasaccharide complex[47] and recent electron micrographs of cholera toxin—model membrane complexes[96] argue for the holotoxin binding to the membrane with the A subunit pointing away from the cell. This orientation is further supported by immunological experiments indicating that the A1 fragment of the surface-bound or surface-assembled holotoxin is accessible to anti-A1 antibodies.[64] It is unlikely that the pentamer dissociates to provide a larger pore for passage of the A1 subunit. Internalization of the toxin must, therefore, be dependent upon some form of endocytosis that is perhaps mediated by the KDEL terminus of the A2 chain.[97] Cholera toxin coupled to horse-radish peroxidase (HRP) was visualized in the Golgi-endoplasmic reticulum-lysosomal system of cultured cells within 1 hour of incubation at 37 °C. The entry of the toxin-HRP conjugate appeared not to occur through coated pits and was negligible at 4 °C.[79,80,84]

From a medical perspective, the strong structural resemblance between CT and LT is puzzling given that the diseases associated with these toxins are quantitatively quite different. Whereas cholera is a fulminant disease that can be fatal within a few hours, the diarrhea caused by enteroxigenic strains of *E. coli* is generally low-grade and transitory. Possible reasons for this disparity include subtle structural differences in the toxins, in toxin processing and secretion, or variation in microbe ecology. Further work will be directed at pinpointing the molecular mechanism(s) and weaknesses of these potent, highly-evolved microbial toxins.

ACKNOWLEDGMENTS

This work was supported by the National Institutes of Health through grant R01 AI28535 and by the U.S. Department of Energy, Office of Health and Environmental Research through contract W-31-109-ENG-38.

REFERENCES
1. Glass RI, Libel M, Brandling-Bennett AD. Epidemic cholera in the Americas. Science 1992; 256:1524-25.
2. Morris JG Jr, Black RE. Cholera and other vibrioses in the United States. New Engl J Med 1985; 312:343-50.
3. Finkelstein RA. Combating epidemic cholera. Science 1992; 257:852.
4. Field M, Rao MC, Chang EB. Intestinal electrolyte and diarrheal disease. New Engl J Med 1989; 321:879-83.
5. Mekalanos JJ, Sadoff JC. Cholera vaccines: fighting an ancient scourge. Science 1994; 265:1387-89.
6. De SN. Enterotoxicity of bacteria-free culture-filtrate of *Vibrio cholerae*. Nature 1959; 183:1533-34.
7. Finkelstein RA, Cholera, the cholera enterotoxins, and the cholera enterotoxin-related enterotoxin family. In: Owen P, Foster TJ, eds. Immunochemical and Molecular Genetic Analysis of Bacterial Pathogens, New York: Elsevier Science Publishers, 1988:85-102.
8. Fishman PH. Mechanism of action of cholera toxin. In: Moss J, Vaughan M, eds. ADP-Ribosylating Toxins and G Proteins: Insights into Signal Transduction. 1990:127-40.
9. Burnette WN. AB₅ ADP-ribosylating toxins: comparative anatomy and physiology. Structure 1994; 2:151-58.
10. Spangler BD. Structure and function of cholera toxin and the related *Escherichia coli* heat-labile enterotoxin. Microbiol Reviews 1992; 56:622-47.
11. Foster JW, Kinney DM. ADP-ribosylating microbial toxins. CRC Critical Review in Microbiology 1985; 11:273-98.
12. Passador L, Iglewski W. ADP-ribosylating toxins. Methods in Enzymology 1994; 235:617-31.
13. Kassis S, Hagmann J, Fishman PH. Mechanism of action of cholera toxin on intact cells. Generation of A1 peptide and activation of adenylate cyclase. J Biol Chem 1982; 257:12148-52.
14. Brunton JL. The shiga toxin family: molecular nature and possible role in disease. In Iglewski B, Clark U, eds. The Bacteria. Vol. 11. New York: Academic Press, 1990:377-97.
15. O'Brien AD, Holmes RK. Shiga and shiga-like toxins. Microbiol Rev 1987; 51:206-20.
16. Cassel D, Pfeuffer T. Mechanism of cholera toxin action: covalent modification of the guanyl nucleotide-binding protein of the adenylate cyclase system. Proc Natl Acad Sci USA 1978; 75:2669-73.
17. Janicot M, Fouque F, Desbuqois B. Activation of rat liver adenylate cyclase by cholera toxin requires toxin internalization and processing in endosomes. J Biol Chem 1991; 266:12858-65.

18. Hansson H-A, Holmgren J, Svennerholm L. Ultrastructural localization of cell membrane GM_1 ganglioside by cholera toxin. Proc Natl Acad Sci USA 1977; 74:3782-86.

19. Critchley DR, Magnani JL, Fishman PH. Interactions of cholera toxin with rat intestinal brush border membranes. J Biol Chem 1981; 256:8724-31.

20. Hardy SJS, Holmgren J, Johansson S et al. Coordinated assembly of multisubunit proteins: oligomerization of bacterial enterotoxins in vivo and in vitro. Proc Natl Acad Sci USA 1988; 85:7109-13.

21. Mekalanos JJ, Collier RJ, Romig WR. Enzymatic activity of cholera toxin. II. Relationships to proteolytic processing, disulfide bond reduction, and subunit composition. J Biol Chem 1979; 254:5855-61.

22. Tomasi M, Battistini A, Araco A et al. The role of the reactive disulfide bond in the interaction of cholera toxin functional regions. Eur J Biochem 1979; 93:621-27.

23. Gill DM, Coburn J. ADP-ribosylation by cholera toxin: Functional analysis of a cellular system that stimulates the enzymic activity of cholera toxin fragment A_1. Biochemistry 1987; 26:6364-71.

24. Gill DM, King CA. The mechanism of action of cholera toxin in pigeon erythrocyte lysates. J Biol Chem 1975; 250:6424-32.

25. Kahn RA, Gilman AG. Purification of a protein cofactor required for ADP-ribosylation of the stimulatory regulatory component of adenylate cyclase by cholera toxin. J Biol Chem 1984; 259:6228-34.

26. Lee C-M, Chang PP, Tsai S-C et al. Activation of heat-labile enterotoxins by native and recombinant adenosine diphosphate-ribosylation factors, 20 kD guanine nucleotide-binding proteins. J Clin Invest 1991; 87:1780-86.

27. Moss J, Tsai SC, Vaughan M. Activation of cholera toxin by ADP-ribosylation factors. Methods in Enzymology 1994; 235:640-47.

28. Moss J, Vaughan M. ADP-ribosylation of guanyl nucleotide-binding regulatory proteins by bacterial toxins. Adv Enzymology 1988; 61:303-79.

29. Moss J, Vaughan M. Activation of cholera toxin and *Escherichia coli* heat-labile enterotoxins by ADP-ribosylation factors, a family of 20 kDa guanine nucleotide-binding proteins. Mol Microbiol 1991; 5:2621-27.

30. Peterson WJ, Ochoa LG. Role of prostaglandins and cAMP in the secretory effects of cholera toxin. Science 1989; 245:857-59.

31. Zhang R-G, Westbrook ML, Westbrook EM et al. The three-dimensional structure of cholera toxin. J Mol Biol 1995; 251:563-73.

32. Zhang R-G, Maulik PR, Westbrook EM et al. The 2.4 Å crystal structure of the cholera toxin B subunit pentamer: choleragenoid. J Mol Biol 1995; 251:550-62.

33. Sixma TK, Pronk SE, Kalk KH et al. Crystal structure of a cholera toxin-related heat-labile enterotoxin from *E. coli*. Nature 1991; 351:371-78.

34. Sixma TK, Kalk KH, van Zanten BAM et al. Refined crystal structure of *Escherichia coli* heat-labile enterotoxin, a close relative of cholera toxin. J Mol Biol 1993; 230:890-918.

35. Wick MJ, Frank DW, Storey DG et al. Structure, function, and regulation of *Pseudomonas aeruginosa* exotoxin A. Ann Rev Microbiol 1990; 44:335-63.

36. Allured, VS, Collier RJ, Carroll SF et al. Structure of exotoxin A of *Pseudomonas aeruginosa* at 3.0 Å resolution. Proc Natl Acad Sci USA 1986; 83:1320-24.

37. Stein PE, Boodhoo A, Tyrell GJ et al. Crystal structure of the cell-binding B oligomer of verotoxin-1 from *E. coli*. Nature 1992; 355:748-50.

38. Choe S, Bennett MJ, Fujii G et al. The crystal structure of diphtheria toxin. Nature 1992; 357:216-22.

39. Fraser ME, Chernaia MM, Kozlov YV et al. Crystal structure of the holotoxin from *Shigella dysenteriae* at 2.5 Å resolution. Nature Structural Biology 1994; 1:59-64.

40. Stein PE, Boodhoo A, Armstrong GD et al. The crystal structure of pertussis toxin. Structure 1994; 2:45-57.

41. Ludwig DS, Ribi HO, Schoolnik GK. Two-dimensional crystals of cholera toxin B-subunit-receptor complexes: Projected structure at 17-Å resolution. Proc Natl Acad Sci USA 1986; 83:8585-88.

42. Mosser G, Mallouh V, Brisson A. A 9 Å two-dimensional projected structure of cholera toxin B-subunit—GM$_1$ complex determined by electron crystallography. J Mol Biol 1992; 226:23-28.

43. Yang J, Tamm LK, Tillack TW et al. New approach for atomic force microscopy of membrane proteins. The imaging of cholera toxin. J Mol Biol 1993; 229:286-90.

44. Schon A, Freire E. Thermodynamics of intersubunit interactions in cholera toxin upon binding to the oligosaccharide portion of its cell-surface receptor, ganglioside GM$_1$. Biochemistry 1989; 28:5019-24.

45. Surewicz WK, Leddy JJ, Mantsch HH. Structure, stability and receptor interaction of cholera toxin as studied by Fourier-transform infrared spectroscopy. Biochemistry 1990; 29:8106-111.

46. Dallas WS, Falkow S. Amino acid sequence homology between cholera toxin and *Escherichia coli* heat-labile toxin. Nature 1980; 288:499-501.

47. Sixma TK, Pronk SE, Kalk KH et al. Lactose binding to heat-labile enterotoxin revealed by X-ray crystallography. Nature 1992; 355:561-64.

48. Sixma TK, Stein PE, Hol WGJ. Comparison of the B pentamers of heat-labile enterotoxin and verotoxin 1: Two structures with remarkable similarity and dissimilarity. Biochemistry 1993; 32:191-98.

49. Galloway TS, van Heyningen S. Binding of NAD$^+$ by cholera toxin. Biochem J 1987; 244:225-30.

50. Murzin AG. OB (oligonucleotide/oligosaccharide-binding) fold: common structural and functional solution for nonhomologous sequences. EMBO J 1993; 12:861-67.

51. Donta ST, Poindexter NJ, Ginsberg BH. Comparison of the binding of cholera and *Escherichia coli* enterotoxins to Y1 adrenal cells. Biochemistry 1982; 21:660-64.

52. Fishman PH, Pacuszka T, Orlandi PA. Gangliosides as receptors for bacterial enterotoxins. Advances in Lipid Research 1993; 25:165-87.

53. Holmgren J, Fredman P, Lindbald M et al. Rabbit intestinal glycoprotein receptor for *Escherichia coli* heat-labile enterotoxin lacking affinity for cholera toxin. Infect Immun 1982; 38:424-33.

54. Fukuta S, Magnani JL, Twiddy EM et al. Comparison of the carbohydrate-binding specificities of cholera toxin and *Escherichia coli* heat-labile enterotoxins LTh-1, LT-IIa, and LT-IIb. Infect Immun 1988; 56:1748-53.

55. Presta LG, Rose GD. Helix signals in proteins. Science 1988; 240:1632-41.

56. De Wolf MJS, Fridkin M, Kohn LD. Tryptophan residues of cholera toxin and its A and B protomers; intrinsic fluorescence and solute quenching upon interacting with the ganglioside GM_1, oligo GM_1, or dansylated oligo GM_1. J Biol Chem 1981; 256:5489-96.

57. Merritt EA, Sarfaty S, van den Akker F et al. Crystal structure of cholera toxin B-pentamer bound to receptor GM1 pentasaccharide. Protein Sci 1994; 3:166-75.

58. Hirst TR, Holmgren J. Conformation of protein secreted across bacterial outer membranes: A study of enterotoxin translocation from *Vibrio cholerae*. Proc Natl Acad Sci USA 1987; 84:7418-22.

59. Hofstra H, Witholt B. Heat-labile enterotoxin in *Escherichia coli*. Kinetics of association of subunits into periplasmic holotoxin. J Biol Chem 1985; 260:16037-44.

60. De Wolf MJS, Van Dessel GAF, Lagrou AR et al. pH-induced transitions in cholera toxin conformation: a fluorescence study. Biochemistry 1987; 26:3799-806.

61. Goins B, Freire E. Thermal stability and intersubunit interactions of cholera toxin in solution and in association with its cell-surface receptor ganglioside GM_1. Biochemistry 1988; 27:2046-52.

62. Streatfield SJ, Sandkvist M, Sixma TK et al. Intermolecular interactions between the A-subunit and B-subunit of heat-labile enterotoxins from *Escherichia coli* promote holotoxin assembly and stability in vitro. Proc Natl Acad Sci USA 1992; 89:12140-44.

63. Bhakuni V, Xie D, Freire E. Thermodynamic identification of stable folding intermediates in the B-subunit of cholera toxin. Biochemistry 1991; 30:5055-60.

64. Orlandi PA, Fishman PH. Orientation of cholera toxin bound to target cells. J Biol Chem 1993; 268:17038-44.

65. Braig K, Otwinowski Z, Hegde R et al. The crystal structure of the bacterial chaperonin GroEL at 2.8 Å. Nature 1994; 371:578-86.

66. Huber R, Romisch J, Paques E-P. The crystal and molecular structure of human annexin V, an anticoagulant protein that binds to calcium and membranes. EMBO 1990; 9:3867-74.

67. Fishman PH, Moss J, Osborne JC Jr. Interaction of choleragen with the oligosaccharide of ganglioside GM_1: Evidence for multiple oligosaccharide binding sites. Biochemistry 1978; 17:711-16.

68. Masserini M, Freire E, Palestini P et al. Fuc-GM1 ganglioside mimics the receptor function of GM1 for cholera toxin. Biochemistry 1992; 31:2422-26.

69. Sattler J, Schwarzmann G, Knack I et al. Studies of ligand binding to cholera toxin III, cooperativity of oligosaccharide binding. Hoppe-Seyler's Z Physiol Chem 1978; 359:719-23.

70. Jobling MG, Holmes RK. Analysis of structure and function of the B subunit of cholera toxin by the use of site-directed mutagenesis. Mol Microbiol 1991; 5:175-67.

71. Saukkonen K, Burnette WN, Mar VL et al. Pertussis toxin has eukaryotic-like carbohydrate recognition domains. Proc Natl Acad Sci USA 1992; 89:118-22.

72. Armstrong GD, Howard LA, Peppler MS. Use of glycosyltransferases to restore pertussis toxin receptor activity to asialogalactofetuin. J Biol Chem 1988; 263:8677-84.

73. Gill DM. Involvement of nicotinamide adenine nucleotide in the action of cholera toxin in vitro. Proc Natl Acad Sci USA 1975; 72:2064-68.

74. Lai C-Y, Xia Q-C, Salotra PT. Location and amino acid sequence around the ADP-ribosylation site in the cholera toxin active subunit A1. Biochem Biophys Res Commun 1983; 116:341-348.

75. Burnette WN, Mar VL, Plater BW et al. Site-directed mutagenesis of the catalytic subunit of cholera toxin: substituting lysine for arginine 7 causes loss of activity. Infect Immun 1991; 59:4266-70.

76. Harford SC, Dykes W, Hobden AN et al. Inactivation of the *Escherichia coli* heat-labile enterotoxin by in vitro mutagenesis of the A-subunit gene. Eur J Biochem 1989; 183:311-16.

77. Domenighini M, Montecucco C, Ripka WC. Computer modeling of the NAD binding site of ADP-ribosylating toxins: active-site structure and mechanism of NAD binding. Mol Microbiol 1991; 5:23-31.

78. Merritt EA, Pronk SE, Sixma TK et al. Structure of partially-activated *E. coli* heat-labile enterotoxin (LT) at 2.6 Å resolution. FEBS Letters 1994; 337:88-92.

79. Joseph KC, Kim SU, Steiber A et al. Endocytosis of cholera toxin into neuronal GERL. Proc Natl Acad Sci USA 1978; 75:2815-19.

80. Joseph KC, Steiber A, Gonatas NK. Endocytosis of cholera toxin in GERL-like structures of murine neuroblastoma cells pretreated with GM_1 ganglioside. J Cell Biol 1979; 81:543-54.

81. Lewis MJ, Pelham HRB. A human homologue of the yeast HDEL receptor. Nature 1990; 348:162-63.

82. Van Heyningen S. Conformational changes in subunit A of cholera toxin following the binding of ganglioside to subunit B. Eur J Biochem 1982; 122:333-37.

83. Gill DM. The arrangement of the subunits of cholera toxin. Biochemistry 1976; 15:1242-48.

84. Tran D, Carpentier J-L, Sawano F et al. Ligands internalized through coated or noncoated invaginations follow a common intracellular pathway. Proc Natl Acad Sci USA 1987; 84:7957-61.

85. Acquotti D, Poppe L, Dabrowski J et al. Three-dimensional structure of the oligosaccharide chain of GM1 ganglioside revealed by a distance-map-

ping procedure: A rotating and laboratory frame nuclear Overhauser enhancement investigation of native glycolipid in dimethyl sulfoxide and in water-dodecylphosphocholine solutions. J Am Chem Soc 1990; 112:7772-78.

86. McDaniel RV, McIntosh TJ. X-ray diffraction studies of the cholera toxin receptor, GM_1. Biophys J 1986; 49:94-96.

87. Thompson TE, Tillack TW. Organization of glycosphingolipids in bilayers and plasma membranes of mammalian cells. Ann Rev Biophys Chem 1985; 14:361-86.

88. Reed RA, Mattai J, Shipley GG. Interaction of cholera toxin with ganglioside GM1 receptors in supported lipid monolayers. Biochemistry 1987; 26:824-32.

89. Ribi HO, Ludwig DS, Mercer KL et al. Three-dimensional structure of cholera toxin penetrating a lipid membrane. Science 1988; 239:1272-76.

90. Wisnieski BJ, Bramhall JS. Photolabelling of cholera toxin subunits during membrane penetration. Nature 1981; 289:319-21.

91. Moss J, Richards RL, Alving CR et al. Effect of the A and B protomers of choleragen on release of trapped glucose from liposomes containing or lacking ganglioside GM_1. J Biol Chem 1977; 252:797-98.

92. Tomasi M, Montecucco C. Lipid insertion of cholera toxin after binding to GM_1 containing liposomes. J Biol Chem 1981; 256:11177-81.

93. Tosteson MD, Tosteson DC. Bilayers containing ganglioside develop channels when exposed to cholera toxin. Nature 1978; 275:142.

94. Krasilnikov OV, Muratkhodjaev JN, Voronov SE et al. The ionic channels formed by cholera toxin in planar bilayer lipid membranes are entirely attributable to its B-subunit. Biochim Biophys Acta 1991; 1067:166-70.

95. Krasilnikov OV, Sabirov RZ, Ternovsky VI et al. A simple method for the determination of the pore radius of ion channels in planar lipid bilayer membrane. FEMS Microbiol Immun 1992; 5:93-100.

96. Cabral-Lilly D, Sosinsky GE, Reed RA et al. Orientation of cholera toxin bound to model membranes. Biophys J 1994; 66:935-41.

97. Houslay MD, Elliott KRF. Is the receptor-mediated endocytosis of cholera toxin a prerequisite for its activation of adenylate cyclase in intact rat hepatocytes? FEBS Lett 1981; 128:289-92.

98. Leong J, Vinal AC, Dallas WS. Nucleotide sequence comparison between heat-labile toxin B subunit cistrons from *Escherichia coli* of human and porcine origin. Infect Immun 1985; 48:73-77.

99. Tsuji T, Honda T, Miwatani T et al. The amino acid sequence of the B-subunit of porcine *Escherichia coli* enterotoxin. FEMS Microbiol Lett 1984; 25:243-46.

100. Yamamoto T, Yokota T. Sequence of heat-labile enterotoxin of *Escherichia coli* pathogenic for humans. J Bacteriol 1983; 155:728-33.

101. Mekalanos JJ, Swartz DJ, Pearson GDN et al. Cholera toxin genes: nucleotide sequence, deletion analysis and vaccine development. Nature 1983; 306:551-57.

102. Dallas WS. Conformity between heat-labile toxin genes from human and porcine enterotoxigenic *Escherichia coli*. Infect Immun 1983; 40:647-52.

103. Kurosky A, Markel DE, Peterson JW. Covalent structure of the B chain of cholera enterotoxin. J Biol Chem 1977; 252:7257-64.

104. Dykes CW, Halliday IJ Hobden AN et al. A comparison of the nucleotide sequence of the A subunit of heat-labile enterotoxin and cholera toxin. FEMS Microbiol Lett 1985; 26:171-74.

105. Lockman H, Kaper JB. Nucleotide sequence analysis of the A2 and the B subunits of *Vibrio cholerae* enterotoxin. J Biol Chem 1983; 258:13722-26.

106. Carroll SF, Collier RJ. Active site of *Pseudomonas aeruginosa* exotoxin A. J Biol Chem 1987; 262:8707-11.

107. Douglas CM, Collier RJ. Exotoxin A of *Pseudomonas aeruginosa*: substitution of glutamic acid 553 with aspartic acid drastically reduces toxicity and enzymatic activity. J Biol Chem 1987; 169:4967-71.

108. Tsuji T, Inoue T, Miyama A et al. Glutamic acid-112 of the A subunit of heat-labile enterotoxin from enterotoxigenic *Escherichia coli* is important for ADP-ribosyltransferase. FEBS Lett 1991; 291:319-21.

109. Olsvik O, Wahlberg J, Petterson B. Use of automated sequencing of polymerase chain reaction-generated amplicons to identify three types of cholera toxin subunit B in *Vibrio cholerae* O1 strains. J Clin Microbiol 1993; 31:22-25.

110. Kraulis PJ. MOLSCRIPT: a program to produce both detailed and schematic plots of protein structures. J Appl Cryst 1991; 24:946-50.

111. Richards FM, Kundrot CE. DEFINE_Structure: a program for specification of secondary and first level supersecondary structure from alpha carbon coordinate list. Proteins: Struc Funct Genetics 1988; 3:71-84.

112. Sixma TK, Aguirre A, Terwisscha van Scheltinga AC et al. Heat-labile enterotoxin crystal forms with variable A/B_5 orientation. FEBS Lett 1992; 305:81-85.

CHAPTER 8

E. COLI HEAT LABILE ENTEROTOXIN AND CHOLERA TOXIN B-PENTAMER— CRYSTALLOGRAPHIC STUDIES OF BIOLOGICAL ACTIVITY

Ethan A. Merritt, Focco van den Akker and Wim G.J. Hol

The heat-labile enterotoxins from *E. coli* are part of a larger class of bacterial toxins whose members are collectively responsible for a variety of diseases known since ancient times and still afflicting the world today. Most notable of these is cholera, but also among them are dysentery, whooping cough, traveler's diarrhea, and the more recently characterized hemolytic uremic syndrome ("hamburger disease"). Although the specific cellular targets, enzymatic activity, and biological mode of action of these toxins vary, their evolutionary kinship may be traced both through sequence homology and a shared protein quaternary structure. Members of this class of toxins are AB_5 hexameric assemblies. Each toxin's catalytic activity is carried on the A subunit, while the pentamer of B subunits mediates receptor recognition and cell surface binding. Atomic resolution structures have been determined for several of the AB_5 toxins. Their structural and functional relationships have been reviewed recently by Burnette[1] and Merritt and Hol.[2]

The cholera toxin family is a subset of the AB_5 toxins consisting specifically of cholera toxin (CT) and the well-characterized *E. coli* heat-labile enterotoxins LT-I, LT-IIa, and LT-IIb. The A subunits of these toxins are homologous in primary sequence, and catalyze the ADP-ribosylation of arginine residues in the intoxicated cell. Probable additional members of the family have been identified from a variety

---

of bacteria on the basis of primary sequence homology and cross-reactivity with anti-CT antibodies.[3] We will focus here primarily on crystallographic studies of *E. coli* heat-labile enterotoxin LT-I, henceforth referred to simply as LT.

LT is 80% sequence identical to cholera toxin in both the A and B subunits, and the two toxins exhibit only minor differences in receptor binding specificity and catalytic activity. The diarrheal disease caused by infection via enterotoxigenic *E. coli* is less severe than that due to infection by *V. cholerae*, but is nevertheless a major health problem because of widespread incidence. Deaths due to *E. coli* infection and mediated by LT have been estimated at 800,000 per year, largely attributable to the death of afflicted infants from dehydration.[4] The difference in severity between cholera and the *E. coli* induced disease is due to virulence factors other than the secreted toxins. LT was the first of the AB₅ toxins to yield an X-ray structure[5,6] and remains a focus for investigation of the molecular details of receptor binding, catalytic mechanism, and immunogenic behavior.

TOXIN ASSEMBLY AND SECRETION

LT is expressed in certain strains of enterotoxigenic *E. coli* (ETEC). The toxin genes *etxA* and *etxB* are encoded on large conjugative plasmids, and have been characterized and cloned from *E. coli* strains isolated from pigs and from humans. The toxin peptide chains are directed to the periplasm of the bacterium by N-terminal signal sequences which are subsequently cleaved. In the periplasm the monomeric A and B chains assemble to form the AB₅ holotoxin. Simultaneous assembly of the A and B subunits is apparently necessary, as isolated A chain will not assemble in vitro with preformed B-pentamer.[7] The B-pentamer will assemble from isolated B chains, but assembly is enhanced by the presence of the A chain. Residues 226-236 of the 240 residue A subunit have in particular been implicated as promoting the oligomerization of the B subunits.[8] The assembled AB₅ toxin is not actively secreted from the *E. coli* periplasm, but rather is released by bacterial lysis. Production of toxin by *E. coli* may therefore be seen as less specialized than that of the homologous toxin by *V. cholerae*, which is encoded chromosomally and is actively secreted from the periplasm of intact cells.

MEMBRANE BINDING AND CELL ENTRY

After release into the lumen of the intestine, the LT holotoxin acts through binding to the surface of intestinal epithelial cells. This process is mediated by the specific recognition and binding of the toxin B-pentamer to the saccharide moiety of ganglioside G_{M1}, a normal component of the cell membrane. Each holotoxin molecule can bind up to five molecules of the receptor. Ganglioside G_{M1} is a glycosphingolipid distinguished by a characteristic branched pentasaccharide protruding from the cell surface and anchored to the cell membrane via a ceramide headgroup. Whereas the specificity of cholera toxin for the G_{M1} saccharide is quite stringent,[9] some binding of the *E. coli* toxin to glycoproteins has been reported.[10-14] LT has also been reported to tolerate more variation than cholera toxin in the specific nature of the ganglioside oligosaccharide; e.g., LT binds weakly to G_{M2} and to asialo-G_{M1}.[15] Weak binding of the toxin to various fragments of the G_{M1} saccharide can also be measured in vitro, and has been explored through crystal structure determinations of the toxin complexed with simpler saccharide moieties (see below).

Translocation of the toxin across the target cell membrane after initial receptor binding is perhaps the least understood stage of intoxication by LT. Early models[16] in which the enzymatically active A subunit was physically inserted into or through the cell membrane as a consequence of receptor binding by the B-pentamer, with or without pore formation by the B-pentamer itself, have been shown to be incorrect. Physical evidence from spectroscopic labeling experiments,[17] atomic force microscopy,[18] electron microscopy,[19] and crystallographic determinations (discussed below) indicate that the catalytic domain of the A subunit is distal to the membrane surface after receptor binding. The toxin, or at least the catalytic domain of the A subunit, must therefore enter the cell in a separate and subsequent stage of the intoxication process. Several lines of evidence presented below suggest that this occurs via the formation of endocytotic vesicles which merge with the Golgi complex. Unlike some other toxins such as diphtheria toxin, there is no evidence for a major pH-induced conformational change or dissociation during cell entry, conformational changes in the cholera AB_5 holotoxin followed by fluorescence quenching are reversible over the range pH 4 to pH 7.[20]

A final activation step which occurs prior to the intracellular enzymatic action of the toxin is the cleavage of the 240 amino acid A subunit into two fragments, A1 and A2. This involves both proteolytic cleavage of a flexible loop between residues 192 and 195,[21] and reduction of a disulfide bridge between Cys 187 and Cys 199. A requirement for equivalent proteolytic cleavage and disulfide reduction is a general feature of the activation of AB_5 bacterial toxins.

ENZYMATIC ACTIVITY

Members of the cholera toxin family, including LT, are ADP-ribosylating toxins which act through covalent modification of regulatory GTP-binding proteins. This enzymatic activity is performed by the A1 domain of the A subunit. The reaction catalyzed is:

$$NAD^+ + ACCEPTOR \rightarrow ADP\text{-ribose-}ACCEPTOR + Nicotinamide + H^+$$

The toxins are capable of modifying arginine residues in a number of proteins, but the physiologically important acceptor for CT and LT is Arg 201 of $G_s\alpha$. This residue lies at the site of GDP/GTP regulation, and the ADP-ribosylated $G_s\alpha$ behaves as if it were constitutively in its GTP-bound form. The modified $G_s\alpha$ therefore leads to elevated intracellular cAMP concentration through continuous stimulation of adenylate cyclase, and eventually to increased Cl$^-$ secretion and a net movement of electrolytes into the lumen of the intestine.

The enzymatic activity of LT (and of cholera toxin) is enhanced in the presence of another G protein known as ARF, or ADP-ribosylation factor. This is believed to occur via direct interaction between ARF and the activated A1 fragment of the toxin, subsequent to or concomitant with separation of the A1 and A2 domains by reduction of the disulfide bond linking them.[22] The ARF:A1 complex can be isolated and shown to be more active than A1 alone.[23] The physiological function of ARF in the cell is believed to be participation in the coatamer-mediated vesicular transport system.[24] It is unknown whether this function is relevant to intoxication by LT and CT.

The precise cascade of intracellular events triggered by elevated cAMP levels after intoxication is not known, but is probably mediated by protein kinase A.[25] An intriguing aspect of this mechanism is the recent observation that transgenic mice expressing a defective homologue to the human cystic fibrosis gene CFTR are resistant to cholera toxin.[26] The CTFR gene codes for a cAMP-regulated Cl$^-$ channel, consistent with this mechanism. Homozygous CTFR-deficient mice did not secrete fluid in response to administered cholera toxin; heterozygotes secretion of Cl$^-$ and fluid was half that of wild-type response.[26] The clear implication is that the relatively high frequency of cystic fibrosis in some human populations may be due to positive selection based on heterozygote resistance to diarrheal diseases caused by bacterial toxins such as LT and CT.

ARCHITECTURE OF THE AB₅ ASSEMBLY

SECONDARY AND TERTIARY STRUCTURE

The AB₅ holotoxin is a remarkable molecular assembly (Fig. 8.1). Five identical B subunits form a regular pentamer surrounding a central pore. The catalytic A1 domain of the A subunit sits entirely on one side of this pentamer. The A2 domain serves to link the two, consisting of a 26-residue long α-helix which lies against one face of the catalytic domain followed by an extended C-terminal region which entirely spans the pore of the B-pentamer and emerges from the other side.

The B-pentamer forms a torus roughly 30 Å in radius and 40 Å thick. The individual B subunits exhibit a mixed α/β topology exhibiting the characteristic features of the "OB-fold." This protein fold was

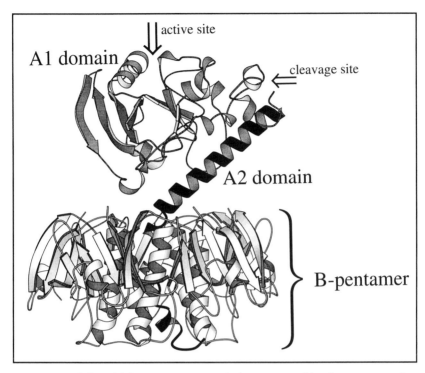

Fig. 8.1. E. coli *heat-labile enterotoxin. AB₅ holotoxin assembly, showing secondary structural elements. Note that the extended tail of the A2 domain spans the central pore of the B-pentamer to emerge at the bottom of the figure. The five identical receptor binding sites are situated around the base of the B-pentamer; two of these comprise the notches visible at both the lower left and right of the figure as drawn. Figure generated from atomic coordinates corresponding to the crystal structure of the LT: galactose complex[30] using the program MOLSCRIPT.[85]*

first observed for staphylococcal nuclease, and proposed to be characteristic of certain oligosaccharide- and oligonucleotide-binding proteins.[27] It has turned out to be shared by a broad set of bacterial toxins, including not only the AB₅ family but also much more distantly related toxins such as Staphylococcal enterotoxin B[28] and the toxic shock syndrome toxin TSST-1.[29] The secondary structural elements making up this fold are two anti-parallel β-sheets each consisting of three strands and two α-helices (Fig. 8.2). In the assembled B-pentamer, the central pore is lined by the five copies of the longer of the two α-helices.

ASSOCIATION OF THE A SUBUNIT WITH THE B-PENTAMER

The AB₅ molecular assembly is stable under physiological conditions. There is not, however, an extensive surface of interaction between the two functional parts. Instead the association of the A subunit

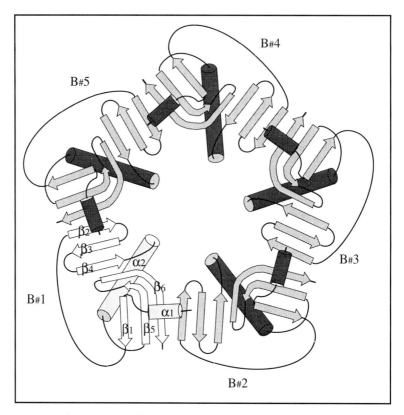

Fig. 8.2. Schematic secondary structure diagram of the five identical B subunits making up the B-pentamer of LT. Each subunit exhibits the characteristic topology of the OB-fold. The secondary structural elements, labeled for a single subunit of the pentamer in the figure, are α_1:5-9, β_1:15-22, β_2:26-30, β_3:47-50, α_2:59-78, β_5:82-88, β_6:94-102. From: Sixma TK, Pronk SE, Kalk KH et al. Nature 1991; 351:371-78.

with the B-pentamer appears to be maintained primarily by the steric constraints arising from protrusion of the C-terminal "tail" of the A2 domain through the central pore of the B-pentamer. Residues 226-232 of this C-terminal region are observed to be in a completely extended conformation which allows them to span the length of the pore while making only minimal direct interaction with the B-subunit residues which form its lining.[5,6] This extended region is capped by a short α-helix comprising residues 232-236 which lies on the far side of the pore from the bulk of the A subunit. The C-terminal four residues (237-240, the RDEL signal sequence) which follow this helix were poorly ordered in the original crystal structure of the LT toxin alone[5,6] but are seen to extend generally parallel to the bottom surface of the B-pentamer in other crystal forms.[30] Inversion or substitution of residues 237-240 does not prevent holotoxin assembly,[31] but complete holotoxin is not recovered for engineered variants in which these residues have been deleted.[8] This may be due to an inability of the holotoxin to assemble at all, but may also be due to drastic destabilization of the AB$_5$ assembly once formed.

One result of this remarkable mode of association is that the attachment of the A subunit to the B-pentamer is rather flexible. The total range of variability observed in crystallographic studies was approximately eight degrees[32,33] but this may underestimate the degree of motion possible in solution. In this context it should be noted that the single difference in three-dimensional structure worthy of note between the crystal structures of *E. coli* LT toxin and of cholera toxin is in the conformation of the C-terminus of the A subunit.[34] Unlike the A2 domain of LT, whose extended region plus C-terminal capping helix spans the B-pentamer pore, the entire A2 domain of cholera toxin forms a continuous single helix broken only by a 52 degree kink at residue 226 where it enters the pore.[34] This more compact conformation reduces the net extent of the C-terminal region sufficiently that it fits entirely within the B-pentamer pore. The amino acid sequences of the two toxins in this region are 68% identical which, while less than the overall sequence identity of 80%, is still substantial. Furthermore, it is possible to assemble stable hybrid toxins from LT A subunits and cholera toxin B subunits or vice versa.[35] Thus while it is possible that the two observed conformations are truly a distinct point of difference between the two toxins, it may also be that the crystal structures have captured a pair of conformations both of which are accessible to both toxins.

IMPLICATIONS FOR MEMBRANE BINDING AND CELL ENTRY

The architecture of the AB$_5$ holotoxin seen in the crystal structures of LT may be combined with the detailed molecular view of the receptor binding geometry provided by the crystal structure of the CTB:receptor oligosaccharide complex.[36] It is clear from these structures that the catalytic A1 domain is positioned at the extreme far end of the toxin assembly from the receptor binding sites. That is, when the toxin binds to the cell membrane the catalytic domain is remote from the membrane surface (Fig. 8.3). This interpretation is supported by low resolution images (20 Å) from atomic force microscopy of cholera toxin B-pentamer and holotoxin bound to synthetic lipid bilayers containing G$_{M1}$.[18] Molecules of the B-pentamer were imaged as a pentamer lying above the plane of the bilayer and exhibiting a central pore or depression. Molecules of the holotoxin were imaged as extending further above the plane of the bilayer and no longer giving evidence of a central pore or depression, i.e., the A subunit lay above the B-pentamer and covered the pore.

An important aspect of this model is that the B-pentamer does not enter the membrane, and the pore in the toxin pentamer does not correspond to or cause a pore in the cell membrane. This is consistent

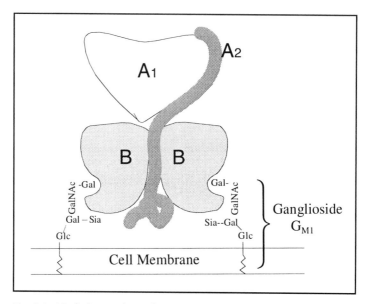

Fig. 8.3. AB₅ holotoxin bound to cell surface receptor, ganglioside GₘₗΙ. *Each toxin molecule contains five receptor binding sites, and binds up to five Gₘₗ molecules. The active site, located on the catalytic domain A1, is positioned above the surface of the B-pentamer facing away from the cell surface. From: Merritt EA, Hol WGJ. Current Opinion in Structural Biology 1995; 5:165-71.*

with previous calorimetric data,[37] with photolabeling studies[17,38] and with surface pressure measurements on monolayers.[39]

The above model implies that the A subunit must enter the cell in a process separate from cell surface receptor recognition and binding. It is most notable that the C-terminal four residues of the LT A subunit mimic a known intracellular trafficking signal (RDEL) which directs intracellular proteins to the endoplasmic reticulum via retrograde transport from the Golgi complex.[40] Intoxication by CT can be inhibited by the brefeldin A,[41,42,43] an antibiotic which disrupts normal Golgi structures. These observations lead to the hypothesis that LT enters the cell through endocytosis, joins the Golgi apparatus, and is directed to the endoplasmic reticulum through RDEL-directed retrograde transport. Cieplak[31] tested this hypothesis by assaying the cytotoxicity of mutant analogs of LT in which the RDEL sequence was inverted or otherwise altered. They found that the altered toxins retained 40% of the wild-type effectiveness in increasing cAMP levels in cell culture, and decreased but did not abolish characteristic elongation of treated cells. These results support a model in which the presence of the RDEL sequences enhances cell entry, but is not absolutely required. This model would also remove the difficulty of explaining why the KDEL peptide observed in the crystal structure of cholera toxin is buried in the pore of the B-pentamer,[34] rather than protruding from the pore to remain receptor-accessible as in LT.

RECEPTOR BINDING SITE

GALACTOSE BINDING SITE

The primary determinant for receptor binding is galactose, the terminal sugar of the G_{M1} branched pentasaccharide. The details of the molecular interactions between the toxin and the galactose upon binding are remarkably consistent in a series of crystal structures of LT complexed with galactose alone,[30] with lactose (Galβ1-4Glc)[44] with the Thomson-Friedenreich antigen (T-antigen: Galβ1-3GalNAc) (van den Akker et al. *Protein Science* 1996; in press) and of the cholera toxin B-pentamer complexed with the full G_{M1} pentasaccharide.[36] In these structures 80% of the accessible surface of the galactose is buried upon binding to a deep cavity in the toxin. Although this cavity is very near the interface between two adjacent monomers of the B-pentamer, all residues contributing directly to sugar binding interactions belong to only one of the monomers (Fig. 8.4). The most notable feature of this interaction is the stacking of the hydrophobic face of the sugar ring over the indole ring of Trp 88. Specificity of the site for galactose arises from extensive hydrogen bonding interactions involving all of the sugar hydroxyl groups other than O1 (the bridging atom to the next sugar residue in the oligosaccharides).

CT B-PENTAMER COMPLEX WITH G_{M1} PENTASACCHARIDE

The entire G_{M1} binding site is revealed by the crystal structure of the cholera toxin B-pentamer bound to five molecules of the G_{M1} pentasaccharide.[36] Both of the branched chains of the saccharide contribute to binding, with approximately 43% of the buried surface of the toxin due to the sialic acid residue, 39% to the terminal galactose residue, and 17% to the N-acetyl galactosamine (Fig. 8.5). The remaining two sugars do not appear to contribute to the binding interaction. The relative rigidity of the saccharide is seen to be due partly to direct and solvent-mediated hydrogen bonds between the two branching chains. As with the complexes of smaller sugars, the binding of each of the five complete G_{M1} pentasaccharide molecules is seen to involve direct interactions with toxin residues belonging to only one of the B subunits in the pentamer.

It is difficult to explain the reported difference in affinity of LT and cholera toxin for either alternative gangliosides or other glycoconjugates on the basis of the receptor binding mode seen in crystal structures for the two toxins. Among all the residues seen to be involved in sugar binding in these structures, there is only a single site

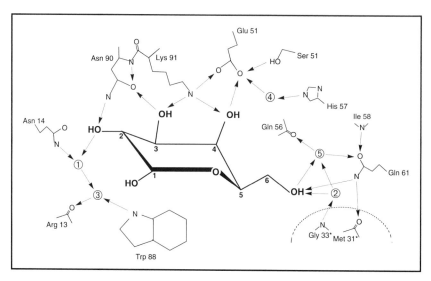

Fig. 8.4. Galactose binding site. Galactose is the terminal sugar on the G_{M1} saccharide, and of the five G_{M1} sugars is inserted most deeply into the receptor binding site. The most notable feature of the galactose:toxin binding interaction is the stacking of the hydrophobic surface of the sugar ring parallel to the conjugated ring system of Trp 88. From: Merritt EA, Sixma TK, Kalk KH et al. Molec Microbiol 1994; 13-745-53.

of sequence variation: residue 13 is a histidine in cholera toxin and in LT derived from human-associated strains of *E. coli*; it is an arginine in the porcine-derived LT corresponding to the available crystal structures. Merritt et al[36] propose that an arginine residue at this position would interact with the two sugar residues of the G_{M1} pentasaccharide not seen to be involved in the CTB:G_{M1} interaction, but this by itself would not account for the variety of differential affinities reported for LT and CT.

STRUCTURAL STUDY OF RECEPTOR-BINDING MUTANTS

The B subunits of LT and of cholera toxin have been subject to mutational studies in order to identify single residue substitutions which alter receptor binding or holotoxin assembly.[46,47] The phenotype of many of the substitutions identified in this way can be explained easily in terms of the observed three-dimensional structures of the toxins, e.g., the critical role of Trp 88 in galactose binding. The role played by other residues, notably 33-35, is more puzzling. Substitution of cholera toxin residue Gly 33 with Glu, Asp, Leu, Ile, or Val results in nontoxic toxin deficient in receptor binding; however, substitution with Lys or Arg leaves receptor binding intact.[46] The Gly 33 → Asp substitution in LT has also been characterized as nontoxic and deficient in

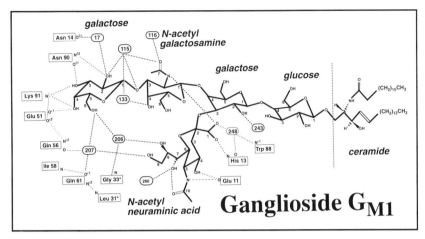

Fig. 8.5. Full G_{M1} oligosaccharide binding site. The cell surface receptor recognized by cholera toxin and E. coli heat-labile enterotoxin is ganglioside G_{M1}, Gal(β1-3)GalNAc(β1-4) {NeuAc(α2-3)} Gal(β1-4)Glc(β1-1)ceramide. This figure shows the direct and solvent-mediated hydrogen bond interactions seen at the receptor binding site in the crystal structure of the G_{M1} oligosaccharide bound to the cholera toxin B-pentamer.[36] Circled numbers correspond to well-ordered water molecules which participate in hydrogen bonding at the binding site. All residues shown are contributed by a single monomer within the toxin B-pentamer with the exception of the two starred residues (Gly33* and Leu31*), which lie in an adjacent monomer. The residues involved in receptor binding are conserved in the amino acid sequences of cholera toxin and E. coli heat-labile enterotoxin with the exception of His13, which is an Arg in some strains of LT, and Leu31*, which is a Met in LT.

receptor binding.[48] While Gly 33 is indirectly involved in the galactose binding interactions seen in the crystal structures described above, this involvement is solely through hydrogen bonding of the peptide backbone (Fig. 8.5), and the nature of the sidechain would not necessarily be predicted to interfere with receptor binding. This issue was partially clarified by the subsequently reported crystal structure of the cholera toxin B-pentamer containing the Gly 33 → Asp substitution.[49] This structure showed that the side chain of aspartic acid, and by extension those of other residues which branch at Cγ, most likely interferes sterically with accommodation of the sialic acid residue of G_{M1}. The structural basis for the critical nature of other residues in this region remains unclear.

CONFORMATIONAL CHANGES DUE TO SACCHARIDE BINDING

One notable difference between the structures of the free LT and CT holotoxins and the structures of LT or CT with sugars bound is in the extent to which the loop consisting of B-pentamer residues 51-60 is well-ordered. This loop protrudes from the bottom surface of the assembled AB_5 holotoxin and contributes residues Glu 51, Gln 56,

His 57, and Gln 61 to the sugar binding site. In the free toxin structures[5,6,34,50] this loop exhibits conformational flexibility, as indicated by the variation seen in conformations of the five separate B monomers and correspondingly high thermal parameters. In the toxin:sugar complexes the loop is much better ordered.[44,36,30] Flexibility of the 51-60 loop in solution probably allows easy entry of the receptor into the binding site. The process of receptor binding may thus be considered as one of a relatively flexible protein binding to a conformationally restricted oligosaccharide. A 15 residue synthetic peptide (CTP3) spanning this loop and corresponding in sequence to residues 50-64 has been shown to elicit antibodies which are cross-reactive with both native LT and native CT. Antibodies to the peptide were partially effective in neutralizing biological activity of both toxins.[51,52] The conformation of the peptide as bound to the antibody is quite different from that observed for the same peptide in any of the toxin structures,[53] however, supporting the idea that the loop exhibits flexibility in solution. The fact that the CTP3 antigen contains residues involved in sugar binding suggests that anti-CTP3 antibodies may directly block the toxin's ability to bind to its receptor.

THE ACTIVE SITE OF LT

IDENTIFICATION OF THE ACTIVE SITE

The NAD$^+$ and substrate binding sites on the toxin A subunit are not understood in nearly as much detail as the receptor binding site. Nevertheless, the location of the active site is clear from mutational studies and from structural and sequence homologies to other ADP-ribosylating toxins.

Both residues Glu 110 and Glu 112 have been suggested as possible catalytic residues. Substitution of either of these glutamic acid residues by an aspartic acid reduces, but does not abolish, activity.[54] Replacement of Glu 112 with a lysine residue inactivates the toxin,[55] although it should be noted that this is a much less conservative substitution. Mutations in the nearby residues Arg 7[56,57] and Ser 61[57] also abolish activity. These residues are clustered at the bottom of a distinct crevice in the A subunit, which thus constitutes at least part of the probable substrate binding site (Fig. 8.1).

This identification of the active site is strongly supported by structural homology of this region to three other ADP-ribosylating toxins. Only a single residue, a glutamic acid (Glu 112 in LT/CT), is strictly conserved among the active sites of these toxins. Nevertheless structural superposition of the LT active site onto those of Pseudomonas exotoxin A,[58,59] diphtheria toxin[60] and pertussis toxin,[61] reveals a consistent geometry for this region.[5,62] Following this consensus geometry one can equate Arg 7 in CT/LT to Arg 9 in pertussis toxin, to His 21 in diphtheria toxin, and to His 440 in exotoxin A; mutations at all of these sites are known to affect activity of the respective toxins. A fairly detailed model for NAD+ binding at the LT active can be built by combining crystallographic studies on the active sites of these bacterial toxins. The crystal structure of the ADP-ribosylation domain of Pseudomonas exotoxin A complexed to AMP and nicotinamide[59] is particularly relevant, as the interactions observed to occur between exotoxin A and the complexed NAD moieties (AMP and nicotinamide) can be extrapolated to homologous interactions in LT. Figure 8.6 shows the AMP and nicotinamide moieties as they would fit in the active site of LT after this superposition. It should be noted that the interactions

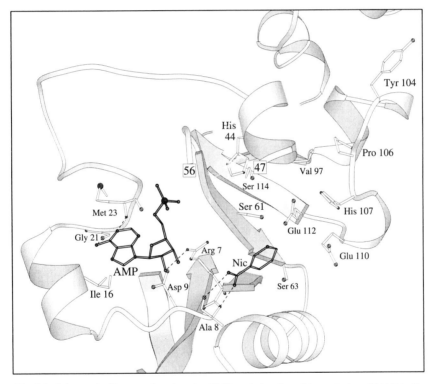

Fig. 8.6. Schematic diagram showing the AMP and nicotinamide moieties of NAD in the active site of LT. The interactions shown are based upon the similar interactions made by the AMP and nicotinamide (Nic) moieties in the crystal structure of Pseudomonas exotoxin A.[59] To generate this model, LT residues 5-9, 14-16, 21-24, 59-72, 83-89, 94-96, 111-116 have been superimposed onto structurally equivalent residues 438-442, 448-450, 454-457, 468-481, 496-502, 541-543, 552-557 in exotoxin A. The resulting superposition yields an r.m.s. deviation of 1.7Å for the 42 Cα atoms. Also highlighted in this figure are residues that have been shown by site-directed mutagenesis to affect ADP-ribosylation as described in the text. Not shown in the figure is the loop consisting of LT residues 47-56, which as described in the text must be displaced from the active site in order to accommodate substrate binding. The endpoints of this loop, however, are indicated by boxed residue numbers.

of the adenosine moiety with exotoxin A are almost identical to the interactions of the adenosine moiety in ApUp diphtheria toxin.[63] This supports the hypothesis of a conserved NAD binding mode in all ADP-ribosylating toxins, including LT.

STRUCTURAL STUDY OF LT ACTIVE SITE MUTANTS

LTA Arg7 → Lys: A noteworthy corollary of this homology model is that the active site cleft as seen in the crystal structure of the wild-type LT is simply not wide enough to accommodate the bound NAD (Fig. 8.6).[5] In crystals of LT containing the Arg7 → Lys substitution, however, the space available in the active site cleft is increased dramatically by displacement of the loop consisting of residues 47-56.[45] The Arg7 → Lys structure may therefore represent an overall view of the active site cleft after NAD binding, although in this case actual NAD binding is prevented by the mutation at residue 7.

The following interactions between the toxin and the adenosine moiety of NAD are likely to be conserved features: (1) Arg 7 forms a hydrogen bond with the O2 hydroxyl group of the ribose moiety analogous to that involving the His440 sidechain of exotoxin A.[45] (2) The carbonyl oxygen of Gly21 forms a hydrogen bond with the N6 atom of the adenine moiety. A glycine at this position is conserved in both the structures of diphtheria toxin and exotoxin A, as well as in the sequences of cholera toxin, pertussis toxin, LT-I, and LT-II. This structural alignment would make Gly 21 a second fully conserved residue known in this family of toxins in addition to the catalytic residue Glu 112. In the structures of both diphtheria toxin and exotoxin A this glycine adopts main chain torsion angles (φ = 105, ψ = 159 in the exotoxin A complex[59]) which are energetically unfavorable for any other amino acid. (3) The backbone nitrogen of Met23 forms a hydrogen bond with the N1 atom of the adenine moiety. (4) Asp 9 forms a hydrogen bond with the O2 hydroxyl group of the ribose moiety. Asp9 has been shown to be important for ADP-ribosylation activity, as substitutions of either Glu or Lys at this position are not tolerated.[64,65] (5) Ile 16 (conserved as either isoleucine or valine in other members of the toxin family) forms a hydrophobic interaction the adenine ring.

Similarly, two interactions involving the nicotinamide moiety of NAD and LT-I are likely to be conserved features. The carbonyl oxygen of Ala8 forms a hydrogen bond with the N7 atom of nicotinamide, while the backbone nitrogen of the same residue forms a hydrogen bond with the O7 atom of nicotinamide. Both of these interactions are mediated via the peptide backbone rather than the sidechain of Ala8, and this residue is not conserved among the ADP-ribosylating toxins.

The modeled NAD binding mode is moreover consistent with mutational analysis carried out at other positions. Glu 112, believed to be the catalytic residue, is in close proximity to N1 of the nicotinamide as expected for catalytic cleavage of the glycosyl bond. Ser61 and Ser63 are also in close proximity to the nicotinamide, consistent with the observation that introduction of large sidechains at these positions disrupts ADP-ribosyl transferase activity.[57,66] His44 is somewhat further from the modeled NAD position, but may be involved in binding the second substrate. Substitution of asparagine at this position prevents the ADP-ribosylation of $G_s\alpha$ by cholera toxin.[64] The homologous residue in pertussis toxin, His35, has been found to be extremely critical for the catalytic rate of ADP-ribosylation but not important to affinity for either of the two substrates.[67] The postulated role of His35 in catalysis is strengthened by the observation of a hydrogen bond between His35 and the catalytic Glu in PT.

LTA Val97 → Lys: Other residues which have been implicated through mutational analysis to be important for activity, including residues 97, 104, 106, 107, and 110[54,66] are clustered in a separate region of the active site cleft (Fig. 8.6). Their importance may well arise from recognition and binding of the native second substrate, $G_s\alpha$. A shift in the conformation in this region during substrate binding may be revealed by the reported structure for another active site mutant of LT in which Val97 is replaced by a lysine.[68] Substitutions at residue 97 of the LT A subunit inactivate the toxin,[66] but this residue does not contribute to the accessible surface of the active site cleft nor does the Val97 → Lys substitution distort the active site conformation seen in the wild-type toxin. Instead, Val 97 lies in the protein interior. It contributes to the surface of a small solvent-filled interior cavity, the opposite side of which is provided by the catalytic residue Glu112. Substitution of valine by the larger lysine sidechain partially fills this interior cavity but does not alter the exterior accessible surface of the protein.[68] The larger, polar sidechain forms a salt bridge to Glu112. The lack of catalytic activity in the mutant may be due to the altered electrostatic character of Glu 112 induced by this interaction, but it may also be that the reduced size of the interior cavity itself explains the loss of activity, i.e., the cavity may accommodate a shift of Glu 112 or a neighboring residue during NAD binding.[68]

POSSIBLE CONFORMATIONAL CHANGES DURING ACTIVATION OF TOXIN

The mechanism of activation of LT and CT remains an interesting question, as the site of activation is more than 20 Å from the catalytic residue Glu112. Van den Akker et al[45] propose a series of conformational changes that propagate from the site of proteolytic cleavage and disulfide reduction to the active site (Fig. 8.7). The final step in this proposed mechanism is a displacement of loop 47-56 from the active site analogous to that observed in the crystal structure of the Arg7 → Lys LT mutant described above. The displaced loop containing residues 47-56 may also contribute a binding surface for the opposite

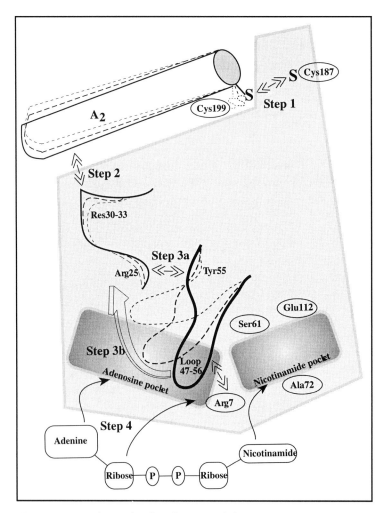

Fig. 8.7. Proposed cascade of conformational changes upon toxin activation. Based upon careful consideration of the conformational differences between the crystal structures of the Arg7 → Lys LT mutant and the wild-type LT structure, van den Akker and coworkers[45] propose the following series of conformation events during the activation of LT: (1) Proteolytic cleavage between residues 192 and 195 and reduction of the disulfide bond between residues 187 and 199 result in dissociation or reorientation of the A2 helix with respect to the A1 domain. (2) Loop 25-36, which is in contact with the A2 helix via residues 30, 31 and 33, will change position. (3) Displacement of loop 25-36 will disrupt the hydrogen bond between the backbone atoms of Arg25 and Tyr55, as well as disrupting hydrophobic interactions between the sidechains of the same two residues. (4) The destabilization of Tyr55, located in loop 47-56, will cause a similar event to that observed due to the destabilization of this loop resulting from substitution of Lys for Arg at residue 7—the displacement of loop 47-56 away from the active site. (5) The latent toxin is now activated and NAD thus has access to the active site. From: van den Akker F, Merritt EA, Pizza MG et al. Biochemistry 1995; 34:10996-11004.

face of the bound NAD, but no details of such an interaction can be deduced from the available structures.

THERAPEUTIC APPLICATIONS

LT, CT AS ADJUVANTS

The ability to cause diarrhea is not the only biological property characteristic of *E. coli* heat-labile enterotoxin. Both LT and CT are also remarkable for their ability to elicit a strong mucosal immune response,[69] notably production of secretory IgA. The stimulation of the mucosal immune system extends not only to a response against the toxin itself, but also to other antigens which may be administered at the same time. This adjuvant effect means that the toxins are of great interest as components of possible vaccines against a variety of diseases. Furthermore, the co-administration of toxin affects the development of tolerance to foreign antigens.[70-72] In particular their mode and site of action make the toxins attractive as components of vaccines targeted against disease organisms whose mode of entry into the body is through the mucosal tissues.

Most early studies of this adjuvant effect claimed efficacy for the toxin B-pentamer alone. Consequently early attempts at toxin-based vaccine design focused on the use or modification of the cholera toxin B-pentamer. Several approaches are possible (Fig. 8.7). The simplest of these is to create a mixture containing both the toxin B-pentamer and the foreign antigen or antigens of interest. Several groups have reported that a stronger immune response results when the foreign antigens are not merely mixed with the B-pentamer, but are instead chemically cross-linked to it.[73] A more sophisticated approach to the same end is to incorporate the foreign epitope into the toxin structure through the use of protein engineering. Dertzbaugh and Elson[74] for example, added foreign antigens to the N-terminus or C-terminus of the toxin B subunit through genetic engineering. Such B-subunit constructs have not proven efficacious, however, in producing the desired immune response.[75]

The crystal structures of the LT holotoxin described above clearly show that the AB_5 assembly may be described as an $A1:A2:B_5$ complex, in which the A1 domain is tethered to the B pentamer only through the intervening A2 domain. This mode of association leaves the A1 domain anchored quite flexibly to the B-pentamer and suggests that the assembly of an AB_5 complex might tolerate replacement of the entire A1 domain with a foreign structure. A hypothetical multitarget vaccine might link entire domains from one or more foreign proteins to the LT B-pentamer in this fashion. Jobling and Holmes[76] have shown that such chimeric toxins based on cholera toxin can be constructed by cloning foreign sequences into plasmids containing the A2 region of the ctxA gene. These fusions products are successfully

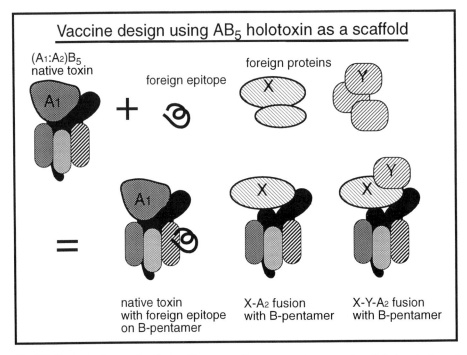

Fig. 8.8. Strategies for vaccine design. The unusual immunogenic properties of cholera toxin and E. coli heat-labile enterotoxin make them attractive vehicles for vaccine design. Optimal efficacy appears to require physical linkage of foreign epitopes, or indeed entire foreign proteins, to the toxin molecule.

assembled into $X:A2:B_5$ hexamers when expressed in *V. cholerae*, and retain the G_{M1} binding ability.

REQUIREMENT FOR ADP-RIBOSYLATION ACTIVITY

It now appears that the immune response provoked by CT and LT is considerably more complex than a simple reaction to the presence of the toxin B subunit. Several studies have reported a requirement for ADP-ribosylation activity in the administered toxin in order for it to act as an adjuvant.[77,78] Other studies, however, have reported adjuvanticity for toxin preparations based on mutant toxins whose A subunit is deficient in ADP-ribosylation.[79,80] There seems to be general agreement that interpretation of early work which claimed efficacy of the B-pentamer alone is confounded by the presence of trace amounts of AB_5 holotoxin, an artifact introduced when the B-pentamer is prepared from holotoxin.

The current picture of immune stimulation by LT and CT may be summarized as follows:

- Binding of the B-pentamer to the mucosal epithelium is required for the induction of an intestinal immune response[71] and may be sufficient to generate a local IgG and IgA response in the Peyer's patch and intestinal lymphoid follicle.[81]
- Administration of B-pentamer is synergistic to the adjuvant effect produced by holotoxin.[82]
- The B-pentamer and the AB_5 holotoxin may have opposite effects on the development of tolerance. Oral administration of antigen conjugated to recombinant cholera toxin B-pentamer suppressed systemic immune response in mice; however, coadministration of AB_5 holotoxin instead abrogated oral tolerance.[72]
- Enhanced production of mucosal IgA is elicited by oral coadministration of antigen and holotoxin.[77] This adjuvant effect is not seen if the holotoxin is replaced by recombinant B-pentamer alone, or by holotoxin lacking ADP-ribosylation activity due to mutation of active site catalytic residue Glu 112 to Lys.[77]
- Engineered holotoxin which is deficient in ADP-ribosylation activity due to substitutions at other sites can nevertheless retain its ability to act as an adjuvant. Orally administered LT holotoxin containing the substitution Arg192 → Gly increased both serum IgG and mucosal IgA production in mice, despite being inactive in ADP-ribosylation assays against the substrate agmatine.[80] Intranasal administration of inactive LT holotoxin containing the substitution Arg7 → Lys was similarly effective in eliciting a mucosal IgA response.[79]

The seemingly contradictory conclusions on a requirement for ADP-ribosylation activity cannot yet be entirely reconciled. It is possible that the adjuvant effect is mediated by nonproductive binding of the A subunit to intracellular target proteins. Some activity-destroying mutations, e.g., Glu112 → Lys, might prevent both productive binding as an ADP-ribosylating enzyme and nonproductive binding to some unknown target which mediates the adjuvant effect. Alternatively, the activity-deficient mutant toxins reported to exhibit adjuvanticity may retain some ADP-ribosylation activity against a minority substrate even though they are inactive against the substrates used in assaying activity. Such variation in substrate specificity is known for homologous ADP-ribosylating toxins. For example, *E. coli* enterotoxin LT-IIb is 62% sequence identical to LT in the catalytic domain; it acts intracellularly to ADP-ribosylate the same target protein G$_s\alpha$, but is inactive against artificial substrates used to assay LT activity in vitro.[83]

PROSPECTS FOR DRUG AND VACCINE DESIGN

It should be evident from the discussion above that structural studies are rapidly expanding our understanding of the biological action of these toxins to the point where receptor binding, catalytic action, and immunogenicity may all be explained and manipulated at the molecular level. Lead compounds for potential anti-diarrheal drugs may soon emerge either from concerted computer modeling linked with crystallographic studies[84] or from serendipitous discoveries such as the adventitious binding of peptides at the receptor binding site in crystal structures of cholera toxin mutants.[49] It is particularly exciting to see that it may be possible to decouple the toxic action of these proteins from their efficacy as immunogens. This encourages us to contemplate the design of effective oral vaccines based on linking foreign antigens to such de-toxified toxins. In this manner a long-standing scourge of humanity would be harnessed instead to combat infectious disease.

ACKNOWLEDGMENTS

This research was supported by the Dutch Chemical Foundation (SON), the Dutch Organization for Scientific Research (NWO), and the National Institutes of Health (AI34501). We would like to thank Ingeborg Feil and Christophe Verlinde for stimulating discussions and our collaborators Tim Hirst, Randall Holmes, Rino Rappuoli, and Jan Wilschut for providing precious protein samples and deep insight into the intriguing properties of the cholera toxin family.

REFERENCES

1. Burnette WN. AB5 ADP-ribosylating toxins: comparative anatomy and physiology. Structure 1994; 2:151-58.
2. Merritt EA, Hol WGJ. AB$_5$ Bacterial Toxins. Current Opinion in Structural Biology 1995; 5:165-71.
3. Holmes RK, Jobling MG, Connell TD. Cholera toxin and related enterotoxins of gram-negative bacteria, In: Moss, Iglewski, Vaughan, Tu, eds. Bacterial toxins and virulence factors in disease, handbook of natural toxins. New York: M. Dekker, 1995;8.
4. Black RE. The epidemiology of cholera and enterotoxigenic *E. coli* diarrheal disease, In: Holmgren J, Lindberg A, Möllby R, eds. Developments of vaccines and drugs against diarrhea, 11th. Nobel Conf. Stockholm 1985, Studentlitteratur, Lund, Sweden, 1986:23-32.
5. Sixma TK, Pronk SE, Kalk KH et al. Crystal structure of a cholera toxin-related heat-labile enterotoxin from *E. coli*. Nature 1991; 351:371-78.
6. Sixma TK, Kalk KH, van Zanten BAM et al. Refined structure of *Escherichia coli* heat-labile enterotoxin, a close relative of cholera toxin. J Mol Biol 1993; 230:890-918.

7. Hardy SJS, Holmgren J, Johansson S et al. Coordinated assembly of multisubunit proteins: oligomerization of bacterial enterotoxins in vivo and in vitro. Proc Nat Acad Sci USA 1988; 85:7109-13.

8. Streatfield SJ, Sandkvist M, Sixma TK et al. Intermolecular interactions between the A and B subunits of heat-labile enterotoxin from *Escherichia coli* promote holotoxin assembly and stability in vivo. Proc Natl Acad Sci USA 1992; 89:12140-44.

9. Fishman PH. Recent advances in identifying the functions of gangliosides. Chem Phys Lipids 1 1986; 42:37-151.

10. Critchley DR, Magnani JL, Fishman PH. Interactions of cholera toxin with rat intestinal brush border membranes. J Biol Chem 1981; 256:8724-31.

11. Holmgren J, Fredman P, Lindblad M et al. Rabbit intestinal glycoprotein receptor for *Escherichia coli* heat-labile enterotoxin lacking affinity for cholera toxin. Infect Immun 1982; 38:424-33.

12. Holmgren J, Lindblad M, Fredman P et al. Comparison of receptors for cholera and *Escherichia coli* enterotoxins in human intestine. Gastroenterology 1985; 89:27-35.

13. Donta ST, Poindexter NJ, Ginsberg BH. Comparison of the binding of cholera and *Escherichia coli* enterotoxins to Y1 adrenal cells. Biochemistry 1982; 21:660-64.

14. Griffiths SL, Critchley DR. Characterisation of the binding sites for *Escherichia coli* heat-labile enterotoxin type I in intestinal brush borders. Biochim Biophys Acta 1991; 1075:154-61.

15. Fukuta S, Magnani JL, Twiddy EM et al. Comparison of the carbohydrate-binding specificities of cholera toxin and *Escherichia coli* heat-labile enterotoxins LTh-I, LT-IIa and LT-IIb. Infect. Immun. 1988; 56:1748-53.

16. Gill DM. The arrangement of subunits in cholera toxin. Biochemistry 1976; 15:1242-48.

17. Tomasi M, Montecucco C. Lipid insertion of cholera toxin after binding to G_{M1}-containing liposomes. J Biol Chem 1981; 256:11177-81

18. Yang J, Tamm LK, Tillack TW et al. New Approach for Atomic Force Microscopy of Membrane Proteins. J Mol Biol 1993; 229:286-90.

19. Cabral-Lilly D, Sosinsky GE, Reed RA et al. Orientation of cholera toxin bound to model membranes. Biophys J 1994; 66:935-41.

20. De Wolf MJS, Van Dessel GAF, Lagrou AR et al. pH-induced transitions in cholera toxin conformation: a fluorescence study. Biochemistry 1987; 26:3799-806.

21. Moss J, Vaughan M. Activation of cholera toxin and *Escherichia coli* heat-labile enterotoxins by ADP-ribosylation factors, a family of 20 kDa guanine nucleotide-binding proteins. Molec Microbiol 1991; 5:2621-27.

22. Welsh CF, Moss J, Vaughan M. ADP-ribosylation factors: a family of guanine nucleotide-binding proteins that activate cholera toxin and regulate vesicular transport. In: Moss, Iglewski, Vaughan, Tu, eds. Bacterial Toxins and Virulence Factors in Disease, Handbook of Natural Toxins. New York: M. Dekker, 1985;8.

23. Tsai S-C, Adamik R, Moss J et al. Guanine nucleotide dependent formation of a complex between choleragen (cholera toxin) A subunit and bovine brain ADP-ribosylation factor. Biochemistry 1991; 30:3697-703.

24. Rothman JE, Orci L. Molecular dissection of the secretory pathway. Nature 1992; 355:409-15.

25. Chang EB, Rao MC. Intracellular mediators of electrolyte transport. In: Field M, ed. Diarrheal Diseases. New York: Elsevier, 1991:49-72.

26. Gabriel SE, Brigman KN, Koller BH et al. Cystic fibrosis heterozygote resistance to cholera toxin in the cystic fibrosis mouse model. Science 1994; 266:107-09.

27. Murzin AG. OB (oligonucleotide/oligosaccharide binding)-fold: common structural and functional solution for nonhomologous sequences. EMBO J 1993; 12:861-67.

28. Swaminathan S, Furey W, Pletcher J et al. Crystal structure of staphylococcal enterotoxin B, a superantigen. Nature 1992; 359:801-06.

29. Acharya KR, Passalacqua EF, Jones EY et al. Structural basis of superantigen action inferred from crystal structure of toxic-shock syndrome toxin-1. Nature 1994; 367:94-97.

30. Merritt EA, Sixma TK, Kalk KH et al. Galactose binding site in *E. coli* heat-labile enterotoxin (LT) and cholera toxin (CT). Molec Microbiol 1994; 13:745-53.

31. Cieplak W Jr, Messer RJ, Konkel ME et al. Role of a potential endoplasmic reticulum retention sequence (RDEL) and the Golgi complex in the cytotonic activity of *Escherichia coli* heat-labile enterotoxin. Molec Microbiol 1995; 16:789-800.

32. Sixma TK, Pronk SE, van Scheltinga T et al. Native nonisomorphism in the structure determination of Heat Labile Enterotoxin (LT) from *E. coli*. In: Wolf W, Evans PR, Leslie AGW, ed. Isomorphous Replacement and Anomalous Scattering. Daresbury, U.K.: SERC CCP4 Study Weekend Proceedings, 1991:133-40.

33. Sixma TK, Aguirre A, Terwisscha van Scheltinga AC et al. Heat-labile enterotoxin crystal forms with variable A/B$_5$ orientation. Febs Lett 1992; 305:81-85.

34. Zhang R-G, Scott DL, Westbrook ML et al. The three-dimensional crystal structure of cholera toxin. J Mol Biol 1995; 251:563-73.

35. Takeda Y, Honda T, Taga S et al. In vitro formation of hybrid toxins between subunits of *Escherichia coli* heat labile enterotoxin and those of cholera enterotoxin. Infect Immun 1981; 34:341-46.

36. Merritt EA, Sarfaty S, van den Akker F et al. Crystal structure of cholera toxin B-pentamer bound to receptor G$_{M1}$ pentasaccharide. Protein Sci 1994; 3:166-75.

37. Goins B, Freire E. Lipid phase separation induced by the association of cholera toxin to phospholipid membranes containing ganglioside G$_{M1}$. Biochemistry 1985; 24:1791-97.

38. Wisnieski BJ, Bramhall JS. Photolabelling of cholera toxin subunits during membrane penetration. Nature 1981; 289:319-21.

39. Mosser G, Brisson A. Conditions of two-dimensional crystallization of cholera toxin B-subunit on lipid films containing ganglioside G_{M1}. J Struct Biol 1991; 106:191-98.

40. Pelham HRB. Recycling of proteins between the endoplasmic reticulum and Golgi complex. Curr Opinion Cell Biol 1991; 3:585-91.

41. Donta ST, Beristain S, Tomicic TK. Inhibition of heat-labile cholera and *Escherichia coli* enterotoxins by brefeldin A. Infect Immun 1993; 61:3282-86.

42. Nambiar MP, Oda T, Chen C et al. Involvement of the Golgi region in the intracellular trafficking of cholera toxin. J Cell Physiol 1993; 154:222-28.

43. Orlandi PA, Curran PK, Fishman PH. Brefeldin A blocks the response of cultured cells to cholera toxin. Implications for intracellular trafficking in toxin action. J Biol Chem 1993; 268:12010-16.

44. Sixma TK, Pronk SE, Kalk KH et al. Lactose binding to heat-labile enterotoxin revealed by X-ray crystallography. Nature 1992; 355:561-64.

45. van den Akker F, Merritt EA, Pizza MG et al. The Arg7Lys mutant of heat-labile enterotoxin exhibits great flexibility of active site loop 47-56 of the A subunit. Biochemistry 1995; 34:10996-11004.

46. Jobling MG, Holmes RK. Analysis of structure and function of the B subunit of cholera toxin by the use of site-directed mutagenesis. Molec Microbiol 1991; 5:1755-67.

47. Nashar TO, Webb HM, Eaglestone S et al. Potent immunogenicity of the B-subunits of *E. coli* heat-labile enterotoxin: receptor binding is essential and induces differential modulation of lymphocyte subsets. Proc Natl Acad Sci USA 1995; 93:226-230.

48. Tsuji T, Honda T, Miwatani T et al. Analysis of receptor-binding site in *Escherichia coli* enterotoxin. J Biol Chem 1985; 260:8552-58.

49. Merritt EA, Sarfaty S, Chang T et al. Surprising leads for a cholera toxin receptor-binding antagonist: crystallographic studies of CTB mutants. Structure 1995; 3:561-70.

50. Zhang R-G, Westbrook ML, Westbrook EM et al. The 2.4 Å crystal structure of cholera toxin B subunit pentamer: choleragenoid. J Mol Biol 1995; 251:550-62.

51. Jacob CO, Pines M, Arnon R. Neutralization of heat-labile toxin of *E. coli* by antibodies to synthetic peptides derived from the B subunit of cholera toxin. EMBO 1984; 3:2889-93.

52. Jacob CO, Arnon R, Finkelstein RA. Immunity to heat-labile enterotoxins of porcine and human *Escherichia coli* strains achieved with synthetic cholera toxin peptides. Infect Immunol 1986; 52:562-67.

53. Shoham M, Scherf T, Anglister J et al. Structural diversity in a conserved cholera toxin epitope involved in ganglioside binding. Protein Sci 1995; 4:841-48.

54. Lobet Y, Cluff CW, Cieplak W Jr. Effect of site-directed mutagenic alterations on ADP-ribosyltransferase activity of the A subunit of *Escherichia coli* heat-labile enterotoxin. Infect Immun 1991; 59:2870-79.

55. Moss J, Stanley S, Vaughan M et al. Interaction of ADP-ribosylation factor with *Escherichia coli* enterotoxin that contains an inactivating lysine 112 substitution. J Biol Chem 1993; 268:6383-87.

56. Burnette WN, Mar VL, Platler BW et al. Site-specific mutagenesis of the catalytic subunit of cholera toxin: substituting lysine for arginine 7 causes loss of activity. Infect Immun 1991; 59:4266-70.

57. Harford S, Dykes CW, Hobden AN et al. Inactivation of the *Escherichia coli* heat-labile enterotoxin by in vitro mutagenesis of the A-subunit gene. Eur .J Biochem 1989; 183:311-16.

58. Allured VS, Collier RJ, Carroll SF et al. Structure of exotoxin A of *Pseudomonas aeruginosa* at 3.0 Ångstrom resolution. Proc Nat Acad Sci USA 83:1320-24.

59. Li M, Dyda F, Benhar I et al. The crystal structure of *Pseudomonas aeruginosa* exotoxin domain III nicotinamide and AMP: conformational differences with the intact exotoxin. Proc Natl Acad Sci USA 1995; 92:9308-12.

60. Choe S, Bennett MJ, Fujii G et al. The crystal structure of diphtheria toxin. Nature 1992; 357:216-22.

61. Stein PE, Boodhoo A, Armstrong GD et al. The crystal structure of pertussis toxin. Structure 1994; 2:45-57.

62. Domenighini M, Magagnoli C, Pizza M et al. Common features of the NAD-binding and catalytic site of ADP-ribosylating toxins. Molec Microbiol 1994; 14:41-50.

63. Bennett MJ, Choe S, Eisenberg D. Refined structure of dimeric diphtheria toxin at 2.0 Å resolution. Protein Sci 1994; 3:1444-63.

64. Kaslow HR, Platler B, Takada T et al. Effects of site-directed mutagenesis on cholera toxin A subunit ADP-ribosyltransferase activity. In: Bacterial Protein Toxins, Zentralblatt fur Bakteriologie Suppl 1992; 23:197-98.

65. Vadheim KL, Singh Y, Keith JM. Expression and mutagenesis of recombinant cholera toxin A subunit. Microb Pathog 1994; 17:339-46.

66. Pizza M, Domenighini M, Hol W et al. Probing the structure-activity relationship of *Escherichia coli* LT-A by site-directed mutagenesis. Molec Microbiol 1994; 14:51-60.

67. Xu Y, Barbancon V, Barbieri JT. Role of histidine 35 of the S1 subunit of pertussis toxin in the ADP-ribosylation of transducin. J Biol Chem 1994; 269:9993-99.

68. Merritt EA, Sarfaty S, Pizza M et al. Mutation of a buried residue causes loss of activity but no conformational change in the heat-labile enterotoxin of *Escherichia coli*. Nature Struct Biol 1995; 2:269-72.

69. Dertzbaugh MT, Elson CO. Cholera Toxin as a Mucosal Adjuvant. In: Spriggs DR, Koff WC, eds. Topics in Vaccine Adjuvant Research Boca Raton: CRC Press, 1991: 119-131.

70. Elson CO, Ealding W. Cholera toxin feeding did not induce oral tolerance in mice and abrogated oral tolerance to an unrelated protein antigen. J Immunol 1984; 133:2892-97.

71. Stok W, van der Heijden PJ, Bianchi ATJ. Conversion of orally induced suppression of the mucosal immune response to ovalbumin into stimulation by conjugating ovalbumin to cholera toxin or its B subunit. Vaccine 1994; 12:521-26.

72. Sun J-B, Holmgren J, Czerkinsky C. Cholera toxin B subunit: an efficient transmucosal carrier-delivery system for induction of peripheral immunological tolerance. Proc Natl Acad Sci USA 1994; 91:10795-99.

73. Wu H.-Y, Russell MW. Induction of Mucosal Immunity by Intranasal Application of a streptococcal Surface Protein Antigen with the Cholera Toxin B Subunit Infect Immun 1993; 61:314-22.

74. Dertzbaugh MT, Elson CO. Reduction in oral immunogenicity of cholera toxin B subunit by N-terminal peptide addition. Infect Immun 1993 61:384-90.

75. Nashar TO, Amin T, Marcello A et al. Current progress in the development of the B subunits of cholera toxin and *Escherichia coli* heat-labile enterotoxin as carriers for the oral delivery of heterologous antigens and epitopes. Vaccine 1993; 11:235-40.

76. Jobling MG, Holmes RK. Fusion proteins containing the A2 domain of cholera toxin assemble with B polypeptides of cholera toxin to form immunoreactive and functional holotoxin-like chimeras. Infect Immun 1993; 61:1168.

77. Lycke N, Tsuji T, Holmgren, J. The adjuvant effect of *Vibrio cholerae* and *Escherichia coli* heat-labile enterotoxin is linked to their ADP-ribosyltransferase activity. Eur J Immunol 1992; 22:2277-81.

78. Lebens M, Holmgren J. Mucosal vaccines based on the use of cholera toxin B subunit as immunogen and antigen carrier. Dev Biol Stand 1994; 82:215-27.

79. Douce G, Turcotte C, Cropley I et al. Mutants of *Escherichia coli* heat-labile toxin lacking ADP-ribosyltransferase activity act as nontoxic, mucosal adjuvants. Proc Natl Acad Sci USA 1995; 92:1644-48.

80. Dickinson BL, Clements JD. Dissociation of *Escherichia coli* heat-labile enterotoxin adjuvanticity from ADP-ribosyltransferase activity. Infect Immun 1995; 63:1617-23.

81. Lindner J, Geczy AF, Russell-Jones GJ. Identification of the site of uptake of the *E. coli* heat-labile enterotoxin, LTB. Scand J Immunol 1994; 40:564-72.

82. Wilson AD, Clarke CJ, Stokes CR. While cholera toxin and B subunit act synergistically as an adjuvant for the mucosal immune response of mice to Keyhole limpet haemocyanin. Scand J Immunol 1990; 31:443-51.

83. Lee C-M, Chang PP, Tsai S-C et al. Activation of *Escherichia coli* heat-labile enterotoxins by native and recombinant adenosine diphosphate-ribosylation factors, 20-kD guanine nucleotide-binding proteins. J Clin Invest 1991; 87:1780-86.

84. Verlinde CLMJ, Merritt EA, van den Akker F et al. Protein crystallography and infectious disease. Protein Science 1993; 3:1670-86.

85. Kraulis P. MOLSCRIPT: a program to produce both detailed and schematic plots of proteins. J Appl Cryst 1991; 24:946-50.

CHAPTER 9

SHIGA TOXIN

Marie E. Fraser, Maia M. Chernaia, Yuri V. Kozlov
and Michael N.G. James

Shiga toxin is named after Kiyoshi Shiga who described the bacterium *Shigella dysenteriae* type 1 in the wake of the 1896 dysentery epidemic in Japan.[1] This bacterium produces a protein toxin, the Shiga toxin. (See ref. 2 for a review.) Certain enterohemorrhagic strains of *Escherichia coli* produce very similar cytotoxins and these are named Shiga-like toxins (SLTs). *S. dysenteriae* causes the severe form of dysentery and the strains of *E. coli* which produce SLTs have been associated with hemorrhagic colitis and the hemolytic uremic syndrome ("hamburger disease") in humans.[3,4] Although the role of Shiga toxin or the Shiga-like toxins in the pathogenesis of *S. dysenteriae* or *E. coli* is not fully understood, the toxin is known to be important in causing the symptoms of these respective diseases.

Shiga toxin is a member of the A-B family of protein toxins. The A subunit has enzymatic activity but is toxic only if it can enter target cells in the host. The B subunit binds to specific receptors on the target cell thereby providing access for the A subunit.

Like cholera toxin and heat-labile enterotoxin, the B part of Shiga toxin consists of five identical subunits.[5] (Pertussis toxin also belongs to the AB$_5$ class of bacterial toxins, although in its case, the five B subunits are dissimilar, consisting of two copies of one subunit (S4) and one copy of each of the other three (S2, S3 and S5).[6]) The B subunit of Shiga toxin is 7.7 kDa in size, comprising 69 amino acid residues.[7] The B subunits bind to glycolipid receptors on the cell surface. The receptor for Shiga toxin is globotriosylceramide.[8,9]

The A subunit of Shiga toxin is 293 amino acid residues (32 kDa) in size.[10,11] The A subunit is easily cleaved proteolytically into two chains, A1 (the amino-terminal part, 27.5 kDa) and A2 (the carboxy-terminal part, 3 kDa).[12] Each part, A1 and A2, has one cysteine residue and in the holotoxin the two chains remain covalently attached by a disulfide bridge between the cysteines. A1 is the enzymatically active part;[13] it is an N-glycosidase.

Protein Toxin Structure, edited by Michael W. Parker. © 1996 R.G. Landes Company.

The N-glycosidase activity common to members of the Shiga toxin family distinguishes them from other bacterial protein toxins. (The other members of the AB_5 class, cholera toxin, heat-labile enterotoxin and pertussis toxin, are ADP-ribosylating toxins.) However several plant toxins, e.g., ricin, have the same activity. These toxins have been named ribosome-inactivating proteins (RIPs). They catalyze the hydrolysis of the N-glycosidic bond between a specific adenine ring (A-4324, rat liver numbering) and the corresponding ribose in eukaryotic 28S rRNA.[14] This highly specific depurination blocks the binding of aminoacyl-tRNA to the ribosome and shuts down protein synthesis in the target cell.[15] The result is cell death.

Shiga toxin has been used as a tool in cell biology to study endocytosis and protein transport. (See ref. 16 for a brief review.) After binding to the cell-surface receptors, Shiga toxin is taken up by the cell via clathrin-coated pits.[17,18] It can travel by retrograde transport from the Golgi apparatus to the endoplasmic reticulum.[19] At some point it, or at least the enzymatic part, must cross the membrane and enter the cytosol to act on the ribosome. The molecular mechanism for crossing the membrane to the cytosol is unknown.

DESCRIPTION OF THE HOLOTOXIN

The five B subunits of Shiga toxin form a regular pentamer (Fig. 9.1). The pentamer is a 25 Å thick disk having a diameter of 55 Å and a central pore of ~10 Å. Five antiparallel β-sheets each composed of six strands are located on the periphery of the B pentamer. The β-strands extend from the top to the bottom of the disk. The five β-sheets surround five parallel α-helices, one from each B subunit. The α-helices are of nearly the same length as the thickness of the disk; they do not extend above or below it. Their amino-termini are at the "bottom" of the disk; their carboxy-termini are at the "top" closest to the A subunit. The loops connecting the elements of secondary structure (α-helices and strands of β-sheet) are quite short at the top of the disk and longer on the bottom. Each B subunit has one internal disulfide bridge connecting Cys 4 (the second residue of the first β-strand of B) and Cys 57 which is in a long loop between the last two strands in the same β-sheet. The interface between one B subunit and the next is located (1) at the center of the β-sheet (three β-strands come from one subunit and three from the neighboring subunit) and (2) where the α-helices pack (Fig. 9.2).

The major part of the A subunit lies "above" the pentamer at the carboxy-termini of the five α-helices of the B pentamer. The exception to this is the carboxy-terminus of A. Although the last six residues at the carboxy-terminus are not visible in the electron density map, Ser 279 starts a short α-helix at the top of the ring of the B helices and the polypeptide extends down into the pore of the B pentamer. The rest of the A subunit rises approximately 30 Å above the B pentamer.

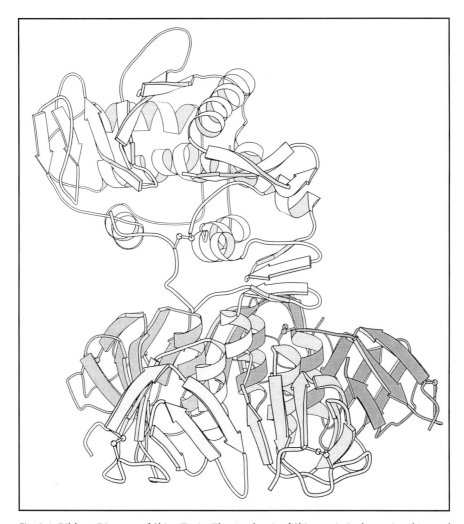

Fig. 9.1. Ribbon Diagram of Shiga Toxin. The A subunit of Shiga toxin is shown in white and the five B subunits with which it associates are shown in varying shades of gray. The ribbons represent α-helices and the β-strands of more than two residues. These elements of secondary structure were determined with DSSP.[44] The residue limits for the elements of secondary structure shown for the A subunit are listed in the legend of Figure 9.3. For the B subunits, the secondary structure is that of a "composite" B subunit. The composite B subunit has β-strands from residue 4 to residue 8, 9 to 14, 20 to 24, 27 to 31, 49 to 53 and 65 to 68; and one α-helix from residue 36 to residue 46. The first β-strand has been shortened from the actual limits of 3 to 8 so that the disulfide bridge connecting residues 4 and 57 is clearly drawn. The disulfide bridges (one in subunit A and one in each of the B subunits) are represented by balls at the positions of the sulfur atoms and sticks connecting them and the α-carbons' positions.

We solved the crystal structure of the holotoxin at 2.5 Å resolution.[45] There are two AB$_5$ hexamers in the asymmetric unit of the crystals. At the current stage of refinement, the structural model includes 69 amino acids for each of the 10 B subunits and 264 and 262 residues of the two A subunits. The A subunits are labeled chains A and L. The B subunits are labeled chains B to K. The hexamer presented here consists of chains A and B to F. Chain B is in front on the right, chain C is in front on the left; continuing around the perimeter are chains D then E then F. Figures 9.1, 9.2, 9.3 and 9.6 were drawn with the program MOLSCRIPT.[46]

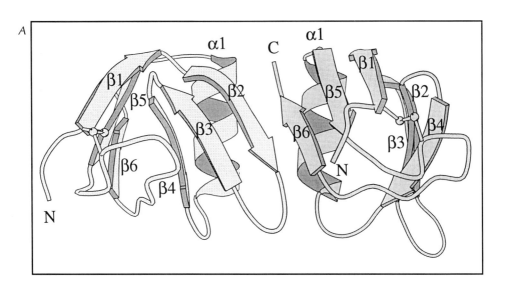

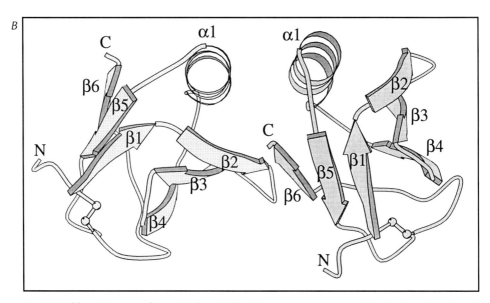

Fig. 9.2. Ribbon Diagram of Two B Subunits. The diagram represents two B subunits, chains B and C, which pack next to each other in the pentamer. Chain B is on the right in darker shading. The view in (a) is looking at one β-sheet. Three β-strands from chain B (β1, β5 and β6) and three from chain C (β2, β3 and β4) form this β-sheet. The view in (b) is from the "top" of the pentamer. It shows how two of the five α-helices (α1) come together at the center of the pentamer. The amino- and carboxy-termini of the chains are labeled N and C respectively.

COLOR FIGURES

Fig. 5.5: Fit of the aerolysin monomers into an image derived from electron microscopy of the aerolysin ion channel.[35] Each monomer is shown as an alpha-carbon backbone and in different colors. The view is looking directly down the channel towards the membrane. This figure was produced using the program O.[59]

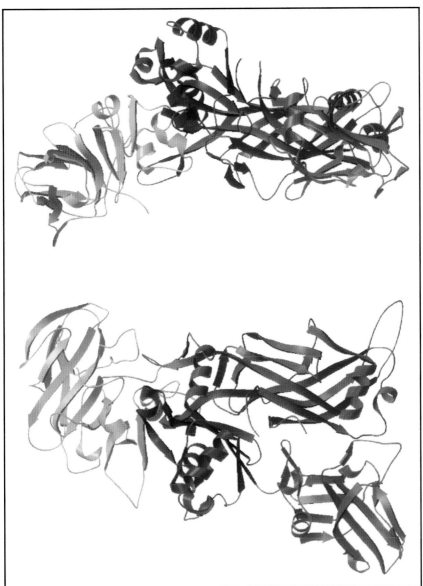

Fig. 6.3. The crystal structure of PA. The domains are colored as follows: 1—yellow, 2—red, 3—blue, 4—green. The molecule has dimensions of roughly 100 Å x 65 Å x 35 Å. Domain 1 contains the residues comprising PA$_{20}$; domain 2 is the only domain long enough to span a membrane; the function of domain 3 is unknown; and domain 4 is believed to bind the cell-surface receptor. See text and Table 6.2 for further details.

Fig. 7.8. The three-dimensional structure of cholera toxin. (view B) A space-filling representation of cholera toxin on an idealized membrane. Although the manner in which the catalytic A subunit passes into the cell remains controversial, choleragen most likely binds to its target membrane with the A subunit facing away. The G_{M1} gangliosides are also shown emerging from the membrane and binding to the ventral surface of the B pentamer.

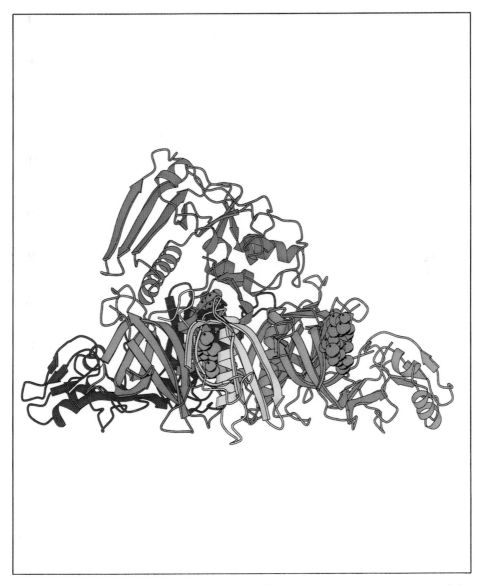

Fig. 10.2. Schematic drawing of pertussis toxin. The S1-subunit (green) sits on top of the B-oligomer, formed of subunits S2 (cyan), S3 (dark blue), two copies of S4 (red), and S5 (yellow). Arrows indicate strands of β-sheet and coiled ribbons indicate α-helices. Bound ligands observed in two separate binding experiments are shown as van der Waals spheres. A sialylated oligosaccharide is shown bound to the sides of subunits S2 and S3 (violet), and ATP is shown bound in the pore of the B-oligomer (orange). This figure, and others showing molecular structures, were drawn with MOLSCRIPT.[104]

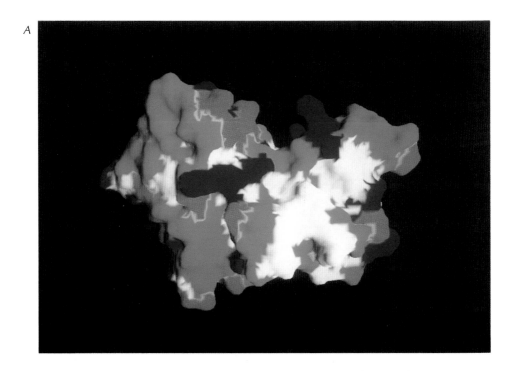

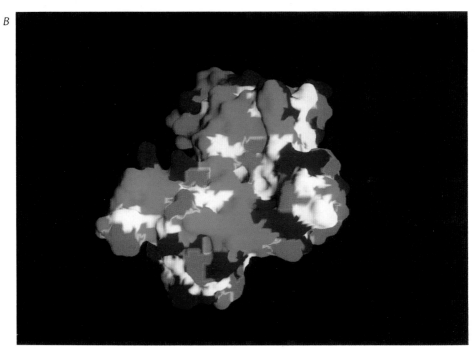

Fig. 11.3. Solvent exposed surface of TSST-1 calculated by GRASP.[55] Acidic residues are red, basic residues are blue, and polar residues are green. (a) Front view. (b) Top rear view.

The globular part of the A subunit is about 30 Å by 35 Å by 55 Å. The longest dimension is in the same direction as the radius of the B pentamer. Since the α-helix that extends into the pore is towards one end of the A subunit, the A subunit overhangs the B pentamer (Fig. 9.1).

The A subunit has a mixed α,β topology (Fig. 9.3). There are nine α-helices (α1 to α9) and three β-sheets. The α-helices all pack on one side of the largest β-sheet. This β-sheet has both parallel and antiparallel β-strands (β1, β4, β5, β6, β7, β8). The second β-sheet is a two-stranded β-hairpin (β2, β3). The third is a four-stranded sheet (β10, β9, β12, β11), made of two β-hairpins packing next to each other with the two central strands running parallel. This sheet is at the bottom of the A subunit and lies just above the B pentamer (Fig. 9.1). Strands β9 and β10 are from the A1 part of the A subunit and strands β11 and β12 are from A2. A1 and A2 are also covalently attached by a disulfide bridge between Cys 242 of A1 and Cys 261 of A2. Cys 242 is the last residue of A1 that is seen in the electron density map. The polypeptide is next visible at Phe 257. The intervening residues include the cleavage sequence and may be in a flexible loop.

B PENTAMER

The structure of the B pentamer of Shiga toxin is known both in complex with the A subunit (the holotoxin) and alone in the structure of the B oligomer of verotoxin-1 from *E. coli*.[20] (Verotoxin is another name for Shiga-like toxin. The toxin kills Vero cells, an African green monkey kidney cell line.[21]) The B subunits of these two toxins are completely identical in sequence. The B oligomer of verotoxin has the same pentameric ring with α-helices surrounding a central pore and five β-sheets surrounding the α-helices. However, the 5-fold symmetry is not perfect. There is a screw component of 1.3 Å along the rotation axis, changing the flat ring of the holotoxin to a "lock-washer." In the verotoxin B pentamer, β-sheet interactions are present between pairs of monomers except at one pair of monomers where there is a gap of ~5 Å. This lack of perfect 5-fold symmetry was attributed to crystallization, due to the binding of a zinc atom at a crystal packing interface. The structure of the B pentamer in the holotoxin agrees with this conjecture. In the holotoxin, there are hydrogen bonding interactions between all neighboring pairs of B subunits. Table 9.1 lists the symmetry operations that relate each B subunit to its neighbor and shows that the symmetry is very close to being perfect 5-fold symmetry.

Since the B pentamer has 5-fold symmetry, it has five copies of any potential receptor-binding site. This makes the holotoxin multivalent provided that the binding sites do not overlap. Binding studies with the B pentamer and the trisaccharide portion of the globotriosylceramide cell-surface receptor indicate that there are five identical noninteracting carbohydrate binding sites per pentamer.[22] Studies also

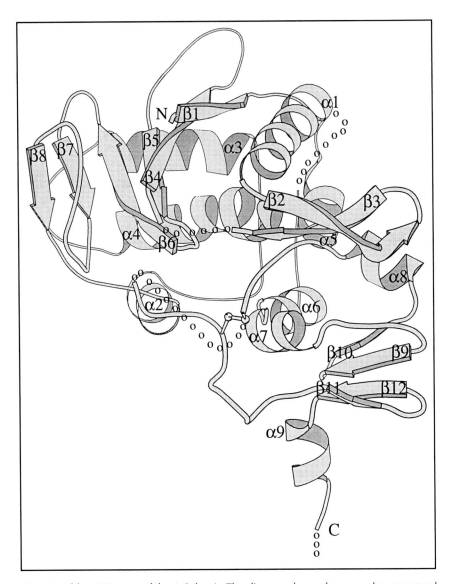

Fig. 9.3. Ribbon Diagram of the A Subunit. The diagram shows the secondary structural elements of chain A. The α-helices run from residues 10 to 24 (α1), 114 to 121 (α2), 132 to143 (α3), 152 to 170 (α4), 172 to 179 (α5), 194 to 200 (α6), 203 to 209 (α7), 229 to 235 (α8) and 279 to 285 (α9). The β-strands are from residues 2 to 6 (β1), 25 to 33 (β2), 36 to 41 (β3), 49 to 55 (β4), 68 to 72 (β5), 77 to 83 (β6), 88 to 91 (β7), 104 to 107 (β8), 218 to 221 (β9), 224 to 226 (β10), 269 to 271 (β11), 275 to 278 (β12). Residues 43 to 46, 184 to 188, 243 to 256 and 288 to 293 of chain A are not visible in the electron density map and have not been included in the current model. They are noted by "o" in the figure. The amino-terminus of the A subunit is labeled N and the carboxy-terminus is labeled C.

Table 9.1 Subunit superpositions

Subunits superposed	root mean square difference (Å) (number of atoms)	Translation along the rotation axis (Å)	Rotation κ (°)	Rotation axis φ (°)	ψ (°)
C on B	0.2 (65)	−0.8	71	331	69
D on C	0.3 (64)	+0.5	70	327	68
E on D	0.4 (68)	−0.2	75	324	71
F on E	0.5 (59)	+2.0	72	333	78
B on F	0.3 (49)	+1.5	74	333	67
H on G	0.3 (61)	+0.3	72	146	108
I on H	0.4 (67)	−0.1	73	146	111
J on I	0.5 (64)	+0.8	72	151	111
K on J	0.3 (56)	−0.4	73	149	108
G on K	0.3 (66)	+0.6	71	149	107
L on A	0.4 (218)	+0.3	175	270	147
KGHIJ on BCDEF	0.4 (286)	+0.2	180	273	149

The two A subunits in the asymmetric unit of the crystal are labeled A and L. The 10 B subunits are B-K. B-F form the pentamer that is associated with A. G-K form the second pentamer that is associated with L. B interacts with A like K interacts with L, C like G, D like H, E like I and F like J. Equivalent α-carbon atoms of the subunits were superposed using the program SUPPOS.[49] Pairs of α-carbon atoms whose distances when superposed differ by more than three times the overall standard deviation were rejected from the superposition. κ, φ and ψ are the polar rotation angles. κ is the rotation about an axis defined by φ and ψ according to the convention of Rossmann and Blow.[50] If the 5-fold noncrystallographic axes were perfect, the translations parallel to the symmetry axes would be 0 Å, the rotation about the symmetry axes would be 72° and there would be a single rotation axis for each pentamer. The derived values are close to this idealized case. A perfect 2-fold axis would have a κ value of 180° and a translation of 0 Å.

indicate that the holotoxin binds to carbohydrates in the same way as the B pentamer does.[23] One molecule of Shiga toxin could bind to more than one glycolipid molecule at the surface of the target cell causing the glycolipids to cluster. This might be the trigger for the endocytosis of the Shiga toxin molecule.

Heat-labile enterotoxin, cholera toxin, pertussis toxin and Shiga toxin all show structural similarity.[24] This structural similarity includes a fold common to the core of the B monomer. The common fold has been named the OB-fold for oligonucleotide/oligosaccharide-binding fold.[25] It includes all of the secondary structural elements of the B subunit of Shiga toxin. The fold is not exclusive to the AB$_5$ class of toxins. It has also been seen in other protein toxins isolated from bacterial pathogens and in other proteins (Table 9.2).[26] The assembly of the five OB-folds with the α-helices to the inside and sheets surrounding them is also part of the structural similarity of the AB$_5$ bacterial protein toxins. Because the amino acid sequences are so dissimilar, it was a

Table 9.2 OB–fold

Protein	Reference
Staphylococcal nuclease	51
toxic shock syndrome toxin–1	52
Staphylococcal enterotoxin B	53
Staphylococcal enterotoxin A	54
aspartyl–tRNA synthetase	55,56,57
lysyl–tRNA synthetase	58
tissue inhibitor of metalloproteinases	59
inorganic pyrophosphatase	60,61,62
cold shock DNA–binding domain	
gene V protein	
ribosomal S17 protein	

These are the other proteins in the SCOP database[26] listed as having the OB-fold. For the cases where no reference is given, the coordinates are available in the Protein Data Bank[63,64] but the paper has not yet been published.

surprise to see that the B pentamer of heat-labile enterotoxin and the B oligomer of verotoxin-1 have the same fold.[20] It was even more surprising to see that one domain of each B subunit of pertussis toxin (the subunits are actually labeled S2, S3, S4 and S5 and there are two copies of S4[6]) fold and assemble in a similar way.[27] Heat-labile enterotoxin, cholera toxin and Shiga toxin all have 5-fold symmetry. They are all potentially multivalent for their receptors on the cell surfaces. Pertussis toxin binds carbohydrates to subunits S2 and S3 so it is potentially bivalent at least.

THE PORE AND A2

The ring of five B subunits leaves a central pore with a diameter of ~10 Å. The symmetric nature of the packing of the B subunits means that the pore has 5-fold symmetry. Since there is room for only a single antiparallel helix from A2, the interactions that residues on this helix make with the pentameric pore are of necessity asymmetric. The top part of the pore is occupied by the carboxy-terminal α-helix of A2 (Fig. 9.1). The sequence of A2 from Thr 281 to Met 287 matches the size and the hydrophobic characteristics of the residues lining the inside of the pore. The side chain oxygen, Oγ1, of Thr 281 interacts with other residues of the A subunit (Oδ1 of Asp 278, Nη1 of Arg 205 and N of Gly 222) but Thr 281 Cγ2 packs down towards Ile 45 of one of the B subunits (chain E). Leu 282 packs under the tryptophan of A2 but also interacts with the Cγ2 atoms of two Thr 46 residues from neighboring B subunits (chains F and B). If Gly 283 were any other type of residue, the Cβ atom would clash with Thr 46 of the next B subunit in the pentamer (chain C). A different rotamer for the threonine would help, but the distance between the two Cβ atoms would still be close, 3.6 Å. This close packing must be a property of the packing of a sixth α-helix antiparallel to the five parallel α-helices that surround it. Residue 284 is an alanine and its Cβ atom packs near the Cγ2 atom of Thr 46 (chain D). The side chain of Ile 285 interacts with the Cγ2 atoms of two Thr 46 residues. One is from the next B subunit (chain E) and the second is the same atom with which Leu 282 interacts (Chain F). Leu 286 interacts with the Cβ atom of Ser 42 (chain B). Met 287 is closest to the Cβ of Thr 38 (chain D). All of these side chain to side chain interactions are hydrophobic.

The last six residues of A2 are Arg-Arg-Thr-Ile-Ser-Ser. Unfortunately, there is no electron density for these residues; therefore we cannot say with which residues of the B subunit they would interact. It is not possible even to know whether they extend through the pore. If they adopted an extended conformation they would. It is likely that they extend at least to the bottom of the B pentamer and that this is why the two AB$_5$ hexamers in the asymmetric unit of the crystal pack with 2-fold noncrystallographic symmetry (Fig. 9.4, Table 9.1). It is possible that the carboxy-termini of the two A subunits interact in the crystal.

The structural similarity of Shiga toxin, heat-labile enterotoxin and pertussis toxin extends to the location of the A subunit with respect to the B oligomer. The A subunit is always at the carboxy-termini of the five α-helices of the pentamer. The A subunit sits on "top" of the B oligomer. In each case the carboxy-terminal sequence of the A subunit interacts with the B oligomer.

A2 of heat-labile enterotoxin extends through the central pore and has a short helix on the other side.[28] The similarity to A2 of Shiga toxin ends there. Most of the residues lining the central pore of heat-labile enterotoxin are charged.[29] The last four residues of A2 are Arg-Asp-Glu-Leu which is an endoplasmic reticulum retention signal. They are thought to interact with the membrane after the enterotoxin is bound.[28]

The pore of pertussis toxin is "more of a pit."[27] The carboxy terminus of the A subunit (S1) extends only part of the way down the pore. On the other hand pertussis toxin has more interactions between S1 and the top of the B oligomer to compensate for having so few interactions in the pore.

How is the interaction between subunit A and the B pentamer important in the mode of action of the Shiga toxin? For delivery into the target cell, the A subunit is dependent on the B subunits. The target cell in the host is determined by the receptor-specificity of the B subunits and binding to the target cell is performed by the B subunits. The B subunits may also play a role in the transport of the A subunit

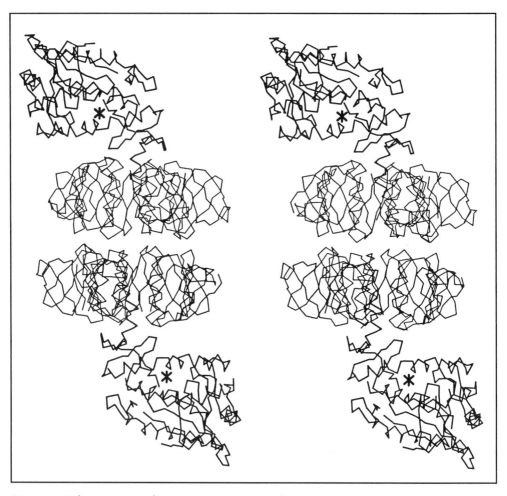

Fig. 9.4. α-Carbon Diagram of Two Hexamers. Two AB₅ hexamers pack in the asymmetric unit of the crystal. The "bottom" of each B pentamer is the surface where the two hexamers interact. This view is down the 2-fold noncrystallographic axis relating the two hexamers. The asterisk shown in each A subunit is at the approximate location of the active site of A1.

and its translocation into the cytosol. A2 is used as a tether to the B pentamer. By making A1 and A2 a single polypeptide, the bacterium ensures that each enzymatic subunit has a tether to the delivery machinery. The disulfide bridge is a necessary link between A1 and A2, since proteolytic cleavage of the polypeptide could happen before the A subunit actually enters the target cell. Reduction of the disulfide would only occur in the reducing environment of the cell after entry. This ensures that the enzyme gets into the cell and perhaps even into the correct compartment in the cell. Once A1 and A2 are cleaved and the disulfide is reduced, A1 is released and there is likely no further need for A2 nor for the B pentamer. The sequence of A2 has to have evolved so that it will bind well to a B monomer or to several B monomers that are in the process of assembling. The fact that B is oligomeric may make the binding take place more easily and then be stronger. A2 never needs to be added to a preformed B pentamer and it never needs to be released from one. The B oligomer may be a pentamer simply because this is favorable for the OB-fold. The OB-folds assemble so that the edge strands of two monomers can form hydrogen bonds. Maybe because of packing considerations, the helices come together. Having five subunits assemble in a pentameric ring leaves a well-packed structure with a pore of the correct size for the A2 tether.

We can only postulate that A2 is not part of the same polypeptide as the B part of the Shiga toxin because this gave the bacterium an advantage. Having the separate polypeptides could have led to a mixing-and-matching among A and B subunits. As an example, the B pentamer of heat-labile enterotoxin is matched with an ADP-ribosylating enzyme in heat-labile enterotoxin. The similar B pentamer of Shiga toxin is matched with an A chain that is an N-glycosidase. Other bacterial toxins also have catalytic subunits that are ADP-ribosylating enzymes, e.g., pertussis toxin, cholera toxin and diphtheria toxin. Only the first two of these three examples are members of the AB$_5$ class.

OTHER A-B INTERACTIONS

Pertussis toxin has more interactions between its A subunit (S1) and the B oligomer than does either heat-labile enterotoxin or the Shiga toxin.[27] A flat surface of S1 lies against the B oligomer. Since the B subunits are not identical they can have evolved independently to optimize interactions with S1. The interactions between the A subunit and the B pentamer of heat-labile enterotoxin are mostly restricted to residues in A2 and the pore.[30] Shiga toxin is the intermediate example. Of the 974 Å^2 of surface area of the A subunit buried by the B pentamer, 385 Å^2 is comprised of the surface area of residues 281 to 287 buried in the pore. The remaining 589 Å^2 is the surface of other residues of the A subunit buried by the B pentamer.

For heat-labile enterotoxin, there is a variability among crystal structures in the orientation of the A subunit with respect to the B pentamer.[31] There is a similar difference between the two AB$_5$ hexamers of Shiga toxin. Table 9.1 shows that the two pentamers in the asymmetric unit of the crystal are related to each other by an exact 180° rotation (a 2-fold rotation). However the rotation necessary to relate the two A subunits is 175°. This means that the two A subunits have slightly different orientations with respect to their B pentamers. This flexibility could be an intrinsic property of the heat-labile enterotoxin and Shiga toxin. It may help in the delivery of the A subunit, or it may simply be a feature of the tethering of the A subunit to the B pentamer that there was no evolutionary pressure to change.

THE A SUBUNIT

The fold of the A subunit of the Shiga toxin is very similar to that of the A chain of ricin[32] (Fig. 9.5), α-trichosanthin,[33,34] α-momorcharin,[35] pokeweed antiviral protein,[36,37] the A chain of abrin and gelonin.[38] These are the members of the family of ribosome-inactivating proteins whose crystal structures are known. They are the only proteins known to have this fold.[26] The fold positions the residues that are absolutely conserved among ribosome-inactivating proteins in a surface cleft. From mutagenesis and from crystallographic studies of substrate analogs bound to ricin,[39] α-momorcharin and α-trichosanthin, this cleft is known to be the active site.

In the Shiga holotoxin, the active site is not directly accessible to substrate (Fig. 9.6a). The side chain of Met 260 from A2 packs into the cleft. It is one of the four residues of the cleavage loop between Cys 242 and Cys 261 that are seen in the electron density map. These four residues are held in position by the disulfide bridge and by a

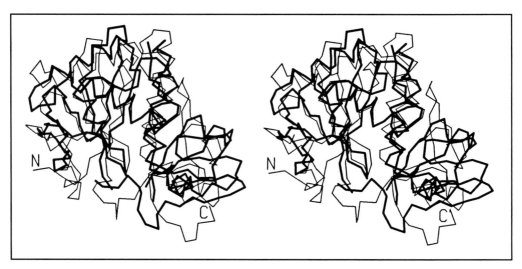

Fig. 9.5. α-Carbon Diagram of Ricin A Superposed on the A Subunit of Shiga Toxin. The α-carbon backbone of ricin A is shown in thin lines with the amino-terminus labeled N and the carboxy-terminus labeled C. The α-carbon backbone of the A subunit of Shiga toxin is shown in thicker lines. The two are oriented so that the carboxy-terminal α-helix of the A subunit of the Shiga toxin in the lower right of the figure is viewed down the axis of the helix (i.e., approximately parallel to the 5-fold axis of the B pentamer).

The two molecules were superposed by first superposing three pairs of equivalent α-carbon coordinates, then allowing the program O[47,48] to do a least squares superposition of all pairs of α-carbon coordinates that matched to within 2.5 Å. 185 pairs of α-carbon atoms matched with a root mean square difference of 1.6 Å.

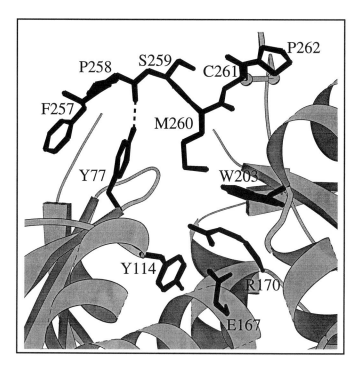

Fig. 9.6. "Active" Sites of Shiga Toxin and Ricin A. For comparison, the "active" sites of (a, shown left) the A subunit of Shiga toxin and (b, shown below) ricin A are drawn. The side chain of Met 260 protrudes into the active site cleft, effectively blocking access to the substrate binding site in the holotoxin. The dotted line between Tyr 77 Oη and Pro 258 O represents a hydrogen bond.

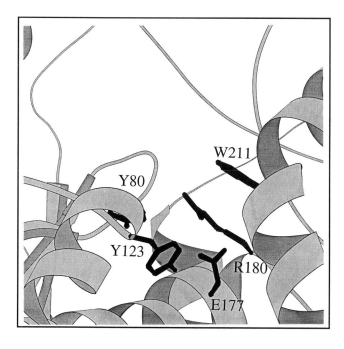

hydrogen bond from Tyr 77 Oη to the carbonyl oxygen atom of Pro 258. In the absence of any polypeptide chain in the active site, Tyr 77 is likely to be in a different position as demonstrated in the structure of ricin A with an empty active site (Fig. 9.6b). With formycin mono-phosphate or adenyl(3' → 5')guanosine in the active site, the equiva-lent tyrosine residue rotates to stack with the purine ring of the sub-strate analogue.[39]

When a sample of Shiga toxin is pretreated with trypsin, followed by dithiothreitol and 8M urea or 1% sodium dodecyl sulfate and then diluted to lower the concentration of the denaturant, its inhibitory activity in cell-free protein synthesis increases.[13] An interpretation of this result is that trypsin cleaves the susceptible loop, dithiothreitol reduces the disulfide bridge still linking A1 and A2 and the denatur-ant releases A1 from A2. A1 then refolds with the active site free of the inhibitory A2 sequence so that it can catalyze the depurination of the ribosome.

The B pentamer, as well as A2, may block the access of 28s rRNA to the active site. Figure 9.4 shows that the active site of A1 faces the B pentamer. This is also true for pertussis toxin[27] but is not true for heat-labile enterotoxin.[30] The analogy of Shiga toxin to pertussis toxin is not simply in the orientation of the active site with respect to the B oligomer. Pertussis toxin also has a disulfide bridge near the active site and it is proposed that this disulfide bridge would have to be reduced to allow a rearrangement of the protein. The rearrangement would open up the active site so that nicotinamide adenine dinucle-otide and the G protein substrate can gain access. This movement would disrupt the interactions between the A subunit (labeled S1 in pertussis toxin) and the B oligomer. Although pertussis toxin does have a loop that a protease readily cleaves, there is no evidence that this step is necessary for activity.

Like Shiga toxin, heat-labile enterotoxin is activated by a proteolytic cleavage of the A subunit and reduction of a disulfide bond between the two parts.[40-42] But the heat-labile enterotoxin has its active site directed away from the B pentamer; the disulfide bridge is at least 25 Å away from the active site;[30] the cleavage loop appears to be flex-ible both before and after cleavage; and cleavage of the loop does not cause conformational changes in the enzymatic part of the A subunit.[43] Why cleavage of the A subunit and a reducing environment are neces-sary to activate the heat-labile enterotoxin is not yet fully understood. So although we can say that Shiga toxin and heat-labile enterotoxin are similar in that both are activated by proteolytic cleavage and re-duction of a disulfide bond, there do not seem to be analogous struc-tural requirements for the activation processes.

FUTURE RESEARCH

The crystal structural results of Shiga toxin provide a more com-plete understanding of the organization of the holotoxin. But the

N-glycosidase portion of the holotoxin is an inactive form. For a more complete understanding of the active site, A1 should be crystallized, both free and in the presence of substrate analogs. As a sidelight it will be interesting to see how the two strands of A1 in the four-stranded β-sheet between the A subunit and the B pentamer rearrange in the absence of A2.

There are still issues in the transport of the holotoxin from receptors on the surface of target cells to its site of action in the cytosol that are not understood. Is the binding of the holotoxin multivalent? What role do the glycolipids play in endocytosis? Do the B subunits play a role in helping A1 to cross the membrane? Does only A1 cross? From what organelle does A1 cross into the cytosol? The crystal structure of the holotoxin may serve as a reference for further experiments in the cell biology of these processes.

REFERENCES

1. Shiga K. Über den Dysenterie-bacillus (*Bacillus dysenteriae*). Zentralbl Bakteriol Orig 1898; 24:913-18.
2. O'Brien AD, Tesh VL, Donohue-Rolfe A et al. Shiga toxin: biochemistry, genetics, mode of action, and role in pathogenesis. Curr Topics in Microbiol Immunol 1992; 180:65-94.
3. Karmali MA, Petric M, Lim C et al. The association between idiopathic hemolytic uremic syndrome and infection by verotoxin-producing *Escherichia coli*. J Infect Dis 1985; 151:775-82.
4. Strockbine NA, Marques LRM, Newland JW et al. Two toxin-converting phages from *Escherichia coli* O157:H7 strain 933 encode antigenically distinct toxins with similar biologic activities. Infect Immun 1986; 53:135-40.
5. Donohue-Rolfe A, Keusch GT, Edson C et al. Pathogenesis of Shigella diarrhea. IX. Simplified high yield purification of Shigella toxin and characterization of subunit composition and function by the use of subunit-specific monoclonal and polyclonal antibodies. J Exp Med 1984; 160:1767-81.
6. Tamura M, Nogimori K, Murai S et al. Subunit structure of islet-activating protein, pertussis toxin, in conformity with the A-B model. Biochemistry 1982; 21:5516-22.
7. Seidah NG, Donohue-Rolfe A, Lazure C et al. Complete amino acid sequence of *Shigella* toxin B-chain. J Biol Chem 1986; 261:13928-31.
8. Jacewicz M, Clausen H, Nudelman E et al. Pathogenesis of Shigella diarrhea. XI. Isolation of a Shigella toxin-binding glycolipid from rabbit jejunum and HeLa cells and its identification as globotriosylceramide. J Exp Med 1986; 163:1391-404.
9. Lindberg AA, Brown JE, Stroemberg N et al. Identification of the carbohydrate receptor for Shiga toxin produced by *Shigella dysenteriae* type 1. J Biol Chem 1987; 262:1779-85.
10. Kozlov Y, Kabishev A, Fedchenko V et al. Cloning and sequencing of Shiga-toxin structural genes. Proc Acad Sci USSR 1987; 295:740-44.

11. Kozlov YV, Kabishev AA, Lukyanov EV et al. The primary structure of the operons coding for *Shigella dysenteriae* toxin and temperate phage H30 Shiga-like toxin. Gene 1988; 67:213-21.

12. Olsnes S, Reisbig R, Eiklid K. Subunit structure of *Shigella* cytotoxin. J Biol Chem 1981; 256:8732-38.

13. Reisbig R, Olsnes S, Eiklid K. The cytotoxic activity of *Shigella* toxin. Evidence for catalytic inactivation of the 60S ribosomal subunit. J Biol Chem 1981; 256:8739-44.

14. Endo Y, Tsurugi K, Yutsudo T et al. Site of action of Vero toxin (VT2) from *Escherichia coli* O157:H7 and of Shiga toxin on eukaryotic ribosomes. RNA N-glycosidase activity of the toxins. Eur J Biochem 1988; 171:45-50.

15. Obrig TG, Moran TP, Brown JE. The mode of action of Shiga toxin on peptide elongation of eukaryotic protein synthesis. Biochem J 1987; 244:287-94.

16. Sandvig K, van Deurs B. Endocytosis and intracellular sorting of ricin and Shiga toxin. FEBS Lett 1994; 346:99-102.

17. Sandvig K, Olsnes S, Brown JE et al. Endocytosis from coated pits of Shiga toxin: a glycolipid-binding protein from *Shigella dysenteriae* 1. J Cell Biol 1989; 108:1331-43.

18. Sandvig K, Prydz K, Ryd M et al. Endocytosis and intracellular transport of the glycolipid-binding ligand Shiga toxin in polarized MDCK cells. J Cell Biol 1991; 113:553-62.

19. Sandvig K, Garred O, Prydz K et al. Retrograde transport of endocytosed Shiga toxin to the endoplasmic reticulum. Nature 1992; 358:510-12.

20. Stein PE, Boodhoo A, Tyrrell GJ et al. Crystal structure of the cell-binding B oligomer of verotoxin-1 from *E. coli*. Nature 1992; 355:748-50.

21. Konowalchuk J, Speirs JI, Stavric S. Vero response to a cytotoxin of *Escherichia coli*. Infection and Immunity 1977; 18:775-79.

22. St. Hilaire PM, Boyd MK, Toone EJ. Interaction of the Shiga-like toxin type 1 B-subunit with its carbohydrate receptor. Biochemistry 1994; 33:14452-63.

23. Donohue-Rolfe A, Jacewicz M, Keusch GT. Isolation and characterization of functional Shiga toxin subunits and renatured holotoxin. Mol Microbiol 1989; 3:1231-36.

24. Merritt EA, Hol WGJ. AB_5 toxins. Curr Opinion in Structural Biology 1995; 5:165-71.

25. Murzin AG. OB(oligonucleotide/oligosaccharide binding)-fold: common structural and functional solution for non-homologous sequences. EMBO J 1993; 12:861-67.

26. Murzin AG, Brenner SE, Hubbard T et al. SCOP: a structural classification of proteins database for the investigation of sequences and structures. J Mol Biol 1995; 247:536-40.

27. Stein PE, Boodhoo A, Armstrong GD et al. The crystal structure of pertussis toxin. Structure 1994; 2:45-57.

28. Sixma TK, Pronk SE, Kalk KH et al. Lactose binding to heat-labile enterotoxin revealed by X-ray crystallography. Nature 1992; 355:561-64.

29. Sixma TK, Stein PE, Hol WGJ et al. Comparison of the B-pentamers of heat-labile enterotoxin and verotoxin-1: two structures with remarkable similarity and dissimilarity. Biochemistry 1993; 32:191-98.

30. Sixma TK, Pronk SE, Kalk KH et al. Crystal structure of a cholera toxin-related heat-labile enterotoxin from *E. coli*. Nature 1991; 351:371-77.

31. Sixma TK, Aguirre A, van Scheltinga ACT et al. Heat-labile enterotoxin crystal forms with variable A/B$_5$ orientation. Analysis of conformational flexibility. FEBS Lett 1992; 305:81-85.

32. Mlsna D, Monzingo AF, Katzin BJ et al. Structure of recombinant ricin A chain at 2.3 Å. Prot Sci 1993; 2:429-35.

33. Xiong JP, Xia ZX, Wang Y. Crystal structure of trichosanthin-NADPH at 1.7 Å resolution reveals active-site architecture. Nature Struct Biol 1994; 1:695-700.

34. Zhou K, Fu Z, Chen M et al. Structure of trichosanthin at 1.88 Å resolution. Proteins 1994; 19:4-13.

35. Ren J, Wang Y, Dong Y et al. The N-glycosidase mechanism of ribosome-inactivating proteins implied by crystal structures of α-momorcharin. Structure 1994; 2:7-16.

36. Monzingo AF, Collins EJ, Ernst SR et al. The 2.5 Å structure of pokeweed antiviral protein. J Mol Biol 1993; 233:705-15.

37. Ago H, Kataoka J, Tsuge H et al. X-ray structure of a pokeweed antiviral protein, coded by a new genomic clone, at 0.23 nm resolution. Eur J Biochem 1994; 225:369-74.

38. Hosur MV, Nair B, Satyamurthy P et al. X-ray structure of gelonin at 1.8 Å resolution. J Mol Biol 1995; 250:368-80.

39. Monzingo AF, Robertus JD. X-ray analysis of substrate analogs in the ricin A-chain active site. J Mol Biol 1992; 227:1136-45.

40. Mekalanos JJ, Collier RJ, Romig WR. Enzymic activity of cholera toxin. II. Relationships to proteolytic processing, disulfide bond reduction, and subunit composition. J Biol Chem 1979; 254:5855-61.

41. Moss J, Osborne JC Jr., Fishman PH et al. *Escherichia coli* heat-labile enterotoxin. Ganglioside specificity and ADP-ribosyltransferase activity. J Biol Chem 1981; 256:12861-65.

42. Moss J, Stanley SJ, Vaughan M et al. Interaction of ADP-ribosylation factor with *Escherichia coli* enterotoxin that contains an inactivating lysine 112 substitution. J Biol Chem 1993; 268:6383-87.

43. Merritt EA, Pronk S, Sixma TK et al. Structure of partially-activated *E. coli* heat-labile enterotoxin (LT) at 2.6 Å resolution. FEBS Lett 1994; 337:88-92.

44. Kabsch W, Sander C. Dictionary of protein secondary structure: pattern recognition of hydrogen bonded and geometrical features. Biopolymers 1983; 22:2577-637.

45. Fraser ME, Chernaia MM, Kozlov YV et al. Crystal structure of the holotoxin from *Shigella dysenteriae* at 2.5 Å resolution. Nature Struct Biol 1994; 1:59-64.

46. Kraulis PJ. MOLSCRIPT: a program to produce both detailed and schematic plots of protein structures. J Appl Cryst 1991; 24:946-50.

47. Jones TA, Zou JY, Cowan SW et al. Improved methods for building protein models in electron density maps and the location of errors in these models. Acta Cryst 1991; A47:110-19.

48. Jones TA, Kjeldgaard M. O—the manual (version 5.10.1) Uppsala, Sweden. 1995; 1-152.

49. SUPPOS is from the BIOMOL package Gröningen, Holland.

50. Rossmann MG, Blow DM. The detection of sub-units within the crystallographic asymmetric unit. Acta Cryst 1962; 15:24-31.

51. Arnone A, Bier CJ, Cotton FA et al. A high resolution structure of an inhibitor complex of the extracellular nuclease of *Staphylococcus aureus* I. Experimental procedures and chain tracing. J Biol Chem 1971; 246:2302-16.

52. Prasad GS, Earhart CA, Murray DL et al. Structure of toxic shock syndrome toxin 1. Biochemistry 1993; 32:13761-66.

53. Swaminathan S, Furey W, Pletcher J et al. Crystal structure of staphylococcal enterotoxin B, a superantigen. Nature 1992; 359:801-06.

54. Schad EM, Zaitseva I, Zaitsev VN et al. Crystal structure of the super-antigen staphylococcal enterotoxin type A. EMBO J 1995; 14:3292-301.

55. Ruff M, Krishnaswamy S, Boeglin M et al. Class II aminoacyl transfer RNA synthetases:crystal structure of yeast aspartyl-tRNA synthetse complexed with tRNA Asp. Science 1991; 252:1682.

56. Delarue M, Poterszman A, Nikonov S et al. Crystal structure of a prokaryotic aspartyl tRNA-synthetase. EMBO J 1994; 13:3219-29.

57. Cavarelli J, Eriani G, Rees B et al. The active site of yeast aspartyl-tRNA synthetase: structural and functional aspects of the aminoacylation reaction. EMBO J 1994; 13:327-37.

58. Onesti S, Miller AD, Brick P. The crystal structure of the lysyl-tRNA synthetase (LysU) from *Escherichia coli*. Structure 1995; 3:163-76.

59. Williamson RA, Martorell G, Carr MD et al. Solution structure of the active domain of tissue inhibitor of metalloproteinases-2. A new member of the OB fold protein family. Biochemistry 1994; 33:11745-59.

60. Arutiunian EG, Terzian SS, Voronova AA et al. X-ray diffraction study of inorganic pyrophosphatase from Baker's yeast at the 3 Å resolution (Russian). Dokl Akad Nauk SSSR 1981; 258:1481.

61. Oganessyan VY, Kurilova SA, Vorobyeva NN et al. X-ray crystallographic studies of recombinant inorganic pyrophosphatase from *Escherichia coli*. FEBS Lett 1994; 348:301-04.

62. Teplyakov A, Obmolova G, Wilson KS et al. Crystal structure of inorganic pyrophosphatase from *Thermus thermophilus*. Prot Sci 1994; 3:1098-107.

63. Bernstein FC, Koetzle TF, Williams GJB et al. The Protein Data Bank: a computer-based archival file for macromolecular structures. J Mol Biol 1977; 112:535-42.

64. Abola EE, Bernstein FC, Bryant SH et al. Protein Data Bank. In: Allen, FH, Bergerhoff, G, Sievers, R, eds. Crystallographic databases—information content, software systems, scientific applications. Bonn/Cambridge/Chester: Data Commission of the International Union of Crystallography, 1987; 107-32.

STRUCTURAL INSIGHTS INTO PERTUSSIS TOXIN ACTION

Penelope E. Stein, Bart Hazes and Randy J. Read

Pertussis toxin is an exotoxin produced by *Bordetella pertussis*, the bacterium that causes whooping cough. It has been estimated that whooping cough still causes about 340,000 deaths of young children in the world each year,[1] although in developed countries the disease has largely been controlled by vaccination. Currently used whooping cough vaccines consist of chemically-killed whole *B. pertussis* cells. However, media concern about the safety of these whole-cell vaccines, while not supported by the medical profession,[2-4] has prompted efforts to develop a new generation of acellular whooping cough vaccines containing pure forms of protective antigens.[5] Pertussis toxin (PT) induces strong protective immunity to whooping cough[6,7] and has therefore been a key component of most acellular vaccines. Since residual activity of PT may have been responsible for occasional adverse reactions associated with whole-cell vaccines,[8] it is best to eliminate all biological activities of the toxin before it is used in component vaccines. Chemical treatment with aldehydes may impair immunogenicity, so site-directed mutagenesis is the preferred method of inactivating the toxin. This approach requires a detailed knowledge of functional determinants.

PT belongs to the class of AB toxins,[9] which consist of a catalytically active (A) subunit associated with a cell-binding (B) component. The A-subunit of PT consists of a single large subunit, S1. The B-component, usually called the B-oligomer, comprises a pentamer with four distinct subunits (S2, S3, two copies of S4, and S5). Pentameric B-components are also found in the cholera toxin (CT) and Shiga toxin (ST) families, although in these toxins all subunits in the pentamer are identical. The PT holotoxin has a molecular weight of 105 kDa (952 amino acids).

Protein Toxin Structure, edited by Michael W. Parker. © 1996 R.G. Landes Company.

The action of PT on a eukaryotic cell is shown schematically in Figure 10.1, although details of the pathway are still speculative. The toxin is released from bacteria in a catalytically inactive (latent) form and attaches to the surface of target cells through binding of the B-oligomer to glycoconjugate receptors. Experimental evidence[10] and the analogy with other AB toxins suggest that PT enters cells by endocytosis, followed by transport to or through the Golgi apparatus. There is growing evidence that PT and a number of other toxins then proceed by retrograde flow through the Golgi stacks to the endoplasmic reticulum (ER), from which they translocate into the cytosol and attack their targets.

A number of unique features of the ER make it a particularly attractive site for toxin translocation.[11,12] Retrograde flow, which has been

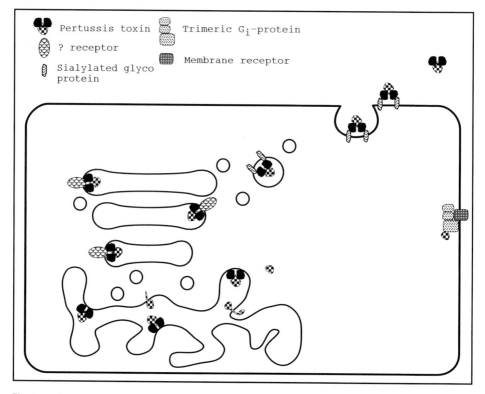

Fig. 10.1. Steps in intoxication by pertussis toxin. Schematic diagram of a eukaryotic cell, illustrating steps that have been shown or inferred to take place in the action of pertussis toxin. A holotoxin molecule binds to glycoproteins on the surface of the target cell and is taken up by endocytosis. It switches (perhaps) to a second receptor that carries it by retrograde flow through the Golgi apparatus to the ER. In the ER, it dissociates, and the activated catalytic S1-subunit is translocated across the membrane into the cytosol. Finally, the catalytic subunit modifies its G protein substrate inside the cell.

recognized recently as a normal process in eukaryotic cells (reviewed in ref. 13), provides an intracellular route to the ER. Protein disulphide isomerases (PDIs) can reduce disulphide bonds in the oxidizing environment of the protein export pathway; disulphide reduction is a common activation mechanism shared by many AB toxins, presumably resulting in exposure of the active site and (except for PT) in dissociation of the A-subunit. While most AB toxins dissociate through a combination of proteolysis and disulphide reduction, the catalytic domain of PT is freed by a unique mechanism. Biochemical data and the crystal structure of a PT-ATP complex suggest that the dissociation of PT is stimulated by ATP, which the holotoxin would encounter for the first time in the ER.[14] Finally, chaperone proteins could help to unfold the catalytic domain prior to membrane translocation, and the protein translocation pore in the ER membrane could provide the back door through which it sneaks into the cytosol.

Pertussis toxin plays a central role in the pathogenesis of whooping cough,[15] probably arising from its profound effects on the immune system and its contribution to the adherence of *B. pertussis* to cilia of the respiratory tract.[16] The activated S1 subunit is an ADP-ribosyltransferase,[17] which catalyses the transfer of ADP-ribose from NAD to a cysteine residue near the carboxy-terminus of the α-subunit of G_i and a few other trimeric GTP binding proteins. The covalently modified $G_i\alpha$-subunit can no longer interact with its receptor, and fails to be activated to its GTP-bound form. This prevents $G_i\alpha$ from inhibiting adenylate cyclase, which in turn leads to an increase in intracellular cAMP. The effects of increased cAMP on various target cells are thought to account for many of the diverse biological activities of PT including effects on the immune system,[15] histamine sensitization and alteration of glucose homeostasis.[18] Furthermore, the B-oligomer can mediate effects that are independent of S1 including proliferation of T lymphocytes,[19] agglutination of erythrocytes[20] and adherence of bacteria to eukaryotic cells.[16] Presumably these effects result from binding of the B-oligomer to cellular receptors, without entry of the toxin into cells.

In this review, we describe how recent structural results have provided an exciting new insight into the pathogenic mechanism of PT. Crystal structures of three major forms of PT have now been determined, representing snap-shots of key events in the toxic pathway. Our new understanding of these events can be exploited in the development of new therapies. The structure of the latent form of PT, on its release from bacteria,[21] has revealed the detailed tertiary structure of the holotoxin and its homologies, not only with other prokaryotic AB toxins, but also with a family of eukaryotic lectins. The structure of the active site in the catalytic S1-subunit has allowed tentative prediction of the activating conformational change induced by disulphide reduction. Knowledge of the active site structure also provides a rational basis for the elimination of catalytic activity in recombinant molecules for use in vaccines. Receptor binding sites on the B-oligomer offer additional targets for the design of inactive vaccine components, since binding to host cells is the initial step in all forms of toxin action. The crystal structure of a complex between PT and a sialylated receptor-like carbohydrate[22] has identified potential target residues in the B-oligomer for creating inactive mutants. Another structure, of ATP bound to PT, has shed light on events inside the target cell prior to dissociation of the toxin.[14] Together, these structures provide a basis for future studies that should reveal additional steps in the toxic mechanism, ultimately enabling reconstruction of the complete picture.

STRUCTURE OF PERTUSSIS TOXIN

Pertussis toxin is released from bacteria as an enzymatically inactive (latent) form of the holotoxin. The crystal structure of this form of PT has been determined at 2.9 Å resolution[21] and is shown schematically in Figure 10.2. The catalytic A-subunit (S1) rests like a cap on the cell-binding B-oligomer made from the remaining five subunits (S2, S3, two copies of S4, S5).

A-SUBUNIT

The A-subunit of PT (S1) comprises 235 amino acids, but studies of carboxy-terminal deletion mutants indicate that residues 1-180 alone are sufficient for a basic level of catalytic activity.[23-25] This amino-terminal portion of S1 shares sequence and structural homology with the enzymatic fragment (A1) of the A-subunit of heat-labile enterotoxin from *Escherichia coli* (LT),[26,27] reflecting their similar catalytic functions. Both toxins ADP-ribosylate the α-subunit of a trimeric G protein, but they attack different amino acids, resulting in different target specificities. Almost all secondary structure elements are conserved within this enzymatic region of the two toxin A-subunits. Pair-wise superposition of 116 Cα atoms gives a root-mean-squared (r.m.s.) difference of 2.0 Å; 33 of the 116 residues are identical. The carboxy-terminal portion of S1 (residues 181-235) forms an α-helix and a small three-stranded antiparallel β-sheet and has no structural homolog in LT or in the other known ADP-ribosylating toxins. This C-terminal portion provides a large part of the interface with the B-oligomer, and the last few residues actually enter a pit in the center of the B-oligomer. Enzymatic studies indicate that the C-terminal portion of S1 plays a role in binding to the $G_i\alpha$ substrate.[28,29] In addition, it is important for the interaction of S1 with lipid bilayers.[30]

Site-directed mutagenesis has defined a cluster of residues on S1 that are crucial for catalytic activity.[23,25,31-36] This active site region shares close structural homology with the active site of LT, and weaker homology with diphtheria toxin (DT)[37] and *Pseudomonas aeruginosa* exotoxin A (ETA).[38] Forty-six Cα atoms in the active site of S1 can be

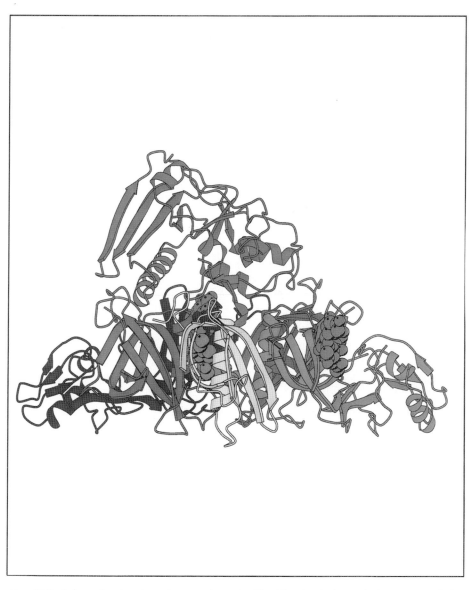

Fig. 10.2. Schematic drawing of pertussis toxin. The S1-subunit (green) sits on top of the B-oligomer, formed of subunits S2 (cyan), S3 (dark blue), two copies of S4 (red), and S5 (yellow). Arrows indicate strands of β-sheet and coiled ribbons indicate α-helices. Bound ligands observed in two separate binding experiments are shown as van der Waals spheres. A sialylated oligosaccharide is shown bound to the sides of subunits S2 and S3 (violet), and ATP is shown bound in the pore of the B-oligomer (orange). This figure, and others showing molecular structures, were drawn with MOLSCRIPT.[104] See color figure in insert.

superimposed on equivalents in the other toxins with an r.m.s. difference of 1.0 Å for LT, 1.5 Å for ETA and 1.6 Å for DT. A sequence alignment based on this structural superposition shows only two residues that are conserved in all four toxins (Fig. 10.3). One of these is a glutamic acid that has been shown to be essential for enzymatic activity in each toxin: residue 129 in PT,[23] 112 in LT,[39] 553 in ETA[40] and 148 in DT.[41] In PT, the side chain of Glu129 is within hydrogen bonding distance of the side chains of His35, Ser52 and Arg9. His35 has been shown, by site-directed mutagenesis, to be important for catalytic activity.[34,42,43] A hydrogen bond between the backbone carbonyl oxygen of Ser52 and the side chain of Arg9 links adjacent β-strands in the active site cleft of S1, and may play a role in maintaining the conformation of the active site. The importance of this hydrogen bond may explain the virtual elimination of catalytic activity associated with mutation of Arg9.[31] Structurally equivalent hydrogen bonds linking a side chain nitrogen atom with a backbone carbonyl group are present in other ADP-ribosylating toxins (Arg7-Ser61 in LT, His440-Tyr470 in ETA, His21-Tyr54 in DT).[27]

The binding sites for NAD and the G protein substrate are still uncertain, and predictions based on the crystal structure of the latent form of PT are complicated by the likelihood that the structure changes upon activation. However, a possible binding site for the adenine part of NAD was suggested by superimposing onto the PT active site the structure of DT bound to the dinucleotide adenylyl-3',5'-uridine monophosphate.[37] The inferred site of NAD binding in PT indeed implies

```
       7               23        35       50
PT    V Y R Y D S       F T A     H L      F V S T S S S R R Y T E
       5               22        44       59
LT    L Y R A D S       L M P     H A      Y V S T S L S L R S A H
      438              455               468
ETA   G Y H G T F       V R A     - -      G F Y I A G D P A L A Y
       19               35        52
DT    S Y H G T K       I Q K     - -      G F Y S T D N K Y D A A


       84              98       128              140
PT    F I G Y I Y E V R    Y G      S E Y L A H R     N I R R V
       81              95       111              123
LT    S T Y Y I Y V I A    F N      Q E V S A L G     Q I Y G W
      494             541      552              565
ETA   R N G A L L R V Y    A I      L E T I L G W     V V I P S
       76             136      147              160
DT    K A G G V V K V T    L S      V E Y I N N W     S V E L E
```

Fig. 10.3. A-subunit sequence alignment. Structurally-based amino acid sequence alignment of the A-subunits of several ADP-ribosyltransferase toxins: pertussis toxin, heat-labile enterotoxin, Pseudomonas exotoxin A and diphtheria toxin. Bold letters indicate sequence identities with pertussis toxin. Adapted from: Stein PE, Boodhoo A, Armstrong GD et al. Structure 1994; 2:45-57.

that a conformational change opening the active site cleft is required to allow entry of catalytic substrates. The PT crystal structure explains how this conformational change may take place following reduction of the disulphide bond (Cys41-Cys201), which is known to be essential for expression of catalytic activity. With this bond broken, the C-terminal portion of S1 that blocks the active site could rotate out of the way. A predicted associated effect would be the exposure of a hydrophobic sequence that might play a role in the translocation of S1 across the cell membrane.

B-OLIGOMER

The PT B-oligomer comprises five subunits: S2, S3, two copies of S4, and S5. Subunits S2 and S3 share 70% sequence identity, but there is no other significant homology among the different B-subunit sequences. Nevertheless, each of the five B-subunits contains a common folding motif of about 100 residues consisting of six antiparallel β-strands forming a closed β-barrel, capped by an α-helix between the fourth and fifth strands. This fold has been identified in several proteins that bind oligosaccharides or oligonucleotides, and has been named the oligomer-binding (OB) fold.[44,45] The OB fold is also found in the cell-binding B-subunits of the CT[26] and ST[46] families, which form pentamers composed of identical subunits. The structural homology among the B-subunits of these three families of AB₅ toxins (PT, the CT family and the ST family) is surprising since their sequences show little homology, they differ in size, and they bind to different cellular receptors. In a comparison of the structures of B-subunits in the three toxin families, the r.m.s. difference in position of 52 structurally equivalent Cα atoms ranges from 1.1 to 1.7Å, with only 4-16% identity in the structurally aligned sequences. A superposition of several OB-fold domains is shown in Figure 10.4. Two disulphide bonds are conserved in all the PT B-subunits, but neither of these is structurally equivalent to the single disulphides in the B-subunits of the CT or ST families. Other differences are found in the length and conformation of loops and termini.

The OB-fold domains of the five PT B-subunits associate through antiparallel β-sheet interactions to form a circular sheet of 30 β-strands surrounding a barrel formed by the five helices. This distorted pentameric structure comprises the central core of the PT B-oligomer, and it is structurally similar to the truly symmetrical B-pentamers of the CT and ST families. The two copies of S4 make equivalent close interactions with S2 (S4a) and S3 (S4b). Subunit S4a interacts on its other side with S3, and S4b with S5.

Although each B-subunit protomer of PT contains an OB-fold domain of about 100 residues, subunits S2 and S3 are significantly larger, each containing an additional amino-terminal domain. These amino-terminal domains, which protrude from the pentameric core of

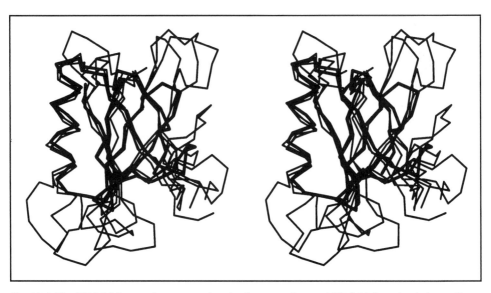

Fig. 10.4. The OB-fold. Stereo view comparing the Cα trace of several OB-fold domains: in heavy lines, the B-subunit of Shiga-like toxin I and, in light lines, pertussis toxin subunits S2 (C-terminal domain), S4 and S5. Adapted from: Stein PE, Boodhoo A, Armstrong GD et al. Structure 1994; 2:45-57.

the B-oligomer, show structural homology to the calcium-dependent (C-type) lectins. The C-type lectin family includes the lectin domain of rat mannose binding protein[47] and E-selectin.[48] Forty-seven Cα atoms of the amino-terminal domains of S2 or S3 can be superimposed onto equivalents in rat mannose binding protein with an r.m.s. difference of 2.2 Å, and a sequence identity of about 20%.[21] Since S2 and S3 have been found to act as cellular adhesins, like the selectin family,[49,50] the structural homology would seem at first sight to reflect a common function. Confusingly, however, S2 and S3 lack the functional region of their lectin homologue. Of 32 conserved amino acids important for carbohydrate recognition in known C-type lectins, only 2 cysteines that form a disulphide bond are also found in S2 and S3. Furthermore, a long loop that was found to be the carbohydrate binding site of mannose binding protein[51] has no structural equivalent in S2 or S3.

A-B Interaction

Although PT and the CT family show structural similarity within both their A and B components, the relative orientation of these components is quite different. The A-subunit of PT is rotated by about 180° relative to the A-subunit of LT, resulting in significant differences in the A-B interaction. The A-subunit of LT is tethered by a carboxy-terminal A2-fragment that extends completely through the central pore of the B-oligomer, where it is tightly held by salt-bridges.[26] However, there are few interactions involving the A1-fragment. In PT, the carboxy-terminus of S1 penetrates only partially through the B-oligomer, resulting in a weaker interaction. Instead there is substantial interaction between a flat surface of S1 and the B-oligomer (Fig. 10.2). Such an extended interface between the A and B components of PT would be much harder to achieve if the B-oligomer were a pentamer of identical monomers, since the A-subunit lacks 5-fold symmetry. The development of this interface may therefore have driven the evolution of the B-oligomer from a symmetrical precursor.

Evolution of the AB₅ Toxins

Structural homologies among the subunits of PT,[21] the CT family,[26] and the ST family[52] suggest that they share common ancestors. In each toxin, a pentameric cell-binding domain is formed by five subunits containing the OB folding motif, but the structurally similar B-oligomers are associated with nonidentical catalytic A-subunits (an ADP-ribosyltransferase in CT and PT, but an N-glycosidase in ST). It seems likely, therefore, that the A and B components have evolved independently. This is further supported by the finding that the A and B components of LT and PT, though individually similar, have different relative orientations in the two holotoxins. The ancestral A-component would have been an enzyme, while the B-oligomer might have been a pentameric protein with a role in cell-surface carbohydrate recognition. Interestingly, the 9Å structure of the nicotinic acetylcholine receptor determined by electron microscopy suggests a superficial resemblance to the toxin B-pentamers.[53]

RECEPTOR BINDING

Identity of the Receptor

Pertussis toxin attaches to its target cells through an interaction between the B-oligomer and glycoconjugate receptors on the cell surface. Terminal sialic acid groups of surface glycoproteins have been shown to be important for toxin recognition,[54] but PT can also recognize molecules bearing various nonsialylated carbohydrate determinants including Lewis a and Lewis X blood group determinants (fucosylated polylactosamines)[50] and lactosylceramide.[49] Experimental evidence suggests that the B-oligomer contains at least two binding domains on subunits S2 and S3, which differ in their specificities for sialylated and nonsialylated carbohydrate moieties.[49,50,55,56] Several receptor candidates have been proposed. The receptor on Chinese hamster ovary (CHO) cells was identified as a glycoprotein of 165 kDa containing a terminal sialyllactosamine moiety.[57] Three possible receptors have been identified on Jurkat cells (a human T cell line): a 70 kDa protein that may be the p73 LPS receptor,[58,59] a 43 kDa unknown protein[60] and the TCR CD3 receptor.[61] The existence of several distinct receptors for PT is not unexpected, especially since glycoproteins with terminal sialic acid residues are quite common. It is therefore likely that there are numerous distinct PT binding proteins on each cell, and the challenge will be to discern which are functionally relevant.

INTERACTION WITH SIALYLATED CARBOHYDRATES

The interaction of PT with sialylated carbohydrates has been investigated crystallographically by soaking crystals in a solution of a sialylated oligosaccharide derived from transferrin. The structure of the complex at 3.5 Å resolution shows two equivalent binding sites for the terminal sialic acid-galactose moiety of the sugar on subunits S2 and S3 of the B-oligomer[22] (Figs. 10.2, 10.5). The binding interaction is essentially identical in S2 and S3, and all residues interacting directly with the sugar are common to both subunits. These interactions involve only the sialic acid group of the sugar, and not the galactose. The sialic acid makes hydrogen bonds with polar or charged groups of Tyr102, Ser104 and Arg125, and its sugar ring makes additional hydrophobic contacts with the aromatic rings of Tyr102 and Tyr103. The crystal structure of this PT-sugar complex is consistent with results of site-directed mutagenesis. Recombinant forms of PT with mutations of Tyr102 and Tyr103 of S2 or S3 show reduced biological

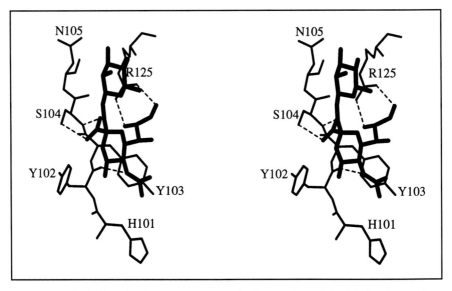

Fig. 10.5. Carbohydrate binding in pertussis toxin. Stereo view of the binding interactions between a sialylated oligosaccharide and the S2-subunit of pertussis toxin. Sialic acid and galactose are drawn in thick lines, and hydrogen bonds are indicated with dashed lines. Subunit S3 contains an essentially identical binding site. Reprinted from: Stein PE, Boodhoo A, Armstrong GD et al. Nature Struct Biol 1994; 1:591-96.

activities, and deletions at residue 105 in S2 or S3, close to the sialic acid binding site, result in reduced cell-binding activity.[62,63] Further crystallographic studies (unpublished) with various soluble sialylated sugars, including 3'- and 6'-sialyllactosamine and sialyllactose, show identical interactions between the sialic acid group and S2/S3. This confirms the original result, and illustrates a tolerance for different sugar linkages beyond the terminal sialic acid residue.

The observed sialic acid binding site lies in the carboxy-terminal domain of S2/S3, which has the OB fold. Although the same fold mediates receptor binding in the CT family, the site of interaction is not conserved. The five receptor binding sites in CT[64] and LT[65] lie on the flat base of the B-oligomer, distant from the A-subunit, while in PT the two binding sites lie on the sides of the B-oligomer (Fig. 10.2). CT and LT bind to the glycolipid G_{M1}, so the sugar-binding results imply that they sit on the cell with their A-subunits pointing away from the membrane surface. Since PT binds to glycoproteins, its orientation is less constrained, so it could bind in any orientation relative to the membrane.

Experimental evidence[49,50,62,63] suggests that residues of the PT B-oligomer that are some distance from the observed sialic acid binding site may also be involved in receptor recognition. Many of these residues lie in the amino-terminal domains of S2 and S3, which have the C-type lectin fold. It seems likely that binding of PT to at least some of its functional receptors extends beyond the observed sialic acid binding site to involve additional parts of the B-oligomer surface.

PHYSIOLOGICAL ROLES OF RECEPTOR BINDING

The primary role of receptor binding by the B-oligomer is believed to be the stimulation of endocytic uptake of the toxic S1-subunit by the host cells. But apart from this function, the mere act of receptor binding appears to trigger several cellular responses that may or may not be functionally relevant. For instance the hemagglutination of red blood cells by PT is most likely just a side effect of its ability to crosslink receptors through its divalent sialic acid binding sites. A more relevant effect may occur when PT crosslinks receptors on a single cell. Crosslinking of receptors by antibodies and/or lectins has been shown in several cases to trigger physiological responses.[61] A particularly interesting example is the rapid down-regulation of TCR CD3 receptors on Jurkat cells, which is accompanied by intracellular second messenger generation.[61] This effect can be induced by the PT B-oligomer, independent of the S1-subunit, or by an anti-TCR monoclonal antibody, but not by the isolated S2/S4 or S3/S4 dimers. Accordingly, the receptor down-regulation and second messenger generation most likely results from receptor crosslinking. For the EGF receptor it has been shown that receptor down-regulation is caused by ligand-induced dimerization followed by transport of the receptor/ligand complex to the lysosome where it is subsequently degraded.[66] It is tempting to speculate that PT stimulates its own uptake by crosslinking CD3 molecules, thereby inducing endocytosis and lysosomal targeting.

Another S1-independent result of receptor binding is the up-regulation of the CD11b/CD18 integrin receptor on leukocytes.[50,67] This effect appears to be mediated by the N-terminal domain of subunits S2 and S3, and mimics the known response of leukocytes to P- and E-selectin binding. Since isolated S2 and S3 subunits are also effective,[67] it appears that receptor crosslinking does not play a role. Up-regulation of the CD11b/CD18 integrin receptor may well be advantageous to *B. pertussis*, since it has been demonstrated that bacterial uptake into and survival in macrophages is enhanced by the interaction between CD11b/CD18 and filamentous hemagglutinin, the major bacterial adhesin.[50] As discussed above, there is some homology between the N-terminal domains of subunits S2 and S3 and the selectins. However, since the crucial carbohydrate-binding loop of the selectins has not been conserved in the S2 and S3 subunits,[21] the relevance of the observed and rather weak homologies is unclear.

Finally, the isolated B-oligomer also inhibits late events in IL-1 signaling in T cells.[68] The effect is obtained even with mutant B-oligomers defective in binding to sialylated carbohydrates, and it cannot be blocked by sialylated glycoproteins, so it presumably involves a secondary receptor-binding site.[69]

INTRACELLULAR TRAFFICKING

In contrast to the rapid effects caused by receptor binding, the effects of ADP-ribosylation of the cytosolic $G_i\alpha$-protein are only observed after a lag phase of at least one hour.[17] This lag phase most likely reflects the time required for endocytosis and intracellular trafficking of the toxin before it can be translocated into the cytosol.

The intracellular route of PT is thought to involve the Golgi apparatus, as brefeldin A, a drug that disrupts the Golgi apparatus, protects cells from PT.[10] Brefeldin A also protects cells against CT,[70-72] LT,[72] ETA and ricin,[73] suggesting that they are all targeted to the Golgi apparatus. Indeed, several toxins have been visualized in the Golgi stacks using electron microscopy.[74] However, the Golgi apparatus may just be a way station rather than the final destination, since two observations suggest that the toxins actually travel as far as the ER. These observations are the successful visualization of ST in the ER by electron microscopy[74] and the presence of a functional ER-retrieval signal in CT, LT and ETA. The ER-retrieval signal, a C-terminal Lys-Asp-Glu-Leu (KDEL) or related tetrapeptide, is known to retrieve soluble proteins from the Golgi back to the ER by means of the KDEL-receptor.[13,75] The importance of the ER-retrieval signal was confirmed by showing that deletion of the signal in ETA reduced toxicity,[76] whereas addition of a KDEL sequence to ricin increased its toxicity.[77] The ER-retrieval signal appears to allow these toxins to hijack the normal host cell mechanisms for recycling proteins.[11,12] We have argued that the known binding of ATP to PT suggests that PT is also transported to the ER (ref. 14, and below), although it does not possess an ER-retrieval signal.

It is interesting to consider whether the toxins remain bound to their cell-surface receptor, or whether they dissociate from that receptor somewhere along the route. The toxins with an ER-retrieval signal might well release their cell-surface receptor in exchange for binding to the KDEL-receptor somewhere in the Golgi apparatus. (In fact, if CT or LT were bound to cell-surface G_{M1} as described above, the ER-retrieval signal would be buried against the membrane, presumably inaccessible to the KDEL-receptor.) For the other toxins it is unclear whether they traverse the Golgi apparatus on their own, by means of their original receptor or by means of another host factor. PT does not contain an ER-retrieval signal but might conceivably bind to a recycling eukaryotic protein, as proposed for ricin and calreticulin.[12] One might even speculate that PT's affinity for nonsialylated carbohydrates is required for binding to a second receptor after initial transport to the Golgi apparatus. The release of the initial receptor could be stimulated by the low pH in late endosomes, a mechanism previously observed for the dissociation of receptor/ligand complexes after endocytosis.[66]

ATP-BINDING TO PERTUSSIS TOXIN

In 1982 Katada and Ui[17] reported that ATP is required to observe PT-induced increases in adenylate cyclase activity. Initially they suggested that ATP stabilized the fragile adenylate cyclase enzyme, but it was later shown that ATP interacts directly with PT.[78] Further experiments showed that ATP binds to a site on the B-oligomer and that ATP binding weakens the interaction between the B-oligomer and the S1-subunit.[79-81] It was further reported that some phospholipids and the detergent CHAPS also activate PT by destabilizing the B-oligomer/S1-subunit interactions, but this time the binding site appeared to reside on the S1-subunit.[79,82] The actual activation of the toxin is believed to be due to the reduction of the Cys41-Cys201 disulphide bond in the S1-subunit.[83] This disulphide bond is very stable in the holotoxin, as 250 mM DTT is required for full activation,[83] whereas it can be reduced by approximately 1 mM DTT in the free S1-subunit.[80] Destabilization of the quaternary structure therefore can lead to activation by facilitating the reduction of the Cys41-Cys201 disulphide bond.

STRUCTURE OF THE ATP COMPLEX

In order to understand the molecular basis of ATP binding to PT and its ability to promote dissociation of the S1-subunit, we have studied the interaction by soaking crystals of PT in ATP.[14] This experiment revealed that an ATP molecule binds in the central pore of the B-oligomer. We have refined a model for the complex to a final R-value of 23.2%, using all reflections between 2.7 and 8.0Å resolution. The final model is very similar to the native structure apart from a few crucial changes in the actual ATP-binding site.

The general location of the ATP-binding site can be seen in Figure 10.2, while Figure 10.6 shows details of the binding interactions. The adenine ring and ribose moiety are deeply buried and stacked between the α-helices of subunits S4b and S5. The ribose moiety makes hydrogen bonds to the side chains of residues Ser61 and Glu65 of S4b, and N7 of the adenine ring forms a hydrogen bond with the side chain hydroxyl of Ser62 of S5. In contrast, the amino group at position 6 and the nitrogen atoms N1 and N3 of the adenine ring are positioned in a hydrophobic pocket without direct hydrogen bonding partners. However, at the 2.7Å resolution limit of our data, we can not exclude the possibility of water-mediated hydrogen bonds. The adenine ring is further stabilized by hydrophobic contacts with the aliphatic portion of the side chain of Arg69 of S4b on one side and the aromatic ring of Phe59 of S5 on the other side. The triphosphate moiety of ATP occupies a more central location in the pore of the B-oligomer on the side opposite the S1-subunit, where it interacts favorably with a cluster of positively charged amino acids.

ATP-Induced Conformational Changes

The binding of ATP to the holotoxin results in only two significant conformational changes (Fig. 10.7). The first change moves the side chain of Arg69 of S4b from a conformation that would block the adenine-binding pocket to one in which it interacts favorably with the

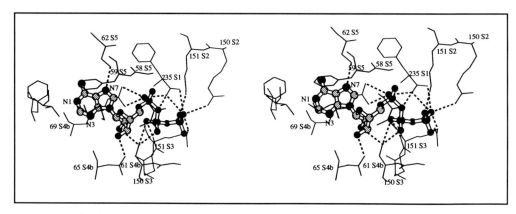

Fig. 10.6. ATP binding in pertussis toxin. Stereo view of the binding interactions between ATP and pertussis toxin. ATP is shown in a ball-and-stick representation, and hydrogen bonds are indicated with dashed lines. Reprinted from: Hazes B, Boodhoo A, Cockle SA, et al. J Mol Biol 1996; 258:661-671.

α-phosphate and the adenine ring. The second change affects the C-terminus of the S1-subunit, which is forced aside by unfavorable steric and electrostatic interactions between its negatively charged C-terminal carboxylate and the triphosphate moiety. Even after movement of the C-terminus by 1.3 Å (measured at the Cα position), the oxygen atoms of the carboxylate remain within 3 Å of the phosphate oxygens. Accordingly it appears that the destabilization of the S1-subunit in the holotoxin is due to the unfavorable electrostatic interactions between the triphosphate moiety and the C-terminal carboxylate.

A PHYSIOLOGICAL ROLE FOR ATP BINDING

Although there is no experimental evidence for a role of ATP binding in vivo, the high binding affinity (about 1 μM), the deeply buried binding pocket and the specific protein interactions argue in favor of a physiological function. This raises the question of where the toxin may encounter ATP. Several authors have suggested that this would be the cytosol.[79,81,84] However, since then it has been established that ATP is also present in the ER.[85,86] This knowledge, combined with the appreciation that several AB-toxins may be transported to the ER, led us to propose that PT might well interact with ATP, and then dissociate, in the ER.[14]

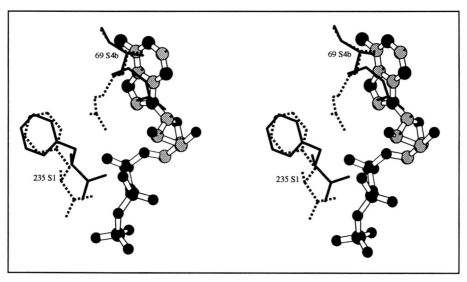

Fig. 10.7. ATP-induced shifts. Stereo view highlighting the conformational shifts in pertussis toxin induced by the binding of ATP. Residues of PT that move significantly are shown in solid lines for the native conformation and dashed lines for the conformation in the complex. The ATP molecule is shown in a ball-and-stick representation.

Precise control of dissociation seems to be central to the mechanism of toxins that enter the ER. Only the catalytic A-subunits are believed to be translocated into the cytosol, so the toxins must dissociate before translocation. However, the timing is crucial; if a toxin dissociates too soon, its A-subunit will never reach the ER. There are common features of these toxins that suggest common mechanisms for the control of dissociation, but the best evidence is for ETA. Its catalytic subunit, but not its binding subunit, can be detected in the cytosol,[87] providing direct evidence for dissociation before translocation. Dissociation of ETA requires cleavage of both a peptide loop and a disulphide bond.[87] Proteolysis does not appear to control the timing of dissociation, because proteolytic cleavage occurs in an early endosomal compartment, whereas reduction of the disulphide occurs later and requires a functional ER-retrieval signal.[87] The control point for dissociation appears to be disulphide reduction, most likely in the ER. The observation that an ETA mutant lacking the disulphide is toxic in vitro but not in vivo[88] is also consistent with a role for the disulphide in timing dissociation.

Disulphide bond reduction is a rather subtle way to target dissociation to the ER by exploiting the subcellular distribution of enzymes. In the oxidizing environment of the protein export pathway, the disulphide would be stable until it encountered the abundant protein disulphide isomerases of the ER. It seems that in the course of evolution this property of the ER has been exploited several times, as several unrelated AB-toxins also require proteolytic cleavage and disulphide reduction for toxicity.[89-91]

PT is an exception to this common mechanism. It too can be proteolytically cleaved,[24,92] but there is no evidence that cleavage is required for toxicity. In addition, reduction of the Cys41-Cys201 disulphide bond in the S1-subunit is required to open the active site to its substrates,[21] but it is not sufficient to allow dissociation of the S1-subunit. PT must use another mechanism, which could be provided by ATP-stimulated dissociation. The subcellular distribution of ATP would ensure that dissociation only takes place after PT arrives in the ER. Accordingly, we have proposed that the ATP-binding site acts in vivo as a sensor that detects arrival of the toxin in the ER and then triggers dissociation to allow the S1-subunit to translocate into the cytosol.[14]

SPECULATIONS ON MEMBRANE TRANSLOCATION

As discussed above, various lines of direct and indirect evidence suggest that a number of toxin A-subunits enter the cytosol through the ER. If one reflects on why evolution has chosen this route repeatedly, one is led to the speculation that it is because the ER has (among other features) a channel through which proteins pass, the protein translocation channel.[11,12]

If the toxins use the protein translocation channel to enter the cytosol, the direction of movement reverses the normal export pathway. There is evidence that peptide chains spanning this channel can in fact move freely in either direction.[93] It has been suggested that protein export is directional because modification or folding of partially-translocated proteins in the lumen of the ER provides a ratchet mechanism: once modified or folded, they become unable to pass back through the channel.[93,94] For a toxin A-subunit to pass freely through this channel, then, it must remain unglycosylated, its disulphide linkages must be broken (perhaps by ER-resident protein disulphide isomerases) and it must unfold while being threaded through the channel (assisted perhaps by chaperone proteins).

One complication is that the translocation channel is only open when a ribosome is docked to it[95] and the elongating chain is long enough to reach the lumen.[96] One might imagine that the toxins have some mechanism to open the channel. It is also possible that, after one protein chain has exited into the lumen, the channel occasionally stays open long enough for a protein chain to enter from the lumenal side. If that were true, then proteins destined for export might occasionally re-enter the cytosol. Such a potential targeting error ought to be accompanied by a quality control mechanism, analogous to those seen in the export pathway.[97] In fact, a characteristic of all toxins that are believed to enter through the ER leads us to propose that there is such a quality control mechanism, which is circumvented by the toxins.

There is an intriguing pattern to the amino acid sequences of toxin A-subunits that are believed to enter the cytosol through the ER: they all have an exceptionally low lysine content. The S1-subunit of PT has no lysines whatsoever,[98] while the translocated components of other toxins (e.g., CT, ETA, ST, ricin) have no more than two lysines, or less than 1% of the amino acid content (results not shown). In contrast, the average globular protein has a lysine content of 6.2%.[99] The unusual composition is not the consequence of overall charge, as the arginine and histidine contents are normal. It is also not general for all AB toxins, as DT, which enters cells by way of acidified endosomes, has a normal lysine content of 8.3% (16 lysines of 193 residues) in its A-subunit.[100]

This pattern has been commented upon previously.[101] A possible explanation was advanced that the presence of lysines would render these toxins susceptible to proteolysis, presumably by trypsin-like proteinases. However, this does not explain why a normal lysine content is tolerated in DT. Instead, we believe that the relevant property of lysines is that they are the residues to which chains of ubiquitin monomers are attached in ubiquitination. Attachment of ubiquitin chains to proteins targets them for degradation by the proteasome.[102] Proteins that completely lack lysines will be insensitive to this degradation

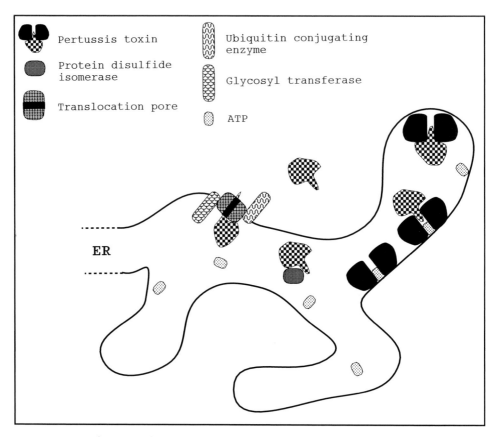

Fig. 10.8. Events during translocation. Schematic diagram of the endoplasmic reticulum, illustrating events postulated to occur during the dissociation and translocation of the S1-subunit of pertussis toxin. The binding of ATP facilitates dissociation of the toxin in the ER. The S1-subunit is then activated by the reduction of its disulphide bridge, possibly through the action of a protein disulphide isomerase. Assisted, perhaps, by chaperone proteins, the S1-subunit passes through the protein translocation pore in an unfolded state where, if it contained any lysine residues, it would be subject to the action of a membrane-bound ubiquitin conjugating enzyme. Finally, it refolds on the cytosolic side of the ER membrane.

system. David J. FitzGerald (personal communication) has also inferred a possible connection to ubiquitination.

A protein traveling in reverse through the translocation channel would exit into the cytosol in an unfolded state. In that state, every lysine residue would be accessible to a ubiquitin conjugating enzyme. (The same vulnerability does not exist in protein export, because the tight connection between ribosome and translocation channel shields the nascent chain from the cytosol.[96]) As it happens, an appropriately positioned ubiquitin conjugating enzyme (UBC6) has been discovered on the cytosolic side of the ER membrane in yeast,[103] and homologues are likely to exist in other eukaryotes. A deletion of the UBC6 conjugating enzyme was found to compensate a temperature sensitive mutation in SEC61, part of the translocation channel in yeast. The authors concluded that the SEC61 mutant is partly unfolded at the nonpermissive temperature, becomes susceptible to ubiquitination by UBC6, and is thus targeted for degradation. In the absence of UBC6 the SEC61 mutant is not ubiquitinated, so it survives.

We propose that the normal role of UBC6 (or some other ubiquitin conjugating enzyme) is to provide a quality control mechanism preventing the occasional leakage of exported proteins back into the cytosol. Under the evolutionary pressure of this quality control mechanism, lysine residues would be replaced in any toxins that enter through the ER, but not in toxins, such as DT, that enter through another compartment. Conversely, eukaryotes would have a selective pressure to maintain a sufficient number of lysine residues in exported proteins (though it would be difficult to distinguish this from the normal lysine content).

Our proposal has a number of testable consequences. First, the replacement of nonessential surface residues of a toxin A-subunit with lysines should reduce cytotoxicity while having no effect on cell-free enzymatic activity. Second, the reduction in cytotoxicity should be reversed in cell lines in which the UBC6 homolog (once identified) was inactivated. (Unfortunately, we do not know of any yeast toxins that undergo retrograde transport, or this experiment could be done in yeast.) Third, attempts to target proteins to the cytosol by exploiting the toxin translocation machinery should be much more successful if the proteins were mutated to contain as few lysine residues as possible.

FUTURE DIRECTIONS

Our structural studies of pertussis toxin have shed light on a number of key steps in toxin action: binding to a target cell, dissociation of the holotoxin and activation of the catalytic subunit. In future work we hope to address the remaining steps that are amenable to crystallography.

As discussed above, there is biochemical evidence of additional binding sites for nonsialylated carbohydrates, which may be relevant to the recognition of some types of target cells or, as we speculate, may even allow the toxin to hijack proteins undergoing retrograde transport to the ER.

However, the biggest gap in our understanding surrounds the activated S1-subunit. The general nature of the activating conformational change that is allowed by the reduction of the Cys41-Cys201 disulphide can be deduced, but knowledge of its precise nature will be essential to a full understanding of substrate binding and the catalytic mechanism. These questions can only be studied with the isolated S1-subunit, in combination with substrates and substrate analogs.

Once the action of pertussis toxin is fully understood at a structural level, we can be confident that site-directed mutagenesis will provide vaccine components devoid of any biological activity. We can also hope to exploit the understanding to devise drug molecules that interfere with one or more steps in the intoxication pathway.

Over millions of years, toxins have evolved to exploit subtleties of eukaryotic cell biology, many of which are still beyond our understanding. Only a few years ago we were unaware of retrograde flow in the protein export pathway, yet this process has long been exploited by a number of toxins. No doubt these toxins still have some tricks to show us, and we still have much to learn about basic science from a study of these fascinating molecules.

ACKNOWLEDGMENTS

This work could not have been carried out without the enthusiastic collaboration of Stephen Cockle (Connaught Laboratories, Toronto) and Glen Armstrong (University of Alberta). Michael Ellison drew our attention to UBC6, and Maxwell D. Cummings took part in early discussions of ubiquitination. Financial support came from the Medical Research Council of Canada, Connaught Laboratories, the Protein Engineering Network of Centres of Excellence, the Alberta Heritage Foundation for Medical Research and the Howard Hughes Medical Institute International Research Scholars Program.

REFERENCES

1. Galazka A. Control of pertussis in the world. World Health Statist Quart 1992; 45:238-47.
2. Cherry JD. Pertussis vaccine encephalopathy: it is time to recognise it as the myth that it is. JAMA 1990; 263:1679-80.
3. Golden GS. Pertussis vaccine and injury to the brain. J Pediatr 1990; 116:854-61.
4. Howson CP, Howe CJ, Fineberg HV. Adverse effects of pertussis and rubella vaccines. Washington: National Academy Press, 1991.
5. Sato Y, Sato H, Tiru M, Brown F. Pertussis: evaluation and research on acellular pertussis vaccines. Basel: Karger, 1991.
6. Munoz JJ, Arai H, Cole RL. Mouse-protecting and histamine-sensitizing activities of pertussigen and fimbrial hemagglutinin from *Bordetella pertussis.* Infect Immun 1981; 32:243-50.
7. Sato H, Sato Y. *Bordetella pertussis* infection in mice: correlation of specific antibodies against two antigens, pertussis toxin, and filamentous haemagglutinin with mouse protectivity in an intracerebral or aerosol challenge system. Infect Immun 1984; 46:415-21.
8. Hewlett EL, Cowell JL. Evaluation of the mouse model for study of encephalopathy in pertussis vaccine recipients. Infect Immun 1981; 57:661-63.

9. Gill DM. Seven toxin peptides that cross cell membranes. In: Jeljaszewica J, Wadstrom T, eds. Bacterial Toxins and Cell Membranes. New York: Academic Press, 1978:291-332.

10. Xu Y, Barbieri JT. Pertussis toxin-mediated ADP-ribosylation of target proteins in Chinese hamster ovary cells involves a vesicle trafficking mechanism. Infect Immun 1995; 63:825-32.

11. Pastan I, Chaudhary V, FitzGerald DJ. Recombinant toxins as novel therapeutic agents. Annu Rev Biochem 1992; 61:331-54.

12. Pelham HRB, Roberts LM, Lord JM. Toxin entry: how reversible is the secretory pathway? Trends Cell Biol 1992; 2:183-85.

13. Nilsson T, Warren G. Retention and retrieval in the endoplasmic reticulum and the Golgi apparatus. Curr Opin Cell Biol 1994; 6:517-21.

14. Hazes B, Boodhoo A, Cockle SA, et al. Crystal structure of the pertussis toxin-ATP complex: a molecular sensor. J Mol Biol 1996; 258:661-671.

15. Pittman M. The concept of pertussis as a toxin-mediated disease. Pediatr Infect Dis 1984; 3:467-86.

16. Tuomanen E, Weiss A. Characterization of two adhesins of *Bordetella pertussis* for human ciliated respiratory epithelial cells. J Infect Dis 1985; 152:118-25.

17. Katada T, Ui M. Direct modification of the membrane adenylate cyclase system by islet activating protein due to ADP-ribosylation of a membrane protein. Proc Natl Acad Sci USA 1982; 79:3129-33.

18. Munoz JJ, Arai H, Bergman RK et al. Biological activities of crystalline pertussigen from *Bordetella pertussis*. Infect Immun 1981; 33:820-26.

19. Tamura M, Nogimori K, Yajima M et al. A role of the B-oligomer moiety of islet-activating protein, pertussis toxin, in development of the biological effects on intact cells. J Biol Chem 1983; 258:6756-61.

20. Arai H, Sato Y. Separation and characterization of two distinct haemagglutinins contained in purified leukocytosis-promoting factor from *Bordetella pertussis*. Biochim Biophys Acta 1976; 444:765-82.

21. Stein PE, Boodhoo A, Armstrong GD et al. The crystal structure of pertussis toxin. Structure 1994; 2:45-57.

22. Stein PE, Boodhoo A, Armstrong GD et al. Structure of a pertussis toxin-sugar complex as a model for receptor binding. Nature Struct Biol 1994; 1:591-96.

23. Pizza M, Bartoloni A, Prugnola A et al. Subunit S1 of pertussis toxin: mapping of the regions essential for ADP-ribosyltransferase activity. Proc Natl Acad Sci USA 1988; 85:7521-25.

24. Burns DL, Hausman SZ, Lindner W et al. Structural characterization of pertussis toxin A subunit. J Biol Chem 1987; 262:17677-92.

25. Barbieri JT, Cortina G. ADP-ribosyltransferase mutations in the catalytic S-1 subunit of pertussis toxin. Infect Immun 1988; 56:1934-41.

26. Sixma TK, Pronk SE, Kalk KH et al. Crystal structure of a cholera toxin-related heat-labile enterotoxin from *E. coli*. Nature 1991; 351:371-77.

27. Sixma TK, Kalk KH, van Zanten BAM et al. Refined structure of *Escherichia coli* heat-labile enterotoxin, a close relative of cholera toxin. J Mol Biol 1993; 230:890-918.

28. Krueger KM, Barbieri JT. Assignment of functional domains involved in ADP-ribosylation and B-oligomer binding within the carboxyl-terminus of the S1 subunit of pertussis toxin. Infect Immun 1994; 62:2071-78.

29. Finck-Barbançon V, Barbieri JT. ADP-ribosylation of αi3 C20 by the S1 subunit and deletion peptides of S1 of pertussis toxin. Biochemistry 1995; 34:1070-75.

30. Hausman SZ, Burns DL. Interaction of pertussis toxin with cells and model membranes. J Biol Chem 1992; 267:13735-39.

31. Burnette WN, Cieplak. W, Mar VL et al. Pertussis toxin S1 mutant with reduced enzyme activity and a conserved protective epitope. Science 1988; 242:72-74.

32. Cockle SA. Identification of an active-site residue in subunit S1 of pertussis toxin by photocrosslinking to NAD. FEBS Lett 1989; 249:329-32.

33. Locht C, Capiau C, Feron C. Identification of amino acid residues essential for the enzymatic activities of pertussis toxin. Proc Natl Acad Sci USA 1989; 86:3075-79.

34. Kaslow HR, Schlotterbeck JD, Mar VL et al. Alkylation of cysteine 41, but not cysteine 200, decreases the ADP-ribosyltransferase activity of the S1 subunit of pertussis toxin. J Biol Chem 1989; 264:6386-90.

35. Burns DL, Manclark CR. Role of cysteine 41 of the A subunit of pertussis toxin. J Biol Chem 1989; 264:564-68.

36. Cortina G, Barbieri JT. Role of tryptophan 26 in the NAD glycohydrolase reaction of the S-1 subunit of pertussis toxin. J Biol Chem 1989; 264:17322-28.

37. Choe S, Bennett MJ, Fujii G et al. The crystal structure of diphtheria toxin. Nature 1992; 357:216-22.

38. Allured VS, Collier RJ, Carroll SF et al. Structure of exotoxin A of *Pseudomonas aeruginosa* at 3.0 Ångstrom resolution. Proc Natl Acad Sci USA 1986; 83:1320-24.

39. Tsujii T, Inoue T, Miyama A et al. A single amino acid substitution in the A subunit of *Escherichia coli* enterotoxin results in a loss of toxic activity. J Biol Chem 1990; 265:22520-25.

40. Douglas CM, Collier RJ. Exotoxin A of *Pseudomonas aeruginosa*: substitution of glutamic acid 553 with aspartic acid drastically reduces toxicity and enzymatic activity. J Bacteriol 1987; 169:4967-71.

41. Tweten RK, Barbieri JT, Collier RJ. Diphtheria toxin: effect of substituting aspartic acid for glutamic acid 148 on ADP-ribosyltransferase activity. J Biol Chem 1985; 260:10392-94.

42. Antoine R, Locht C. The NAD-glycohydrolase activity of pertussis toxin S1 subunit: involvement of the catalytic His-35 residue. J Biol Chem 1994; 269:6450-57.

43. Xu Y, Barbançon-Finck V, Barbieri JT. Role of histidine 35 of the S1 subunit of pertussis toxin in the ADP-ribosylation of transducin. J Biol Chem 1994; 269:9993-99.

44. Murzin AG, Chothia C. Protein architecture: new superfamilies. Curr Opin Struct Biol 1992; 2:895-903.

45. Murzin AG. OB (oligonucleotide/oligosaccharide binding)-fold: common

structural and functional solution for nonhomologous sequences. EMBO J 1993; 12:861-67.

46. Stein PE, Boodhoo A, Tyrrell GJ et al. Crystal structure of the cell-binding B oligomer of verotoxin-1 from *E. coli*. Nature 1992; 355:748-50.

47. Weis WI, Kahn R, Fourme R et al. Structure of the calcium-dependent lectin domain from a rat mannose-binding protein determined by MAD phasing. Science 1991; 254:1608-15.

48. Graves BJ, Crowther RL, Chandran C et al. Insight into E-selectin ligand interaction from the crystal structure and mutagenesis of the lec EGF domains. Nature 1994; 367:532-38.

49. Saukkonen K, Burnette WN, Mar VL et al. Pertussis toxin has eukaryotic-like carbohydrate recognition domains. Proc Natl Acad Sci USA 1992; 89:118-22.

50. van't Wout J, Burnette WN, Mar VL et al. Role of carbohydrate recognition domains of pertussis toxin in adherence of *Bordetella pertussis* to human macrophages. Infect Immun 1992; 60:3303-08.

51. Weis WI, Drickamer K, Hendrickson WA. Structure of a C-type mannose-binding protein complexed with an oligosaccharide. Nature 1992; 360:127-34.

52. Fraser ME, Chernaia MM, Kozlov YV et al. Crystal structure of the holotoxin from *Shigella dysenteriae* at 2.5 Å resolution. Nature Struct Biol 1994; 1:59-64.

53. Unwin N. Nicotinic acetylcholine receptor at 9 Å resolution. J Mol Biol 1993; 229:1101-24.

54. Armstrong GD, Howard LA, Peppler MS. Use of glycosyltransferases to restore pertussis toxin receptor activity to asialogalactofetuin. J Biol Chem 1988; 263:8677-84.

55. Witvliet MH, Burns DL, Brennan MJ et al. Binding of pertussis toxin to eukaryotic cells and glycoproteins. Infect Immun 1989; 57:3324-30.

56. Heerze LD, Chong PCS, Armstrong GD. Investigation of the lectin-like binding domains in pertussis toxin using synthetic peptide sequences. J Biol Chem 1992; 267:25810-15.

57. Brennan MJ, David JL, Kenimer JG et al. Lectin-like binding of pertussis toxin to a 165-kilodalton Chinese hamster ovary cell glycoprotein. J Biol Chem 1988; 263:4895-99.

58. Clark CG, Armstrong GD. Lymphocyte receptors for pertussis toxin. Infect Immun 1990; 58:3840-46.

59. Armstrong GD, Clark CG, Heerze LD. The 70 kilodalton pertussis toxin-binding protein in Jurkat cells. Infect Immun 1994; 62:2236-43.

60. Rogers TS, Corey SJ, Rosoff PM. Identification of a 43 kilodalton human T lymphocyte membrane protein as a receptor for pertussis toxin. J Immunol 1990; 145:678-83.

61. Witvliet MH, Vogel ML, Wiertz EJHJ et al. Interaction of pertussis toxin with human T lymphocytes. Infect Immun 1992; 60:5085-90.

62. Lobet Y, Feron C, Dequesne G et al. Site-specific alterations in the B-oligomer that affect receptor-binding activities and mitogenicity of pertussis toxin. J Exp Med 1993; 177:79-87.

63. Loosmore S, Zealey G, Cockle S, et al. Characterization of pertussis toxin analogs containing mutations in B-oligomer subunits. Infect Immun 1993; 61:2316-24.

64. Merritt EA, Sarfaty S, Vandenakker F et al. Crystal structure of cholera toxin B-pentamer bound to receptor GM1 pentasaccharide. Protein Sci 1994; 18499:17830-21057.

65. Sixma TK, Pronk SE, Kalk KH et al. Lactose binding to heat-labile enterotoxin revealed by X-ray crystallography. Nature 1992; 355:561-64.

66. French AR, Tadaki DK, Niyogi SK et al. Intracellular trafficking of epidermal growth factor family ligands is directly influenced by the pH sensitivity of the receptor/ligand interaction. J Biol Chem 1995; 270:4334-40.

67. Rozdzinski E, Burnette WN, Jones T et al. Prokaryotic peptides that block leukocyte adherence to selectins. J Exp Med 1993; 178:917-24.

68. O'Neill LAJ, Ikebe T, Sarsfield SJ et al. The binding subunit of pertussis toxin inhibits IL-1 induction of IL-2 and prostaglandin production. J Immunol 1992; 148:474-79.

69. McCarthy LM, O'Neill LAJ. Effects of pertussis toxin and its binding subunit on IL-1 signalling in T cells. Biochem Soc Trans 1995; 23:112S.

70. Nambiar MP, Oda T, Chen C et al. Involvement of the Golgi region in the intracellular trafficking of cholera toxin. J Cell Physiol 1993; 154:222-28.

71. Orlandi PA, Curran PK, Fishman PH. Brefeldin A blocks the response of cultured cells to cholera toxin. J Biol Chem 1993; 268:12010-16.

72. Donta ST, Beristain S, Tomicic TK. Inhibition of heat-labile cholera and *Escherichia coli* enterotoxins by brefeldin A. Infect Immun 1993; 61:3282-86.

73. Yoshida T, Chen C, Zhang M et al. Disruption of the Golgi apparatus by brefeldin A inhibits the cytotoxicity of ricin, modeccin, and *Pseudomonas* toxin. Exp Cell Res 1991; 192:389-95.

74. Sandvig K, Ryd M, Garred O et al. Retrograde transport from the Golgi complex to the ER of both Shiga toxin and the nontoxic Shiga B-fragment is regulated by butyric acid and cAMP. J Cell Biol 1994; 126:53-64.

75. Pelham HRB. The retention signal for soluble proteins of the endoplasmic reticulum. TIBS 1990; 15:483-86.

76. Chaudhary VK, Jinno Y, FitzGerald D et al. *Pseudomonas* exotoxin contains a specific sequence at the carboxyl terminus that is required for cytotoxicity. Proc Natl Acad Sci USA 1990; 87:308-12.

77. Wales R, Chaddock JA, Roberts LM et al. Addition of an ER retention signal to the ricin A chain increases the cytotoxicity of the holotoxin. Exp Cell Res 1992; 203:1-4.

78. Lim L-K, Sekura DD, Kaslow HR. Adenine nucleotides directly stimulate pertussis toxin. J Biol Chem 1985; 260:2585-88.

79. Burns DL, Manclark CR. Adenine nucleotides promote dissociation of pertussis toxin subunits. J Biol Chem 1986; 261:4324-27.

80. Moss J, Stanley SJ, Watkins PA et al. Stimulation of the thiol-dependent

ADP-ribosyltransferase and NAD glycohydrolase activities of *Bordetella pertussis* toxin by adenine nucleotides, phospholipids and detergents. Biochemistry 1986; 25:2720-25.

81. Hausman SZ, Manclark CR, Burns DL. Binding of ATP by pertussis toxin and isolated toxin subunits. Biochemistry 1990; 29:6128-31.

82. Krueger KM, Barbieri JT. Molecular characterization of the in vitro activation of pertussis toxin by ATP. J Biol Chem 1993; 268:12570-78.

83. Moss J, Stanley SJ, Burns DL et al. Activation by thiol of the latent NAD glycohydrolase and ADP-ribosyltransferase activities of *Bordetella pertussis* toxin (islet-activating protein). J Biol Chem 1983; 258:11879-82.

84. Kaslow HR, Lim L-K, Moss J et al. Structure-activity analysis of the activation of pertussis toxin. Biochemistry 1987; 26:123-27.

85. Braakman I, Helenius J, Helenius A. Role of ATP and disulphide bonds during protein folding in the endoplasmic reticulum. Nature 1992; 356:260-62.

86. Clairmont C, Demaio A, Hirschberg C. Translocation of ATP into the lumen of rat liver and canine pancreas rough endoplasmic reticulum derived vesicles and its binding to lumenal glucose regulated proteins. J Cell Biol 1991; 115:255a.

87. Ogata M, Chaudhary VK, Pastan I et al. Processing of *Pseudomonas* exotoxin by a cellular protease results in the generation of a 37,000-Da toxin fragment that is translocated to the cytosol. J Biol Chem 1990; 265:20678-85.

88. Madshus IG, Collier RJ. Effects of eliminating a disulfide bridge within domain II of *Pseudomonas aeruginosa* exotoxin A. Infect Immun 1989; 57:1873-78.

89. Gordon VM, Leppla SH. Proteolytic activation of bacterial toxins: role of bacterial and host cell proteases. Infect Immun 1994; 62:333-40.

90. Grant CCR, Messner RJ, Cieplak W, Jr. Role of trypsin-like cleavage at arginine 192 in the enzymatic and cytotonic activities of *Escherichia coli* heat-labile enterotoxin. Infect Immun 1994; 62:4270-78.

91. Freedman RB, Hirst TR, Tuite MF. Protein disulphide isomerase: building bridges in protein folding. TIBS 1994; 19:331-36.

92. Krueger KM, Mende-Mueller LM, Barbieri JT. Protease treatment of pertussis toxin identifies the preferential cleavage of the S1 subunit. J Biol Chem 1991; 266:8122-28.

93. Ooi CE, Weiss J. Bidirectional movement of a nascent polypeptide across microsomal membranes reveals requirements for vectorial translocation of proteins. Cell 1992; 71:87-96.

94. Simon SM, Peskin CS, Oster GF. What drives the translocation of proteins? Proc Natl Acad Sci USA 1992; 89:3770-74.

95. Simon SM, Blobel G. A protein-conducting channel in the endoplasmic reticulum. Cell 1991; 65:371-80.

96. Crowley KS, Liao SR, Worrell VE et al. Secretory proteins move through the endoplasmic reticulum membrane via an aqueous, gated pore. Cell 1994; 78:461-71.

97. Hammond C, Helenius A. Quality control in the secretory pathway. Curr Opin Cell Biol 1995; 7:523-29.

98. Loosmore SM, Cunningham JD, Bradley WR et al. A unique sequence of the *Bordetella pertussis* toxin operon. Nucleic Acids Res 1989; 17:8365.

99. Wishart DS, Boyko RF, Willard L et al. SEQSEE: a comprehensive program suite for protein sequence analysis. Comp Appl Biosci 1994; 10:121-32.

100. Greenfield L, Bjorn MJ, Horn G et al. Nucleotide sequence of the structural gene for diphtheria toxin carried by corynebacteriophage beta. Proc Natl Acad Sci USA 1983; 80:6853-57.

101. London E, Luongo CL. Domain-specific bias in arginine/lysine usage by protein toxins. Bioch Bioph Res Commun 1989; 160:333-39.

102. Ciechanover A. The ubiquitin-proteasome proteolytic pathway. Cell 1994; 79:13-21.

103. Sommer T, Jentsch S. A protein translocation defect linked to ubiquitin conjugation at the endoplasmic reticulum. Nature 1993; 365:176-79.

104. Kraulis P. MOLSCRIPT: a program to produce both detailed and schematic plots of proteins. J Appl Cryst 1991; 24:946-50.

============CHAPTER 11============

STRUCTURE OF TOXIC SHOCK SYNDROME TOXIN-1

Douglas H. Ohlendorf, David T. Mitchell, G. Sridhar Prasad,
R. Radhakrishnan, Cathleen A. Earhart and Patrick M. Schlievert

Toxic shock syndrome toxin-1 (TSST-1) is a member of a group of toxins known as pyrogenic toxin superantigens (PTSAgs). These toxins are produced by *Staphylococcus aureus* and a number of Group A streptococci. Included in the family of PTSAgs are staphylococcal enterotoxins A through H excluding F (SEA, SEB, SEC1, SEC2, SEC3, SED, SEE, SEG and SEH), exotoxins A through C from *Streptococcus pyogenes* (SPEA, SPEB and SPEC), and TSST-1 from *S. aureus*.

PTSAgs are nonglycosylated proteins with molecular weight ranging from 22 kDa to 30 kDa. They are stable from pH 2.5 to pH 11 and to temperatures of 60°C or higher.[1] For example, TSST-1 shows no loss of activity after a 1 hr incubation at 100°C. PTSAgs are all synthesized with an amino terminal signal sequence that is removed during secretion. The genes encoding these toxins have been cloned and sequenced. All the staphylococcal enterotoxins have a disulfide loop. The level of sequence homology, i.e., the number of identical amino acid residues, varies substantially among these proteins. SEA, SED and SEE form one group of toxins with homologies between 53 and 81%. SEB, the SECs and SPEA form another group with homologies between 50 and 66%. SPEC is 30% homologous with SPEA but shows no statistically significant homology with SEB or the SECs. TSST-1 and SPEB show no statistically significant homology to any other toxin or to each other.[2] The most recent addition to the PTSAgs, SEH, is 37% homologous with SEA, SED and SEE, 33% homologous with SEB and 27% homologous with SECs.[3]

BIOLOGICAL RELEVANCE

PTSAgs have been shown to be involved in a number of acute conditions. The enterotoxins are the most common cause of food

---

Protein Toxin Structure, edited by Michael W. Parker. © 1996 R.G. Landes Company.

poisoning in humans.[4] As its name suggests, TSST-1 is the key agent in toxic shock syndrome (for a review, see ref. 5). TSST-1 alone or in combination with enterotoxins is associated with over 75% of the reported cases of TSS. The remaining cases are associated with either SEB or SECs.[6] TSS is a severe multisystem condition characterized by high fever, rash, skin desquamation (peeling) upon recovery, and hypotension. This sometimes fatal disorder can be induced by as little as 1 µg of TSST-1. TSS has often been associated with tampon usage in menstruating women. However today 40-50% of reported cases of TSS occur from nonmenstrual infections. A related condition, RED (recalcitrant, erythematous, desquamating) disorder, which affects AIDS patients, appears to be due to the staphylococcal toxins.[7] TSST-1 has been found in the kidneys of 18% of children who have died of sudden infant death syndrome.[8] Finally, a recent study[9] has shown that organisms producing TSST-1 are present in 60% to 70% of cases of Kawasaki syndrome, a childhood illness similar to TSS except that hypotension is absent; 5% to 25% of children who recover Kawasaki syndrome are found to have developed coronary artery aneurysms. Kawasaki syndrome is now "the leading cause of acquired heart disease in the United States."[10] The lethal properties of the PTSAgs are associated with their ability to potentiate endotoxin shock.

The term superantigen describing this family of toxins refers to their ability to profoundly expand CD4[+] T cells that contain particular amino acid sequence in the variable region of the chain (Vβ) of the T cell receptor.[11] In a normal antigenic response to an infection, peptide fragments (antigens) from the foreign agent are displayed by class II major histocompatibility complex (MHC) molecules on the outside of antigen presenting cells such as B lymphocytes, dendritic cells and macrophages. The peptide is bound between two α helices in a domain formed by portions of the α and β chains of the class II MHC.[12] A very small percentage (0.0001% to 0.001%) of helper CD4[+] T cells containing receptors having a surface complementary to the MHC:antigen complex are activated. The proliferation of the T cells and the release of cytokines such as interleukins stimulate the growth of other T cells and of B cells and their differentiation into plasma cells which secrete antibodies. In contrast, superantigens bind as intact molecules at locations outside the antigen-binding groove.[13,14] They serve as a bivalent bridge between the class II MHC α chain and the T cell receptor with the appropriate Vβ chain with little regard for what is bound in the antigen-binding groove.[15-17] This leads to massive mobilization of T cells with large release of interleukins and tumor necrosis factors α and β. For example, during episodes of acute toxic shock syndrome, Vβ2[+] T cells may ultimately account for half or more of all T cells in the host.

STRUCTURE OF TSST-1

TSST-1 was discovered in 1981.[18,19] The gene for TSST-1 codes for a 234 residue protein, the first 40 residues of which are cleaved during secretion. The mature 194 residue protein has a molecular weight of 22,049 Da and an isoelectric point of 7.2. TSST-1 has been crystallized in five crystal forms that diffract to resolutions ranging from 4.2 to 1.9 Å[20,21] (W.J. Cook, personal communication). We have crystallized several mutants of TSST-1 including a tetramutant (TSST-1H = TSST-1 [T69I, Y80W, E132K, I140T]) produced from a fusion of the first half of the gene present in *S. aureus* isolates from human sources and the second half of the genes present in isolates from ovine mastitis.[22] The structure of the *C222*$_1$ crystal form of TSST-1 was reported by our group in 1993[23] and by Acharya et al in 1994.[24] Both groups reported models refined to 2.5 Å resolution with R-factors of 0.226 (F > 2σ) and 0.213 (F > 3σ), respectively. The structure of TSST-1 in this crystal form has been subsequently refined to an R-factor of 0.154 (F > 1σ) to 2.05 Å resolution. The structural model contains every amino acid in the three TSST-1 molecules in the crystallographic asymmetric unit plus 405 solvent molecules (Prasad GS, Radhakrishnan R, Mitchell DT et al, in preparation). These three molecules combine with crystallographic symmetry to form a pseudo-*P6*$_1$*22* assembly. The r.m.s. difference among the three molecules is about 0.4 Å for Cαs and 1 Å for all atoms.

Ribbon drawings showing the secondary structural elements of TSST-1 are presented in Figure 11.1. Figure 11.1a presents the standard orientation of TSST-1 to which orientation descriptors such as front and top used in this discussion refer. Helices are labeled with letters; β strands are labeled with numbers. The view presented in Figure 11.1b is from the top and rear, looking down into a prominent groove at the bottom of which is helix B.

TSST-1 is folded into two domains. Domain A is composed of residues 1-17 and 90-194. Overall the architecture of Domain A can be likened to a baby in a crib. The mattress is the 5-strand mixed topology β sheet on the left of Domain A (β floor in Figure 11.1a). Helix B, composed of residues 125-141, is the baby lying on the mattress. Front and back flaps cover and restrict access to helix B. Across the top of the Domain A is helix A formed by residues 3-14. Helix A hangs above helix B further restricting access. The β floor and helix B together form a motif known as a β grasp.[26] Other proteins that contain this motif include the immunoglobulin-binding domain of streptococcal protein G, ubiquitin, and ferredoxin.[26,27]

The smaller Domain B is composed of residues 18-89 which fold into a 5-strand mixed topology β barrel. All pairs of adjacent strands are parallel except for strands 3 and 5. This fold has been called the OB-fold[28] and is found in staphylococcal nuclease,[29] in B-subunits of heat-labile enterotoxin[30] and verotoxin-1, in the anticodon-binding

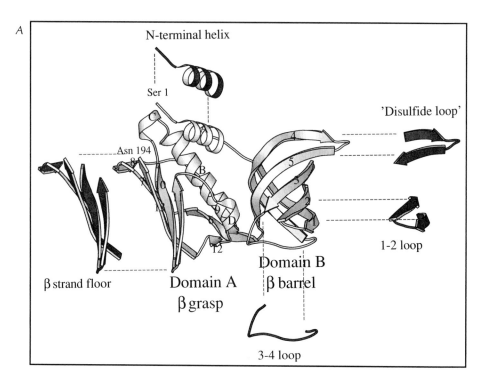

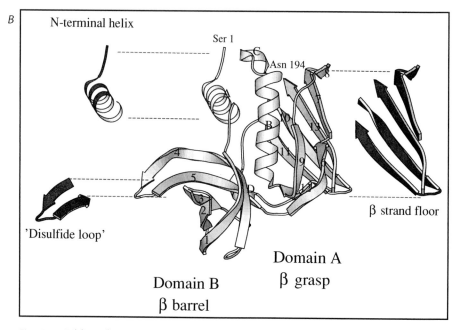

Fig. 11.1. Ribbon drawing using MOLSCRIPT[54] of TSST-1 with individual structural components pulled away from molecule. (a) Front view. (b) Top rear view.

domain of aspartyl-tRNA synthetase,[28] in the major cold-shock protein from *Bacillus subtilis*,[31] and in the active domain of tissue inhibitor of metalloproteinase-2.[32]

The interface between Domain A and Domain B is unusual in its composition. Normally intermolecular interfaces are hydrophobic or neutral. When present, charges are typically paired with opposing charged residues forming buried salt bridges since burying an unsatisfied charge can be very destabilizing.[33] In TSST-1 Lys 67 and Glu 177 are buried in separate regions of the interdomain interface without any neutralizing counterions. This interface is no more hydrophobic than the rest of the molecule. In addition Edwin and Kass[34] found that upon treatment of TSST-1 with pepsin a 12 kDa fragment corresponding to residues 88-194 could be isolated. An examination of the structure of TSST-1 shows that this cleavage site is buried near the center of the domain interface. Taken together these data suggest that separation of the two domains may be a low energy breathing mode of TSST-1.

The structure of the enterotoxin SEB has also been determined.[35,36] Despite only 20% sequence homology, the overall fold of TSST-1 and SEB are quite similar. Considering the secondary structural elements presented in Figure 11.1, there are six significant differences between the two structures:

(1)   The amino terminus of SEB contains 20 additional residues which stretch over the β floor. Alignments of the amino acid sequences of the enterotoxins (see Fig. 11.2) suggest that the enterotoxins and the streptococcal pyrogenic exotoxins all have more residues in their amino termini than does TSST-1. Studies of SEA suggest that a binding site for zinc required for MHC-binding is formed by residues of the amino terminus and the β floor.[37]

(2)   TSST-1 has a regular β duplex at the top of Domain B where SEB has an irregular loop containing a disulfide bridge. Although no cysteines are found in TSST-1, this region in Figure 11.1a is labeled the "disulfide loop" because most other PTSAgs contain a disulfide bridge in this region.

(3)   The loop between β strands 1 and 2 in TSST-1 is short with the duplex smoothly wrapping around the barrel. In SEB the end of this loop curls away from the barrel.

(4)   The connector between β strands 3 and 4 in TSST-1 forms an extended loop. In SEB there is a 10 residue helix connecting these strands. The presence of three prolines in this segment from TSST-1 may make helix formation impossible. The proximity of this helix in SEB may play a role in the curling of the β duplex just described above.

```
                                  aaaaaaaa   aaaaa bb   bbbbbbbbbb b bbbbbbb   bbb        bbbb       bbbb
TSST-1  .........  .........  STNDNIKDLL DWYSSG.SDT FTNSEVLDNS L.GSMRIKNT DG.....SIS LIIFPSPYYS PAFTKGEKVD
SEB     ....ESQPD  PKPDELHKSS KFTGLMENMK VLYDDN.HVS AINVKSIDQF LYFDLIYSIK DTKLGNYDNV RVEFKNKDLA DKY.KDKYVD
SEC3    ....ESQPD  PMPDDLHKSS EFTGTMGNMK YLYDD.HYVS ATKVKSVDKF LAHDLIYNIN DKKLNNYDKV KTELLNEDLA NKY.KDEVVD
SEA     ...SEKSEEI NEKALRKKSE LQGTALGNLK QIYYYNEKAK TENKESHDQF LQHTILFKGF FTDHSWYNDL LVDFDSKDIV DKY.KGKKVD
SED     .......SV  KEKELHKKSE LSSTALNNMK HSYADKNPII GENKSTGDQF LENTLLYKKF FTDLINFEDL LINFNSKEMA QHF.KSKNVD
SEE     ....SEEI   NEKDLRKKSE LQRNALSNLR QIYYYNEKAI TENKESDDQF LENTLLFKGF FTGHPWYNDL LVDLGSKDAT NKY.KGKKVD
SEH     .........  ..EDLHDKSE LTDLALANAY GQY.NHPFI  KENIKSDEIS GEKDLIFRNQ ..GDSGNDL  RVKFATADLA QKF.KNKNVD
SPEA    .....QQD   PDPSQLHRSS L.VKNLQNIY FLYE.GDPVT HENVKSVDQL LSHDLIYNVS ...GPNYDKL KTELKNQEMA TLF.KDKNVD
SPEC    .........  ..DSKKDISN V.KSDLLYAY TITPYDYKDC RVN.FSTTHT LNIDTQKYR. ...GKDYIS  SEMSYEASQK ..FKRDDHVD
                                           *  *                                                        **
```

```
        bbbbbbbbbb bb         bbb        bbbbbb     bbbbbbbb   bbbbbbbb b  bb         bbbbbbb    aaaaaaaaa
TSST-1  LNTKRTKKSQ HTSEGT...  .....YIHF  QISGVTN..T EKLPTPI..E LPLKVKVHGK DSPLKY.GPK FDKKQLAIST LDFEIRHQLT
SEB     VFGANYYYQC TFSKKTNDIN SHQTDKRKTC MYGGVTEHNG NQLDKY..RS ITVRVFEDGK NL.LSF.DVQ TNKKKVTAQE LDYLTRHYLV
SEC3    VYGSNYYVNC YFSSKDNVGK VTS...GKTC MYGGITKHEG NHFDNGNLQN VLIRVYENKR NT.ISF.EVQ TDKKSVTAQE LDIKARNFLI
SEA     LYGAYYGYQC AGGTPN.... ....KTAC   MYGGVTLHDN NRLTEE..KK VPINLWLDGK QNTVPLETVK TNKKNVTVQE LDLKARRYLQ
SED     VYPIRYSINC YGGEID.... .......RTAC TYGGVTPHEG NKLKER..KK IPINLWINGV QKEVSLDKVQ TDKKNVTVQE LDAQARRYLQ
SEE     LYGAYYGYQC AGGTPN.... ....KTAC   MYGGVTLHDN NRLTEE..KK VPINLWIDGK QTTVPIDKVK TSKKEVTVQE LDLQARHYLH
SEH     IYGASFYYKC EKISEN.... ....ISEC   LYGGTTL.NS EKLAQE..RV IGANVWVDGI QKETEL..IR TNKKNVTLQE LDIKIRKILS
SPEA    IYGVEYYHLC YLCENAE... ....RSAC   IYGGVTNHEG NHLEIP..KK IVVKVSIDGI QS..LSFDIE TNKKMVTAQE LDYKVRKYLT
SPEC    VFGLFYILNS H......... .....TGEY  IYGGITPAQN NKVNHKLLGN LFISGESQQN ....LNNKII LEKDIVTFQE IDFKIRKYLM
        * *                             *** **                 *                     * * * ***
```

```
        aaa  aaaa   a  bbb     bbbbb bbb  bbbb       bb         b bbbbbbbb  bb
TSST-1  QIHGLYRSSD KTG...GYW  KITMNDGSTY QSDLS...K  .YNTEKPPIN IDEIKTIEAE IN
SEB     KNKKLYEFNN SP..YETGYI KFIE.FENSF WTDMMPAPGD KFDSQSKYLM MYNDNKMVDS KD..VKIEVY LTTKKK
SEC3    NKKNLYEFNS SP..YETGYI KFIESNGNTF WYDMMPAPGD KFD.QSKYLM IYKDNKMVDS KS..VKIEVH LTTKNG
SEA     EKYNLYNSDV FDGKVQRGLI VFHTSTEPSV NYDLFGAQGQ YSNTLL...R IYRDNKSINS EN..MHIDIY LYTS
SED     KDLKLYNNDT LGGKIQRGKI EFDSSDGSKV SYDLFDVKGD FPEKQL...R IYSDNKTLST EH..LHIDIY LYEK
SEE     GKFGLYNSDS FGGKVQRGLI VFHSSEGSTV SYDLFDAQGQ YPDTLL...R IYRDNKTINS EN..LHIDLY LYTT
SEG     DKYKIYYKDS EISK..GLI  EFDMKTPRDY SFDIYDLKGE NDYEID...K IYEDNKTLKS DD.ISHIDVN LYTKKKV
SPEA    DNKQLYTNGP SK..YETGYI KFIPKNKESF WFDFFPEPEF T..QSKYLM  IYKDETL.DS NT..SQIEVY LTTK
SPEC    DNYKIYDAT. SP..YVSGRI EIGTKDGKHE QIDLFDSPNE G..TRSDIFA KYKDNRIINM KN.FSHFDIY LEK
        * *                                        *****
```

Fig. 11.2. Amino acid sequences of PTSAgs aligned using primary and tertiary structural homology. The letters a and b above the TSST-1 sequence indicate the presence of α helices and β strands, respectively, in TSST-1. Residues in the interface between domains in TSST-1 are marked with asterisks.

(5) The front flap covering helix B in TSST-1 extends only up as far as the central axis of helix B. In SEB this front flap extends nearly up to helix A completely blocking access from the front to helix B.

(6) The rear flap covering helix B in TSST-1 is similar in size to the front flap and again extends as far as the central axis of helix B. In SEB the rear flap is much smaller than the front flap allowing more extensive access to helix B from the back.

By using the significant sequence homology present among some PTSAgs and the structural homology seen between SEB and TSST-1 it is possible to align their amino acid sequences (see Fig. 11.2). Residues that form the interface between Domain A and Domain B in TSST-1 are marked. Many of these positions are highly conserved among the PTSAgs listed. These residues are found at the start of β strand 4, at the end of β strand 5, and at the junction of β strand 9 and helix B. This conservation gives evidence to the critical role these residues play in connecting the domains.

FUNCTIONALLY IMPORTANT FEATURES

Figure 11.3 shows that polar residues are widely distributed on the surface of TSST-1 with the exception of the front of Domain B. Mutational analysis[38] and the structures of the TSST-1:MHC complex[39] and SEB:MHC complex[17] indicate that this surface forms the MHC-binding site. Of the residues of TSST-1 that have been shown to interact with the class II MHC molecule, the uncharged nature of the surface is maintained in SEB except for residues equivalent to Leu 44 and Ile 46. In SEB the equivalent residues are Arg 65 and Glu 67 which are adjacent to each other. Glu 67 is involved in a buried salt link with Lys 39a of the MHC molecule HLA-DR1.[17] In the TSST-1:MHC complex Lys 39a is buried but without a neutralizing charge.

There are several questions concerning the TSST-1:MHC interaction which deserve attention in the future. The first question deals with the covering by TSST-1 of the bound antigen and the top of the MHC molecule. The structure reported by Kim et al[39] suggests that there is little contact between the T cell receptor and the MHC molecule in the TSST-1:MHC:T cell receptor ternary complex. If this is the case then one wonders why the MHC molecule is needed at all. For the SEB:MHC:T cell receptor ternary complex there are data which indicate interactions between the MHC molecule and T cell receptor.[40,41] Is TSST-1 different? The second question is the specificity of TSST-1 and other superantigens for class II MHC molecules. Given the substantial structural similarity between the class I and class II MHCs,[42] one wonders why no superantigen exists for the class I MHC molecules. The third question is the significance of the structural

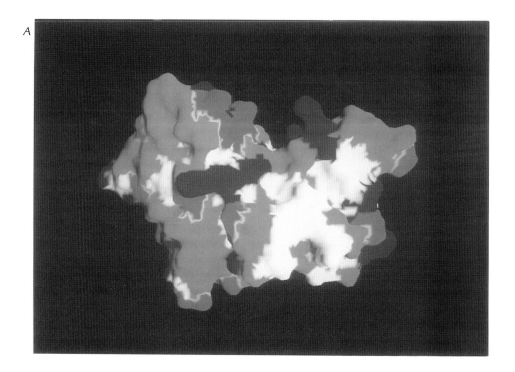

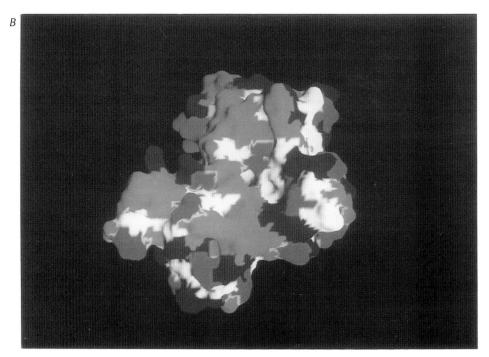

Fig. 11.3. Solvent exposed surface of TSST-1 calculated by GRASP.[55] Acidic residues are red, basic residues are blue, and polar residues are green. (a) Front view. (b) Top rear view. See color figure in insert.

homology between Domain B and other OB-fold proteins. Superposition of the peptide backbones of Domain B with that of staphylococcal nuclease overlays the active site of the latter protein with the right half of putative MHC-binding site. Since the other OB-fold proteins bind either oligonucleotides or oligosaccharides, one wonders whether a similar function exists for this region of PTSAgs. The fourth question is the significance of the structural homology between Domain A and the other β grasp proteins. Both PTSAgs and protein G bind immunoglobulins. Is this a coincidence or is there an evolutionary relationship that may extend to their modes of interaction?

The binding site for the T cell receptor is thought to be formed by residues on the top of Domain A. This is supported by a large number of mutations that reduce or eliminate superantigenicity.[38,43-5] This putative T cell receptor binding site coincides with a surface used in forming intermolecular contacts in TSST-1H, in three crystal forms of TSST-1, and in the TSST-1:MHC complex (Mitchell DT, Kim J, Urban RG et al, in preparation). The structural homology with the immunoglobulin-binding domain of staphylococcal protein G suggests that the interaction with the T cell receptor β chain might involve the edge of the β floor as seen in the complex between protein G and an immunoglobulin Fab fragment complex.[47] Structural analysis of a superantigen:T cell receptor complex will be the best way to address this question.

Although the precise mechanism of their gastrointestinal effects remain unclear, the disulfide loop at the top of Domain B seems to play a role in the ability of the enterotoxins to induce vomiting. TSST-1 has no disulfide loop and is not an enterotoxin. It does seem clear that the emetic properties of the enterotoxins is independent of their superantigenic properties.[48,49] Further mutational analysis of this region is required to ascertain the functional properties of this region.

The ability of the PTSAgs to kill the host is largely due to their ability to potentiate endotoxin shock. Over the past few years evidence has been obtained that the host lethality of TSST-1 is independent of superantigenicity[22,50,51] These data point to residues comprising the middle of helix B under the putative T cell receptor-binding site as being critical in this biological activity. In addition to further mutational studies, structural characterization of complexes of TSST-1 with portions of endotoxin would be extremely useful.

The ability of TSST-1 to produce hypotension in toxic shock syndrome and possibly aneurysms in Kawasaki syndrome suggests an effect on endothelial cells in blood vessels. TSST-1 has been shown to bind to cultured aortic endothelial cells with dissociation constants ranging from 6×10^{-7} M to 6×10^{-10} M.[52,53] Since these cells contain neither MHC molecules or T cell receptors, this ability seems to be unrelated to the ability to bind either of these molecules. The region of TSST-1 involved in binding endothelial cells is unknown. The mapping of this interaction through the screening of a library of TSST-1 mutants would provide a start in understanding this interaction.

What is clear is that the family of PTSAgs has a number of properties that govern its biological functions. The recent structural analyses of these molecules are first steps toward deciphering the structural foundation for these activities. New studies combining mutagenesis, structure determination and biological analysis will allow other steps down the road toward the design of new pharmacological agents against diseases produced by these molecules and of proteins that can act as immunomodulators in the treatment of cancer and AIDS.

REFERENCES

1. Tranter HS. Foodborne staphylococcal illness. Lancet 1990; 336:1044-46.
2. Betley MJ, Borst DW, Regassa LB. Staphylococcal enterotoxins, toxic shock syndrome toxin and streptococcal pyrogenic exotoxins: A comparative study of their molecular biology. Chem Immunol 1992; 55:1-35.
3. Ren K, Bannan JD, Pancholi V et al. Characterization and biological properties of a new staphylococcal exotoxin. J Exp Med 1994; 180:1675-83.
4. Bergdoll MS. Enterotoxins. In: Easmon CSF, Adlam C, eds. Staphylococci and Staphylococcal Infections. London: Academic Press, 1983: 187-248.
5. Bohach GA, Fast DJ, Nelson RD et al. Staphylococcal and streptococcal pyrogenic toxins involved in toxic shock syndrome and related illnesses. Crit Rev Microbiol 1990; 17:251-72.
6. Schlievert PM. Role of toxic shock syndrome toxin 1 in toxic shock syndrome: Overview. Rev Infect Dis 1989; 11:S107-09.
7. Cone LA, Woodard DR, Byrd RG et al. A recalcitrant, erythematous, desquamating disorder associated with toxin-producing staphylococci in patients with AIDS. J Infect Dis 1992; 165:638-43.
8. Newbould MJ, Malam J, McIllmurray JM et al. Immunohistological localisation of staphylococcal toxic shock syndrome (TSST-1) antigen in sudden infant death syndrome. J Clin Pathol 1989; 42:935-39.
9. Leung DYM, Meissner HC, Fulton DR et al. Toxic shock syndrome toxin-secreting *Staphylococcus aureus* in Kawasaki syndrome. Lancet 1993; 342:1385-88.
10. Rowley AH, Gonzalez-Crussi F, Shulman ST. Kawasaki syndrome. Curr Prob Pediatr 1991; 21:387-405.
11. Marrack P, Kappler JW. The staphylococcal enterotoxins and their relatives. Science 1990; 248:705-11.
12. Brown JH, Jardetsky TS, Gorga JC et al. Three-dimensional structure of the human class II histocompatibility antigen HLA-DR1. Nature 1993; 364:33-39.
13. Dellabona P, Peccoud J, Kappler J et al. Superantigens interact with MHC class II molecules outside of the antigen groove. Cell 1990; 62:1115-21.
14. Karp DR, Teletski CL, Scholl P et al. The a1 domain of the HLA-DR molecule is essential for high-affinity binding of the toxic shock syndrome toxin-1. Nature 1990; 346:474-76.

15. White J, Herman A, Pullen AM et al. The Vb-specific superantigen staphylococcal enterotoxin B: stimulation of mature T cells and clonal deletion in neonatal mice. Cell 1989; 56:27-35.

16. Herman A, Kappler JW, Marrack P et al. Superantigens: Mechanism of T-cell stimulation and role in immune responses. Ann Rev Immunol 1991; 9:745-72.

17. Jardetsky TS, Brown JH, Gorga JC et al. Three-dimensional structure of a human class II histocompatibility molecule complexed with superantigen. Nature 1994; 368:711-18.

18. Bergdoll MS, Crass BA, Reiser RF et al. A new staphylococcal enterotoxin, enterotoxin F, associated with toxic-shock-syndrome *Staphylococcal aureus* isolates. Lancet 1981; 2:1017-21.

19. Schlievert PM, Shands KN, Dan BB et al. Identification and characterization of an exotoxin from *Staphylococcus aureus* associated with toxic shock syndrome. J Infect Dis 1981; 96:14-18.

20. Passalacqua EF, Brehm RD, Acharya KR et al. Purification, crystallization and preliminary x-ray analysis of toxic shock syndrome toxin-1 from *Staphylococcus aureus*. J Mol Biol 1992; 228:983-86.

21. Earhart CA, Prasad GS, Murray DL et al. Growth and analysis of crystal forms of toxic shock syndrome toxin 1. Proteins 1993; 17:329-34.

22. Lee PK, Kreiswirth BN, Deringer JR et al. Nucleotide sequences and biological properties of toxic shock syndrome toxin-1 from ovine and bovine-associated *Staphylococcus aureus*. J Infect Dis 1992; 1056-63.

23. Prasad GS, Earhart CA, Murray D et al. Structure of toxic shock syndrome toxin-1. Biochemistry 1993; 32:13761-66.

24. Acharya KR, Passalacqua EF, Jones EY et al. Structural basis of superantigen action inferred from crystal structure of toxic-shock syndrome toxin-1. Nature 1994; 367:94-97.

26. Overington JP. Comparison of three-dimensional structures of homologous proteins. Curr Opin Struct Biol 1992; 2:394-401.

27. Kraulis PJ. Similarity of protein G and ubiquitin. Science 1991; 254:581-82.

28. Murzin AG. OB(oligonucleotide/oligosaccharide binding)-fold: Common structural and functional solution for nonhomologous sequences. EMBO J 1993; 12:861-67.

29. Hynes T, Fox RO. The crystal structure of staphylococcal nuclease refined at 1.7 Å resolution. Proteins 1991; 10:92-105.

30. Hol WJ, et al. Chapter 8 in this book. 1996.

31. Schindelin H, Marahiel MA, Heinemann U. Universal nucleic acid-binding domain revealed by the crystal structure of the *B. subtilis* major cold-shock protein. Nature 1993; 364:164-68.

32. Williamson RA, Martorell G, Carr MD et al. Solution structure of the active domain of tissue inhibitor of metalloproteinase-2. A new member of the OB fold protein family. Biochemistry 1994; 33:11745-59.

33. Dao-pin S, Anderson DE, Baase WA et al. Structural and thermodynamic consequences of burying a charged residue within the hydrophobic core of T4 lysozyme. Biochemistry 1991; 30:11521-29.

34. Edwin C, Kass EH. Identification of functional antigenic segments of toxic shock syndrome toxin 1 by differential immunoreactivity and by differential mitogenic responses of human peripheral blood mononuclear cells, using active toxin fragments. Infect Immun 1989; 57:2230-36.

35. Swaminathan S, Furey W, Pletcher J et al. Crystal structure of staphylococcal enterotoxin B, a superantigen. Nature 1992; 359:801-06.

36. Swaminathan S et al. Chapter 12 in this book. 1995.

37. Fraser JD, Urban RG, Strominger J et al. Zinc regulates the function of two superantigens. Proc Natl Acad Sci USA 1992; 89:5507-11.

38. Hurley JM, Shimonkevitz R, Hanagan A et al. Identification of class II MHC and T cell receptor binding sites in the superantigen toxic shock syndrome toxin-1. J Exp Med 1995; 181:2229-35.

39. Kim JS, Urban RG, Strominger JL et al. Toxic shock syndrome toxin-1 complexed with a class II major histocompatibility molecule HLA-DR1. Science 1994; 266:1870-74.

40. Labrecque N, Thibodeau J, Mourad W et al. T cell receptor-major histocompatibility complex class II interaction is required for the T cell response to bacterial superantigens. J Exp Med 1994; 180:1921-29.

41. Deckhut AM, Chien Y-H, Blackman MA et al. Evidence for a functional interaction between the b chain of major histocompatibility complex class II and the T cell receptor a chain during recognition of a bacterial superantigen. J Exp Med 1994; 180:1931-35.

42. Stern LJ, Wiley DC. Antigenic peptide binding by class I and class II histocompatibility proteins. Structure 1994; 2:245-51.

43. Blanco L, Choi EM, Connolly K et al. Mutants of staphylococcal toxic shock syndrome toxin 1: Mitogenicity and recognition by a neutralizing monoclonal antibody. Infect Immun 1990; 58:3020-28.

44. Irwin MJ, Hudson KR, Fraser JD et al. Enterotoxin residues determining T cell receptor Vb binding specificity. Nature 1992; 359:841-43.

45. Bonventre PF, Heeg H, Cullen C et al. Toxicity of recombinant toxic shock syndrome toxin 1 and mutant toxins produced by *Staphylococcus aureus* in a rabbit infection model of toxic shock syndrome. Infect Immun 1993; 61:793-99.

47. Derrick JP, Wigley DB. Crystal structure of a streptococcal protein G domain bound to an Fab fragment. Nature 1992; 359:752-54.

48. Harris TO, Grossman D, Kappler JW et al. Lack of complete correlation between emetic and T cell stimulatory activities of staphylococcal enterotoxins. Infect Immun 1993; 61:3175-83.

49. Hovde CJ, Marr JC, Hoffmann ML et al. Investigation of the role of the disulphide bond in the activity and structure of staphylococcal enterotoxin C1. Molec Microbiol 1994; 13:897-909.

50. Murray DL, Prasad GS, Earhart CA et al. Immunological and biochemical properties of mutants of toxic shock syndrome toxin-1. J Immunol 1994; 152:87-95.

51. Murray DL, Earhart CA, Mitchell DT et al. Localization of biologically important regions on toxic shock syndrome toxin-1. Infect Immun 1995; in press.

52. Lee PK, Vercellotti GM, Deringer JR et al. Effects of staphylococcal toxic shock syndrome toxin 1 on aortic endothelial cells. J Infect Dis 1991; 164:711-19.

53. Kushnaryov VM, MacDonald HS, Reiser RF et al. Reaction of toxic shock syndrome toxin 1 with endothelium of human umbilical cord vein. Rev Infect Dis 1989; 11:S282-87.

54. Kraulis PJ. MOLSCRIPT: A program to produce both detailed and schematic plots of protein structures. J Appl Cryst 1991; 24:946-50.

55. Nichols A, Bharadwaj R, Honig B. GRASP: Graphical representation and analysis of surface properties. Biophys J 1993; 64:A166.

CHAPTER 12

STRUCTURE
OF STAPHYLOCOCCAL
ENTEROTOXINS

S. Swaminathan, William Furey and Martin Sax

Staphylococcal enterotoxins are microbial agents causing food poisoning.[1] They induce vomiting and diarrhea in humans,[2] and they were ranked in the 1970s as the second most common cause of food poisoning in the United States of America. In 1970, it was shown that even though they are enteric, their primary targets are T lymphocytes.[3] For a long time the actual mechanism of action of these toxins was not properly understood. Studies over the past decade have enabled us to understand these toxins and their functions and mechanism better.

Staphylococcus aureus produces a series of low molecular weight (25,000-30,000 Daltons) enterotoxic proteins, which are similar in composition and activity and are classified according to their antigenic differences. There are five distinct types of *Staphylococcal* enterotoxins labeled SEA (staphylococcal enterotoxin A) through SEE.[4] SEC is further divided into SEC1, SEC2 and SEC3 due to minor epitope variation. All these toxins share significant sequence homology, some more than the others.[6] Toxic shock syndrome toxin 1 (TSST1) and exfoliative toxins which are not enteric are also produced by *S. aureus*. All these toxins use similar mechanisms of pathogenesis.

Recently, these bacterial toxins have come to the forefront of research because of their ability to induce massive, yet selective T cell proliferation (see ref. 7 and the references therein). When presented by a major histocompatibility complex class II molecule (MHCII), they are able to bind to T cell receptors (TCR) and stimulate all T cells bearing particular types of Vβ chains. This stimulation of T cells is orders of magnitude greater than that due to regular antigenic peptides.

Protein Toxin Structure, edited by Michael W. Parker. © 1996 R.G. Landes Company.

It is believed that the binding site for these toxins on TCR is on Vβ chains only and that other variable chains of TCR contribute little to this binding. The current consensus is that the formation of an MHCII-SE-TCR ternary complex triggers the proliferation of all T cells bearing particular types of Vβ elements. Because of their ability to cause this immense T cell proliferation they are called superantigens.[8] Although the mechanisms of actions of both antigenic peptides and superantigens require that they both be presented by an MHCII molecule for T cell activation to occur,[9,10] ordinary antigens have to be preprocessed by the antigen presenting cells into small peptides (13 to 17 residues in length) before they can bind to MHCII and activate T cells.[11] These antigenic peptides bind to a cleft formed by two α helices of α and β chains of the MHCII molecule.[12,13] On the other hand, superantigens need no such preprocessing, for they bind as intact molecules but to a different part of the MHCII molecule.[12,14] This difference in the mode of binding gives rise to differences in the interactions between the antigen/superantigen and MHCII, and TCR. Antigenic peptides make contacts with all variable chains of TCR and the fraction of T cells stimulated is limited by these interactions. On the other hand, superantigens make contacts only with the Vβ chain of the TCR. Since there are only a limited number of Vβ elements in humans, a very large fraction of all TCRs is activated.[15] When an MHCII-SE-TCR complex is formed three kinds of interactions occur viz. MHCII—SE, MHCII—TCR and SE—TCR. Even though soluble TCR Vβ chains bind to SEs, the latter alone are not able to stimulate T cells suggesting that MHCII—TCR interaction is a necessary condition for T cell stimulation.[16]

Different modes of MHCII—SE association have been proposed which are believed to depend on the kind of SEs. For example, while zinc is required to mediate MHCII—SEA complex formation,[17] it is not required for SEB to bind to MHCII. SEA competes with both SEB and TSST1, but SEB and TSST1 do not compete with each other.[18-21] These results suggest that the binding sites for SEs on MHCII may be separate but overlapping.

Although superantigenicity may affect the host adversely through clonal deletion,[8] it nevertheless may have therapeutic possibilities also. Injecting mice with SEB can reduce severity of lupus nephritis, an autoimmune disease; treating animals with SEB also can protect them against multiple sclerosis.[15]

Three dimensional structures help in understanding the stereochemistry of different binding sites in a protein molecule. The first three dimensional structure of a superantigen, SEB, was determined in our laboratory and helped in defining and understanding the stereochemistry of binding sites in SEB.[22] The correlation of secondary structural elements in SEB with regions in the sequence homologous to other SE sequences also led us to predict that all SEs would have similar structures. Subsequently, crystal structures of TSST1,[23,24] SEC2,[47] SEC3[25], MHCII—SEB complex[26] and MHCII—TSST1 complex[27] have been reported and the prediction confirmed. Three dimensional models for SEA and SEE based on the similarity of their sequence to those of other SEs also have been reported. In this chapter the crystal structures of SEB and SEC2 are described. The stereochemistry of different binding sites are discussed in the light of the available structures and complexes. An attempt is also made here to identify the Vβ specificity defining residues in *Staph* enterotoxins.

STRUCTURE OF SEB

The three dimensional structure of SEB has been determined to 2.0 Å resolution. The SEB molecule consists of two domains. Domain 1 consists of residues 1-120 and residues 127-239 compose domain 2. Residues 1-29 of domain 1 are more closely associated with domain 2 in three dimensional space. A schematic diagram of the folding and a ribbon drawing of the SEB molecule are shown in Figures 12.1a and 12.1b. Domain 1 consists of two β sheets, one formed by strands β1, β4, and β5 and the other by β2, β3, β4 and β5. It also contains three α- helices viz. α1, α2 and α3. Strands β4 and β5 form part of both sheets; the two β sheets together form a cylindrical barrel whose inner wall is lined with hydrophobic residues forming a hydrophobic core. One end of this cylinder is capped by a helix, α3 and the other is covered partly by a loop formed by residues 54-60 (β3β4 loop). This folding pattern of domain 1, also seen in a few other toxin structures, is now identified as the oligomer binding or the OB fold.[28]

Domain 2 mainly consists of two α helices, α4 and α5 and a twisted β sheet formed by β strands β6, β7, β9, β10 and β12 resembling a modified β grasp motif; it also contains two very short β strands, β10 and β11. The two α helices form the central core of the SEB molecule, and are sandwiched between the twisted β sheet of domain 2 and the β cylinder of domain 1. Each of the α helices α4 and α5 has one side exposed to the solvent and forms a groove with adjoining residues forming two faces of the SEB molecule.

STRUCTURE OF SEC2

We have also determined the crystal structure of SEC2 at 2.7 Å resolution. SEC2 also consists of two domains similar to SEB. The N-terminus domain (residues 1-120) consists of three α helices and five β strands (see Fig. 12.2). The C-terminus domain is similar to its counterpart in SEB. While the general topology is the same as SEB, there are some minor differences. The conformation of the first three N-terminal residues are different in SEB and SEC2. There are only 18 residues (inclusive of two cystines) in the disulfide loop of SEC2 which is three less than in SEB. Strand β7 is also short in SEC2. In SEC2 the helix formed by residues 210 to 219 is made up of two

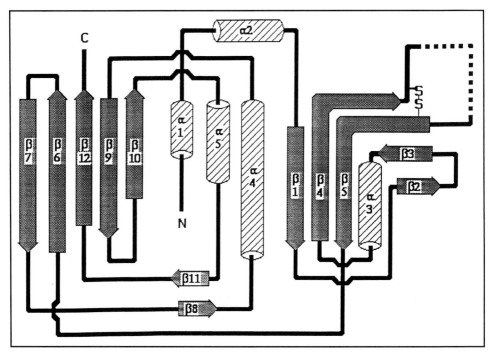

Fig. 12.1a. A schematic representation of the secondary structure of SEB. Cylinders and arrows represent α-helices and β-strands. The discontinuous line in the disulfide loop indicates seven residues not seen in the electron density map. The residues forming the secondary structural elements are: 13-17 (α1), 21-29 (α2), 33-39 (β1), 48-52 (β2), 63-68 (β3), 70-78 (α3), 81-89 (β4), 112-120 (β5), 127-138 (β6), 141-151 (β7), 154-156 (β8), 157-172 (α4), 182-190 (β9), 195-200 (β10), 210-217 (α5), 222-224 (β11), 229-236 (β12). Reprinted with permission from: Swaminathan S et al. Nature 1992; 359:801-806. © 1992 Macmillan Magazines Ltd.

distinct helices, an α helix (210-214) and a 3_{10} helix (215-219). The secondary structural elements in SEB, SEC2, and TSST1[23] are compared in Table 12.1. TSST1 has about 18 residues less at the N terminus than SEB or SEC2 and helix α3 of SEB or SEC2 is absent in TSST1 and is replaced by an extended loop.

In both crystal structures the disulfide loop is disordered and a few residues could not be located in the electron density maps during X-ray structure determination. The high thermal factors for residues forming the disulfide loops are consistent with either dynamic or static disorder. Since incubation with trypsin cleaves[29] the peptide bond between 98 and 99 of SEB, it was conjectured that the missing residues might possibly have been the result of proteolysis taking place in the native protein. However, a reduced SDS-PAGE run on the native crystals did not reveal two bands indicating that there is one continuous polypeptide chain (unpublished results).

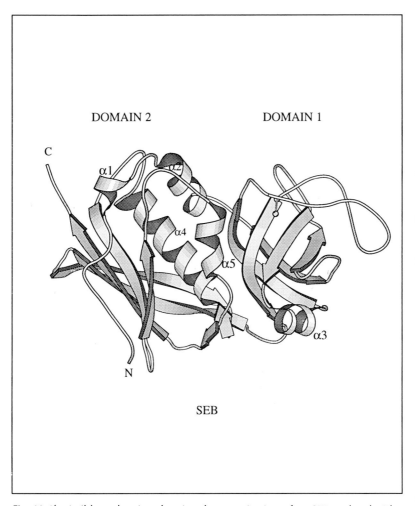

Fig. 12.1b. A ribbon drawing showing the organization of an SEB molecule. The conformation of the disulfide loop should be viewed with caution since part of it was not seen in the electron density maps. The seven missing residues were included in the model for the sake of continuity.

MODELING OF SEA AND SEE

The crystal structure determinations of SEB and SEC2 showed that the conformation and folding pattern of these two SEs are very similar. Crystal structures of SEC3 and a distantly related toxin, TSST1 also revealed that all these molecules possess a common folding as had been suggested by us[22] and we now call this common fold the SE-fold. Based on this fact and the sequence homology of all SEs, the three dimensional structures of SEA and SEE have been modeled.[47] Unlike SEB and SEC, SEA requires Zn to bind to MHCII molecules and

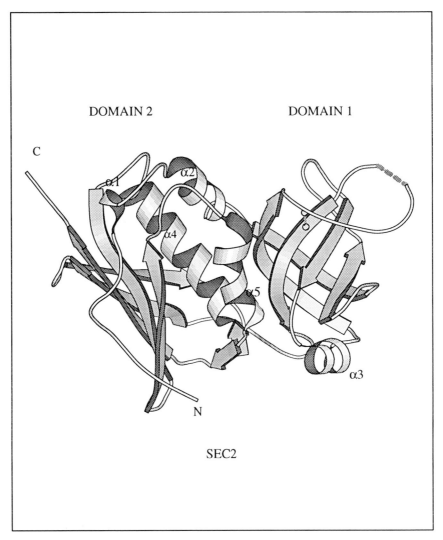

Fig. 12.2. A ribbon drawing of an SEC2 molecule. The dashed line represents the four missing residues in the disulfide loop.

accordingly a Zn atom was included at this putative site in the model. This binding site is shown in the ribbon drawing of SEA (Fig. 12.3).

MUTATIONAL ANALYSIS ON SEB AND STRUCTURAL FEATURES

Mutational analysis has been carried out on SEB to identify residues or regions responsible for superantigenic activity.[30] Three such regions were first examined and in each region specific amino acids responsible for MHCII binding or T cell activation were identified.

Table 12.1. Comparison of secondary structures of SEB, SEC2 and TSST1

SEB		SEC2		TSST1†	
Residues	**Secondary structure**	**Residues**	**Secondary structure**	**Residues**	**Secondary structure**
13 –17	α1	13 – 16	α1		
21 – 29	α2	21 – 28	α2	7 – 15	α1
33 – 39	β1	33 – 38	β1	18 – 20	β1
48 – 52	β2	48 – 52	β2	32 – 37	β2
63 – 68	β3	62 – 67	β3	41 – 47	β3
70 – 78	α3	70 – 77	α3		
81 – 89	β4	82 – 90	β4	60 – 75	β4
112 – 120	β5	108 – 117	β5	79 – 89	β5
127 – 138	β6	130 – 137	β6	101 – 106	β6
141 – 151	β7	140 – 147	β7	109 – 111	β7
154 – 156	β8	153 – 155	β8	119 – 124	β8
157 – 172	α4	156 – 171	α4	125 – 140	α2
182 – 190	β9	181 – 190	β9	152 – 158	β9
195 – 200	β10	194 – 200	β10	161 – 166	β10
210 – 217	α5	210 – 214	α5	173 – 177	3_{10}
		215 – 219	3_{10}	181 – 182	β11
222 – 224	β11	222 – 224	β11	183 – 185	3_{10}
229 – 236	β12	229 – 236	β12	186 – 193	β12

† From reference 23.

Some of the residues affected both MHCII and TCR binding; however, since MHCII binding is a prerequisite for TCR activation, these residues provide no useful information about TCR activation, but they do provide information about MHCII binding. On the other hand residues that influence T cell activation but not MHCII binding are likely to be in the T cell binding site on SEB. This information in conjunction with the three dimensional structure of SEB revealed features about the stereochemistry of the MHCII and TCR binding sites.

Residues 9 to 23 form region 1 and affect both TCR and MHCII binding. Residues 9, 14, 17 and 23 in this region were mutated either alone or in combination with others. Residue 9 is in a random coil before α1. Asn23 which is conserved in all *S. enterotoxins* is located in a helix, α2 and its side chain is exposed to solvent. Mutation of this residue affects TCR activation, mainly. Residues 14 and 17 were found to be bifunctional, affecting both MHCII and TCR binding. While Ser14 is on α1, Phe17 is located just at the C terminal portion of α1. Asp9, Ser14 and Asn23 are exposed to solvent, but Phe17 lies in a loop sandwiched between two other loops and is buried. This

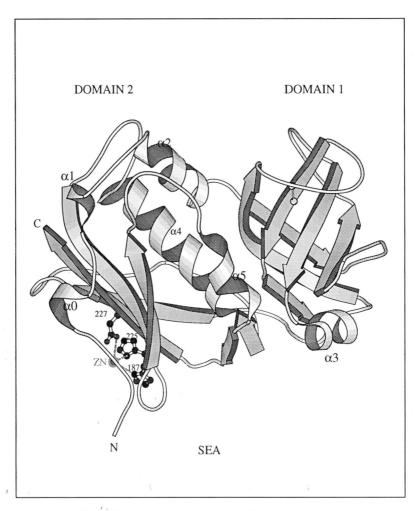

Fig. 12.3. A ribbon drawing of an SEA molecule. The putative zinc binding site is shown as a ball and stick model. The zinc ion is coordinated by His187, His225 and Asp227 of SEA. The other coordination to this zinc ion is presumably His81 of the β chain of the MHCII molecule.

suggests that mutation of Phe17 with Ser (F → S) might have affected the structural integrity of the three loops at this region and changed the conformation of the molecule resulting in the disruption of MHCII binding. The association of Ser14 with MHCII binding and Asn23 with TCR activity, reveals the structural basis of the bifunctional role of region 1. The proximity of these two residues also suggests that the MHCII and the TCR are close to each other in the ternary complex, MHCII-SEB-TCR.

Residues 40 to 53 form region 2 and affect MHCII binding in all S. enterotoxins. Most of the mutational studies in this region involved Phe44 while some of them involved 41, 45, 46, 48, 52 and 53 also. These mutations affected the MHCII binding and consequently the TCR activation. Phe44, lying on a turn connecting strands β1 and β2, is exposed to solvent and is in a position to make hydrophobic contacts with MHCII. Residues 48 to 52 are in strand β2.

Asn60 and Tyr61 form region 3 and mutation of either one of them affects TCR binding but not MHCII binding. These residues are located in a turn connecting β2 and β3 and are exposed to solvent.

TCR BINDING SITE

The TCR binding site in SEB was identified from the three dimensional structure and from the results based on mutational analysis. There is a shallow cavity in the molecular surface in the interface of the two domains and is formed by residues from both domains (see Figs. 12.4a and 12.4b). Residues forming this shallow cavity are Glu22, **Asn23**, Val26, **Leu27**, Asp29, Asp30, Asn31, Val33, Asp55, Leu58, Asn60, **Tyr61**, Ala87, Asn88, **Tyr89**, Tyr90, **Tyr91**, **Gln92**, **Thr112**, Phe208, Asp209, Gln210 and **Leu214**. Residues in bold type are conserved in all S. enterotoxins. Residues 22-23 of region 1 and residue 60-61 of region 2 from the mutational studies[30] form part of this cavity. The molecular topology as well as mutational evidence, clearly indicated that this is the TCR binding site. A similar cavity has been identified in the three dimensional structures of SEC2, SEC3 and to a lesser extent in TSST1 in which the first 18 N terminal residues corresponding to SEB are missing.

Further evidence from other mutational studies also indicate that residues in the three stacked loops adjacent to this cavity are functionally important. The amino acid sequences of SEA and SEE are 82% identical, but SEA binds to mouse Vβ3 whereas SEE binds to mouse Vβ11. Using a hybrid molecule of SEA-SEE, Irwin et al[31] have shown that the Vβ specificity defining residues are in the C terminal half of these two toxins. They have further localized the region to residues 207 and 208 of SEA. By changing residues 207 and 208 to correspond to SEE (S → P and N → D), the Vβ specificity could be changed to correspond to SEE. Similar results were obtained when SEE residues were changed to correspond to SEA and it was concluded that these residues defined the Vβ specificity for SEA and SEE. A similar mutational study was done by Mollick et al[32] independently. Even though the specificity changed, the level of activity (Vβ binding) was always less than that of wild type. To bring the level of activity of these chimeric molecules to the wild type level, changes in the sequence at the N-terminal region are also required. Changes in the N-terminal

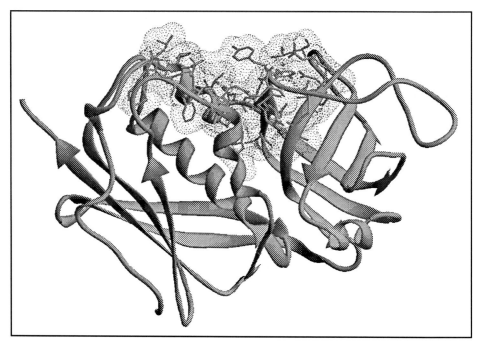

Fig. 12.4a. The TCR binding site of SEB shown as a dot surface viewed as in Figure 12.2. The residues implicated in TCR binding are shown as a stick model.

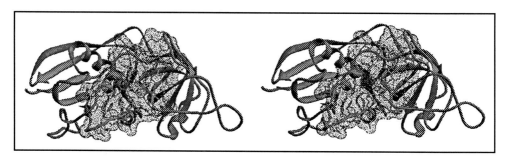

Fig. 12.4b. A stereo view of the TCR binding site as viewed from the top of Figure 12.4a.

region were localized to residues 20-27 of SEA. Further, when residues 20-24 and the C-terminal residues 207-208 corresponded to the sequence of the target toxin the level of activity (TCR binding) was equal to that of the wild type.

In TSST1, the regions corresponding to 111-118 and 135-158 have been implicated in TCR binding.[23] These residues are situated respectively in the β7β8 and α2β9 loops of the TSST1 structure. The α2β9 loop corresponds to the α4β9 loop of SEB, which forms one of the three stacked loops in SEB.

All these results taken together confirm the original finding of the TCR binding site in SEB. The three stacked loops (17-21, 173-181, and 201-209 of SEB) and particularly, the slope of the ridge on the side of the interface of the two domains seem to be important for TCR binding. Although the Vβ chain of the TCR might make contacts with more than one residue in this cavity, Vβ specificity may be controlled by one or two residues or by a combination of them as discussed later. This Vβ binding cavity in the interface of the domains may be a common feature of all bacterial superantigens, at least it is for the *Staph* enterotoxins.

MHCII BINDING SITE IN S. ENTEROTOXINS

Though the TCR binding site could be defined with reasonable confidence from the topology of the molecule and from mutational studies, defining the MHCII binding site was more difficult. Residues implicated in MHCII binding by mutational studies are situated at the two ends of the molecule and do not coalesce into a single site although they occur on the same side of the molecule. The available mutational data indicated that the MHCII binding site is on the α5 face of the SEB molecule (see Fig. 12.5). This groove is well formed and runs along the length of the helix. Residues 13-17 and 44-52 are on either side of this groove. The residues lining this groove are His12, Phe17, Met24, Tyr28, Tyr89, Gln92, Tyr115, Asp161, Arg165, Glu191, Phe196, Trp197, Tyr198, Asp199, Met201, Phe208, Lys212, Tyr213, Met215, Tyr217, Asp219, Lys221, Val223 and Val228. This groove and the TCR binding site are adjacent to one another making it possible for TCR and MHCII to be in close proximity. Further, residue Phe208 falls both in the TCR binding site and in this groove attributing a dual functionality for this residue. In TSST1, a portion of the region between 155-194 in particular has been implicated in MHCII binding.[33] On the basis of this, Acharya et al[23] had predicted residues between 170 and 180 which are closely packed to the N-terminal β cylinder to be the MHCII binding site. Residues 170-180 are located in the 3_{10} helix and in the loop connecting this helix and strand β11. Even though the mode of binding of SEB and TSST1 to MHCII may be different, the general binding site seems to be the same.

In the crystal structure of an MHCII-SEB complex, SEB residues Phe44, Leu45 and Phe47 make hydrophobic contacts with the MHCII molecule. Mutation of these residues also was found to disrupt MHCII binding. In addition, residues 63-68 and 92-96 are also lining the interface of SEB and MHCII. Three residues in the region 210-217 also make contacts with MHCII. These three residues are in helix α5. These observations also conform to the suggestion that the MHCII binding is on the α5 face of the SEB molecule. However, residue Phe17 of

SEB implicated in MHCII binding[30] makes no contact with MHCII. Mutation of this residue which is buried in the molecule might disrupt the conformation of the SEB molecule and thus indirectly affect MHCII binding.

MHCII-SEA binding is intriguing. Residues Phe44, Leu45, Tyr89 and Tyr115 (numbering corresponds to SEB) are conserved in SEB and SEA. The residue corresponding to Glu67 of SEB is Asp in SEA. All these residues line the MHCII-SEB interface.[26] Substitution of SEA Leu48 (corresponding to Leu45 of SEB) reduces MHCII binding but does not abrogate it completely. The similarity of the primary sequence and the secondary structures suggest that the MHCII-SEA binding will be similar to MHCII-SEB binding. However, SEB and TSST1 do not inhibit SEA binding to MHCII completely. But, SEA competes with SEB and TSST1 suggesting some overlap in binding sites. Moreover, it has been shown that His81 of MHCII β chain is important for SEA binding. Fraser et al[17] have shown that zinc mediation is required for SEA, SED or SEE to bind to MHCII. It was also suggested that this zinc might act as a bridge between His81 of MHCII β chain and an SEA molecule.[34] It is also argued that SEA might have two different binding sites for MHCII and one SEA molecule may crosslink two MHCII molecules.[26] A crystal structure determination of an MHCII-SEA complex should answer these questions.

Hoffmann et al[25] have suggested a different binding site for MHCII on the SEC3 molecule. This site is on the other face of the molecule (α4 face). The present experimental and structural evidence does not support this mode of binding. Jett et al[35] have identified a sequence of residues as important for inducing T lymphocyte proliferation in SEB. This sequence, viz. KKKVTAQEL (152-160), is in β8 and α4 of SEB, with five of these residues pointing into the α5 face.

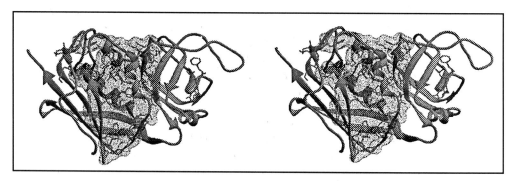

Fig. 12.5. A stereo view of the α5 face of the SEB molecule. The α5 groove is shown as a dot surface and the residues forming this groove as a stick model. Residues 13-17, 44, 45 and 46 implicated in MHCII binding are also shown but not as dot surface.

POSSIBLE BIOLOGICAL ROLE OF THE α4 FACE OF SEB

The groove involving α4 lies on the side opposite the α5 face (Fig. 12.6) and is lined with residues Asp30, His32, Ser34, Ile36, Asp83, Phe85, Thr119, His121, Gln125, Tyr129, Phe146, Asp147, Val148, Gln149, Thr150, Asn151, Lys152, Glu159, Tyr162, Leu163, His166 and Tyr167. It contains residues 113-166 of SEB, which is homologous to the human and mouse invariant chain. The invariant chain associates with nascent MHCII molecules inside the cell and disassociates from them as the complexes reach the surface of the cell.[6] However, the role of this sequence in SEB is not clear. Carboxymethylation of histidines prevents all activities of SEB.[36] There are five histidines in SEB viz., His12, His32, His115, His121 and His166, of which three are in the groove formed by α4 which suggests that this groove may have important biological functions (Fig. 12.7).

EMETIC SITE

This site is thought to reside in the region 113-126 in SEB.[37] Half of this region is in the loop connecting domains 1 and 2; moreover, of the 14 residues, 11 are conserved in all S. enterotoxins. His 121 which is conserved in all S. enterotoxin has been suggested as the most important residue for emetic activity. In the crystal structure His121 lies on the loop connecting domains 1 and 2 and is exposed to solvent.

Vβ SPECIFICITY

Staphylococcal enterotoxins may be classified into two groups based on their sequence homology. SEA, SED and SEE form one group, while SEB and SECs form the other. There is high sequence homology between members of the same group. However, *Staph.* enterotoxins within the same group may bind to different Vβ chains of T cells although there may be overlap of Vβ specificity.[7] For example, SEA and SEE share 90% sequence homology (82% identity), but have different Vβ specificities. SEA stimulates human Vβ5.3 (hVβ5.3) or mouse Vβ3

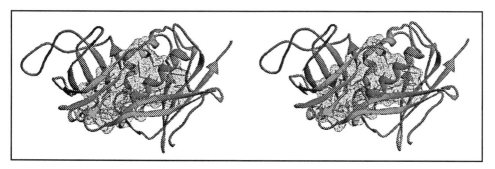

Fig. 12.6. A stereo view of the dot surface of the α4 groove which is on the opposite side of α5 face.

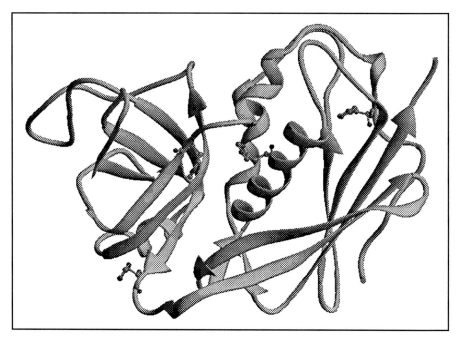

Fig. 12.7. Location of histidines 12, 32, 121 and 166 in SEB are shown. The histidine in the disulfide loop is not shown. RIBBONS[45] was used to make Figures 12.4, 12.5, 12.6 and 12.7.

(mVβ3) but not hVβ8. On the other hand SEE stimulates hVβ5.1 and not hVβ8; SED stimulates hVβ5 and hVβ12 but not hVβ8. Similarly, both SEB and SEC2 stimulate hVβ12, 14, 15, 17 and 20; but while SEB activates hVβ3 and not hVβ13.2, SEC2 does the reverse. Mutational studies have been carried out on SEs to pinpoint the Vβ specificity defining residues. We have defined Vβ specificity determining residues by comparing the stereochemistry of the TCR binding sites of SEB, SEC2, SEA and SEE.

A stereo view of the residues comprising the TCR binding site in SEC2 is shown in Figure 12.8. Residues forming the various TCR binding sites are compared and given in Table 12.2. The list here includes some adjacent residues which lie on the three stacked loops, characterized as an important structural unit of SEB with respect to the interaction with the TCR. Eight residues differ between SEB and SEC2 in this site. Three of these residues, 31, 32 and 87 are located either deep inside the cavity or have their side chains pointing to the interior of the molecule and apparently do not contribute to Vβ specificity. Of the five remaining residues, Oγ of Thr20 in SEC2 makes a hydrogen bonding contact of 3.24 Å with Nδ2 of Asn23; in SEB residue 20 is Leu and this contact is missing (Table 12.2). In SEB, Oδ1 and Oδ2 of Asp209 make hydrogen bonds (2.85 and 2.50 Å respec-

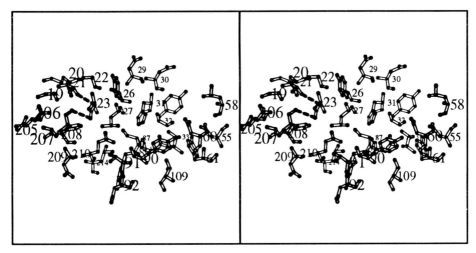

Fig. 12.8. A stereo view of the residues forming the TCR binding site in SEC2.

tively) with Nε2 of Gln92 whereas in SEC2, Gln92 is replaced by Asn92 and these hydrogen bond interactions are absent; this may be because the side chain of Asn is one carbon atom shorter than Gln. The specificity could also be a cooperative property of residues 20 and Asn 23; while in SEC2 Nδ2 and Oδ1 of Asn23 are engaged in hydrogen bonding interactions 3.43 Å and 3.26 Å respectively with Oγ1 of Thr20 and Nζ of Lys207, in SEB only Oδ1 of Asn23 is involved in hydrogen bonding (3.29 Å) with Lys207 leaving Nδ2 open. In addition, in SEB residues 22 and 26 are Glu and Val while they are Gly and Tyr in SEC2. SEB and SEC2 have common Vβ specificity for hVβ12, 14, 15, 17 and 20. But while SEB stimulates hVβ3, SEC2 stimulates hVβ13.2. Since residue 23 has been identified as a critical residue for TCR binding,[30] it is reasonable to conclude that the specificity defining residues in SEB and SEC2 are at positions 20, 22, and 26 in cooperation with residue 92. Residues 20, 22 and 26 have been shown to define the subtype specificity of SEC1 and SEC2 from monoclonal antibody binding studies.[38]

A similar comparison is made for SEA, SED and SEE. In SEA and SED the residue at position 21 (corresponding to 20 in SEB or SECs) is Thr while it is Asn in SEE. The side chain atoms of residues at this position are close to the side chain atoms of Asn25 (equivalent

Note: Recent high resolution structure determination of SEC2 reveals a zinc atom bound to the molecule. It is coordinated by residues Asp83, His118 and His122. The fourth coordination is from Asp9 of a symmetry related molecule. This zinc site is different from the zinc site modeled in SEA. It is not yet clear whether this zinc contacts MHCII molecule or not.

Table 12.2. Comparison of residues forming the TCR binding site

SEB	G19	**L20**	M21	**E22**	N23	**V26**	L27	D29	D30	**N31**
SEC1	G19	**L20**	M21	**E22**	N23	**V26**	L27	D29	D30	**H31**
SEC2	G19	**T20**	M21	**G22**	N23	**Y26**	L27	D29	D30	**H31**
SEC3	G19	**T20**	M21	**G22**	N23	**Y26**	L27	D29	D30	**H31**
SEA	*G20*	*T21*	L23	*G24*	N25	Q28	I29	Y31	Y32	N33
SEE	*R17*	*N18*	L20	*S21*	N22	Q25	I26	Y28	Y29	N30
SEB	**H32**	V33	D55	L58	N60	Y61	**A87**	N88	Y89	Y90
SEC1	**Y32**	V33	D55	L58	N60	Y61	**S87**	N88	Y89	Y90
SEC2	**Y32**	V33	D55	L58	N60	Y61	**S87**	N88	Y89	Y90
SEC3	**Y32**	V33	D55	L58	N60	Y61	**S87**	N88	Y89	Y90
SEA	E34	K35	F58	H61	W63	Y64	A90	Y91	Y92	G93
SEE	E31	K32	F55	H58	W60	Y61	A87	Y88	Y89	G90
SEB	**Y91**	**Q92**	T112	G205	D206	K207	F208	D209	Q210	L214
SEC1	**V91**	**N92**	T109	G205	D206	K207	F208	D209	Q210	L214
SEC2	**V91**	**N92**	T109	G205	D206	K207	F208	D209	Q210	L214
SEC3	**V91**	**N92**	T109	G205	D206	K207	F208	D209	Q210	L214
SEA	Y94	Q95	T104	G203	Q204	Y205	*S206*	*N207*	T208	L210
SEE	Y91	Q92	T101	G200	Q201	Y202	*P203*	*D204*	T205	L207

Residues forming the TCR binding site along with some adjacent residues which lie on the three stacked loops are given based on the analysis of SEB.[22] A common numbering scheme has not been chosen in this table; each SE is numbered sequentially. Sequences for SEB, SEC1, SEC2 and SEC3 are from references 41, 42, 43 and 44 and for SEA and SEE from 6. Differences in residue type in each group are in boldface.

to Asn23 in SEB or SECs) in the three dimensional models. While Asn 22 has two possible interactions with Asn 18 in SEE, Asn25 has only one possible interaction with Thr 21 in SEA. SEA stimulates mVβ3 while SEE stimulates mVβ11.[31] In SEA, Nε2 and Oε1 of Asn207 are available to make hydrogen bonds with Oδ1 and Nδ2 of Gln95 while in SEE Asp204 will have both hydrogen bonding interactions with Nε2 of Gln92 leaving Oε1 open similar to SEB. Apparently this difference will also contribute to the Vβ specificity. As discussed earlier, it has been shown[31,32] that when residues 206 and 207 in SEA were mutated to correspond to SEE (S → P, N → D, respectively) the specificity of SEA was converted to that of SEE but the level of Vβ binding was smaller compared to the wild type. It was also seen that when both the N and C terminal residues are from the same serotype the binding was restored to that of the wild type.[32] This along with our results show that the residue at position 21 in SEA (18 in SEE) is the specificity defining residue along with residues Ser206 and Asn207 in SEA (or Pro203 and Asp204 of SEE). However, the role of Ser206

(SEA) or Pro203 (SEE) is not clear from the three dimensional structures of enterotoxins alone. Recently, it was shown that residues 20-24 and 200-207 of SEA should be changed to correspond to SEE to have SEE wild type levels of activity (Lamphear J and Rich RR, personal communication).

OLIGOMER BINDING FOLD AND THE BINDING OF GLYCOSPHINGOLIPIDS TO SEB

The folding pattern of domain 1 (Fig. 12.9) has been observed in other toxins and has been identified as the OB fold.[28] It was suggested that a carbohydrate moiety or a nucleotide might bind to a specific site in this domain which is located between the N terminal portion of $\alpha 3$ and the beginning of the loop connecting $\beta 4$ and $\beta 5$. But it was shown subsequently that while this fold is conserved in most of the toxins the binding site of the carbohydrate moiety is not conserved.[39] Chatterjee and Jett[40] have investigated the binding of SEB to human proximal tubular (PT) cells. Their studies indicated that SEB binds to a neutral glycosphingolipid in kidney PT cells. Since lactose forms the head group of lactosylceramide, the crystal structure of a SEB- lactose complex was investigated to determine the binding site of lactose in SEB. It was found that lactose makes one hydrogen bond with Asp127 and another with the same residue via a water molecule. This residue is located in the beginning of domain 2 and is close to the loop connecting domains 1 and 2. Chatterjee and Jett also showed that SEB was unable to displace the binding of SEA or TSST1 to PT cells. Similarly, neither SEA nor TSST1 competed with SEB suggesting that SEB had a different binding mode than SEA or TSST1 to PT cells. Furthermore pre-incubation of PT cells with endoglycoceramidase completely inhibited the binding of labeled SEB. Since this enzyme is known to cleave the glycosyl moiety from glycosphingolipids, it is inferred that SEB binds to glycosphingolipids. Our studies (unpublished) on the crystal structure of SEB complexed with 3'-sialyl-lactose show that this sugar moiety binds to a different region of SEB than lactose. It appears from these studies that the binding site in SEB for glycosphingolipids depends on the nature of the lipid.

CONCLUSION

All *Staph* enterotoxins have a common fold and the binding sites are generally the same in all of them except for a zinc binding site which is present in some but not in others. The mode of binding may differ for these toxins notwithstanding their common structure-function relationships. The TCR Vβ chain may make several contacts with SEs but the Vβ specificity is defined by one or two residues and could be a cooperative property of variable and conserved residues. The TCR combining sites of different SEs are distinguishable by differences in

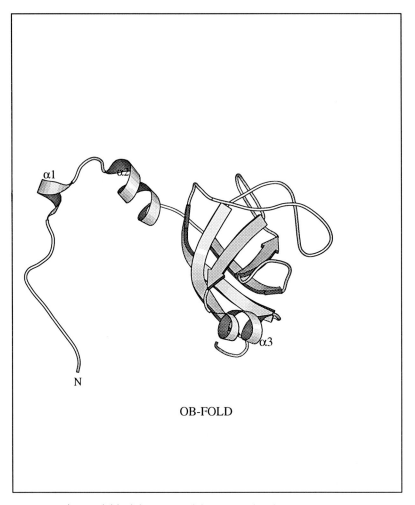

Fig. 12.9. The OB-fold of domain 1 of the SEB molecule. MOLSCRIPT[46] was used
to make Figures 12.1a, 12.2, 12.3, 12.8 and 12.9.

the hydrogen bonding motifs of the specificity determining residues.
These variations in stereochemical interactions can be used to identify
the specificity determining sites. SEs bind to the gycosyl moiety of
glycosphingolipids suggesting that at least in PT cells the receptors are
glycosphingolipids for SEB.

ACKNOWLEDGMENTS

This work was supported by the Department of Veterans Affairs
and by the US Army Medical Research and Development Command.

REFERENCES

1. Bergdoll MS. Staphylococcal intoxications. In: Reimann H, Bryan FL, eds. Food borne infections and intoxications. New York: Academic Press, 1979:443-94.
2. Spero L, Johnson-Winegar A, Schmidt JJ. Enterotoxins of staphylococci. In: Hardgree MC, Tu AT, eds. Handbook of Natural Toxins. New York: Dekker, 1988:131-63.
3. Peavy DL, Adler WH, Smith RT. The mitogenic effects of endotoxin and staphylococcal enterotoxin B on mouse spleen cells and human peripheral lymphocytes. J Immunol 1970; 105:1453-58.
4. Bergdoll MS. The staphylococcal enterotoxins—an update. In: Jeljazowicz, ed. The Staphylococci. New York: Guster Fischer Verlag, 1988:247-66.
5. Lee AC-M, Robbins RN, Reiser RF et al. Isolation of specific and common antibodies to staphylococcal enterotoxins B, C1 and C2. Infect Immun 1980; 27:431-34.
6. Marrack P, Kappler J. The staphylococcal enterotoxins and their relatives. Science 1990; 248:705-11.
7. Kotzin BL, Leung DY, Kappler J et al. Superantigens and their potential role in human disease. Adv Immunol 1993; 54:99-166.
8. White J, Herman A, Pullen AM et al. The Vβ-specific superantigen staphylococcal enterotoxin B: stimulation of mature T cells and clonal deletion in neonatal mice. Cell 1989; 56:27-35.
9. Fleisher B, Schrezenmeier H. T cell stimulation by staphylococcal enterotoxins. Clonally variable response and requirement for major histocompatibility complex class II molecules on accessory or target cells. J Exp Med 1988; 167:1697-707.
10. Carlsson R, Fisher H, Sjögren HO. Binding of staphylococcal enterotoxin A to accessory cells is a requirement for its ability to activate human T cells. J Immun 1988; 140:2484-88.
11. Rudensky AY, Preston-Hurlburt P, Hong S-C et al. Sequence analysis of peptides bound to MHC class II molecules. Nature 1991; 353:622-27.
12. Brown JH, Jardetzky T, Saper MA et al. A hypothetical model of the foreign antigen binding site of class II histocompatibility molecules. Nature 1988; 332:845-50.
13. Bjorkman PJ, Saper MA, Samraoui B et al. The foreign antigen binding site and T cell recognition regions of class I histocompatibility antigens. Nature 1987; 329:521-28.
14. Janeway Jr CA, Yagi J, Conrad PJ et al. In: Göran Müller ed. Immunological Reviews. Copenhagen: Munsgaard 1989; 107:61-88.
15. Johnson HM, Russell JF, Pontzer CH. Superantigens in Human Disease. In Scientific American 1992:92-101.
16. Fraser JD. Structural model of staphylococcal enterotoxin A interaction with MHC class II antigens. In: Huber BT, Palmer E, eds. Superantigens: a pathogen's view of the immune system. Current Communications. Cold Spring Harbor Press 1993; 7:7-29.

17. Fraser JD, Urban RG, Strominger JL et al. Zinc regulates the function of two superantigens. Proc Natl Acad Sci USA 1992; 89:5507-11.

18. Fraser JD. High affinity binding of staphylococcal enterotoxins A and B to HLA-DR. Nature 1989; 339:221-23

19. Pontzer CH, Russell JK, Johnson HM. Localization of an immune functional site on staphylococcal enterotoxin A using the synthetic peptide approach. J Immunol 1989; 143:280-84.

20. Purdie K, Hudson KR, Fraser JD. Bacterial superantigens. In: MacCluskey J, ed. Antigen processing and presentation. Boca Raton: CRC Press, 1991:193-214.

21. Scholl PR, Diez A, Geha RS. Staphylococcal enterotoxin B and toxic shock syndrome toxin-1 bind to distinct sites on HLA-DR and HLA-DQ molecules. J Immunol 1989; 143:2583-88.

22. Swaminathan S, Furey W, Pletcher J et al. Crystal structure of staphylococcal enterotoxin B, a superantigen. Nature 1992; 359:801-06.

23. Acharya KR, Passalacqua EF, Jones EY et al. Structural basis of superantigen action inferred from crystal structure of toxic-shock syndrome toxin-1. Nature 1994; 367:94-97.

24. Prasad GS, Earhart CA, Murray DL et al. Structure of toxic shock syndrome toxin 1. Biochemistry 1993; 32:13761-66.

25. Hoffmann ML, Jablonski LM, Crum KK et al. Predictions of T cell receptor and major histocompatibility complex binding sites on staphylococcal enterotoxin C1. Infect Immun 1994; 62:3396-407.

26. Jardetzky TS, Brown JH, Gorga JC et al. Three-dimensional structure of a human class II histocompatibilty molecule complexed with superantigen. Nature 1994; 368:711-18.

27. Kim JS, Urban RG, Strominger JL et al. Toxic shock syndrome toxin-1 complexed with a class II major histocompatibility molecule HLA-DR1. Science 1994; 266:1870-74.

28. Murzin AG. OB (oligonucleotide/oligosaccharide binding)-fold: common structural and functional solution for nonhomologous sequences. EMBO J 1993; 12:861-67.

29. Spero L, Warren, JR, Metzger JF. Effect of a single peptide bond scission by trypsin on the structure and activity of staphylococcal enterotoxin B. J Biol Chem 1973; 248:7289-94.

30. Kappler JW, Herman A, Clements J et al. Mutations defining functional regions of the superantigen staphylococcal enterotoxin B. J Exp Med 1992; 175:387-96.

31. Irwin MJ, Hudson KR, Fraser JD et al. Enterotoxin residues determining T cell receptor Vβ binding specificity. Nature 1992; 359:841-43.

32. Mollick JA, McMasters RL, Grossman D et al. Localization of a site on bacterial superantigens that determine T cell receptor β chain specificity. J Exp Med 1993; 177:283-93.

33. Soos JM, Russel JK, Jarpe MA et al. Identification of binding domains on the superantigen, toxic shock syndrome-1, for class II MHC molecules. Biochem Biophys Res Commun 1993; 191:1211-17.

34. Karp DR, Long EO. Identification of HLA-DR1 β chain residues critical for binding staphylococcal enterotoxins A and E. J Exp Med 1992; 175:415-24.

35. Jett M, Neill R, Welch C et al. Identification of staphylococcal enterotoxin B sequences important for induction of lymphocyte proliferation by using synthetic peptide fragments of the toxin. Infect Immun 1994; 62:3408-15.

36. Scheuber PH, Golecki JR, Kickhofen B et al. Skin reactivity of unsensitized monkeys upon challenge with staphylococcal enterotoxin B: A new approach for investigating the site of toxic action. Infect Immun 1985; 50:869-73.

37. Spero L, Marlock BA. Biological activity of the peptides of staphylococcal enterotoxin C formed by limited tryptic hydrolysis. J Biol Chem 1978; 253:8787-91.

38. Turner TN, Smith CL, Bohach GA. Residues 20, 22 & 26 determine the subtype specificities of staphylococcal enterotoxin C1 & C2. Infect Immun 1992; 60:694-97.

39. Stein PE, Boodhoo A, Armstrong GD et al. Structure of a Pertussis toxin-sugar complex as a model for receptor binding. Nature Struct Biol. 1994; 1:591-96.

40. Chateerjee S, Jett M. Glycosphingolipids: The putative receptor for staphylococcus aureus enterotoxin-B in human kidney proximal tubular cells. Mol Cellular Biochem 1992; 113:25-31.

41. Jones CL, Khan SA. Nucleotide sequence of the enterotoxin B gene from staphylococcus aureus. J Bacteriol 1986; 166:29-33.

42. Bohach GA, Schlievert PM. Nucleotide sequence of the staphylococcal enterotoxin C1 gene and relatedness to other pyrogenic toxins. Mol Gen Genet 1987; 209:15-20.

43. Bohach GA, Schlievert PM. Conservation of the biologically active portions of staphylococcal enterotoxins C1 and C2. Infect Immun 1989; 57:2249-52.

44. Hovde CJ, Hackett SP, Bohach GA. Nucleotide sequence of the staphylococcal enterotoxin C3 gene: Sequence comparison of all three type C staphylococcal enterotoxins. Mol Gen Genet 1990; 220:329-33.

45 Carson, M. RIBBONS 2.0. J Appl Cryst 1991; 24:958-61.

46. Kraulis, P.J. MOLSCRIPT: a program to produce both detailed and schematic plots of protein structures. J Appl Cryst 1991; 24:946-50.

47. Swaminathan S, Furey W, Pletcher J, Sax M. Residues defining Vβ specificity in staphylococcal enterotoxins. Nature Structural Biology 1995; 2:680-686.

CHAPTER 13

THE STRUCTURE OF RIBOSOME INACTIVATING PROTEINS FROM PLANTS

Jon D. Robertus and Arthur F. Monzingo

Higher plants lack an immune system and have evolved a range of defenses against fungal, bacterial and viral pathogens. These include: synthesis of antimicrobial phytoalexins;[1] proteinase inhibitors;[2] membrane disrupting polypeptides;[3] and lytic enzymes such as β-1, 3 glucanases[4] and chitinases.[5]

Frequently plants also contain enzymes which inhibit eukaryotic ribosomes giving rise to the name ribosome inactivating proteins, or RIPs. Numerous reviews of this class of proteins have been written.[6,7] The RIPs are divided into two classes. Type 1 RIPs are single chain proteins of Mr 30,000. Examples are pokeweed antiviral protein (PAP), trichosanthin, gelonin, and others.[8] The notion that these RIPs serve a defensive function in plants is supported by the observations that expression of PAP in transgenic tobacco plants increases resistance to certain viruses,[9] while expression of barley RIP is correlated with resistance to fungal infection.[10]

Type 2 RIPs are heterodimers consisting of an active, or A chain linked by a disulfide bond to a binding, or B chain. Each chain has a Mr 32,000 and together they form a potent cytotoxin. The B chain binds to cell surface glycoproteins and is taken into the host cell by endocytosis. The toxin may escape from the endosome or may traffic to other membrane bound compartments in the cell. Eventually, the disulfide bond between A and B chains is reduced, and the A chain is released into the cytoplasm. The A chains of the heterodimeric type 2 RIPs are homologous with the type 1 RIPs.[11]

Biochemists and medical researchers have made extensive use of both forms of RIPs in the search for therapeutic agents. The toxins

Protein Toxin Structure, edited by Michael W. Parker. © 1996 R.G. Landes Company.

have been conjugated to targeting molecules to create "magic bullets" which should be selective for the target cell line. The most common targeting proteins are antibodies used to create a class of agents called immunotoxins. After endocytosis, the toxin attacks the target cell ribosomes and kills it. This aspect of toxin use has been extensively reviewed in Frankel;[12] articles therein describe a number of immunotoxin systems aimed at cancer and other target cells.

BIOCHEMISTRY BACKGROUND OF RIPS

The archetypal example of the RIP family is the type 2 protein ricin. Ricin is isolated as a heterodimer, consisting of a 32,000 dalton A chain (RTA) linked by a disulfide bond to a 32,000 dalton B chain (RTB). The amino acid sequence of ricin has been determined chemically,[13,14] as well as from the nucleotide sequences of both the cDNA[15] and the genomic DNA.[16] The latter studies revealed that ricin is synthesized as a proenzyme with a leader peptide upstream of the A chain and a 12 residue linking peptide which joins the A and B chains. The leader and linking peptides are processed away to form the native ricin heterodimer. Both chains of native ricin are glycoproteins, possessing high mannose carbohydrate groups.[17]

It appears that A and B chains have a strong affinity for one another, mediated by hydrophobic forces, and that association of the chains is necessary for toxicity. The disulfide link between the chains does not appear to be critical for toxicity, except in maintaining protein-protein interactions at very low toxin concentration.[18]

The B chain lectin has a binding preference for galactosides,[19] although it binds much more strongly to complex galactosides from cell surface carbohydrates than to simple sugars.[20] Once the B chain attaches to cell surface receptors it appears to mediate endocytotic uptake of the protein.[21] It was also shown that, in addition to the galactoside binding uptake mechanism, ricin can be taken up by and intoxicate those cells which possess mannose receptors. In this case, the mannose residues of ricin are themselves recognized by a cellular receptor.[22] The B chain is known to bind two galactosides in a noncooperative manner.[23,24] In one study the dissociation constants of the two sites were found to be 0.03 mM and 0.4 mM, while another study using fluorescent galactose analogs showed more nearly equal affinities for the two sites. Chemical modification studies have implicated Tyr 248 in the "strong" galactose binding site,[25] and tryptophan in the "weak" site.[26]

After endocytosis, RTA reaches the cytoplasm of the target cell by unknown mechanisms. In the cytoplasm of a typical eukaryotic cell, it enzymatically attacks the 60S ribosomal subunit and disrupts protein synthesis. Kinetic parameters for the ricin reaction vary with the source of ribosomes and with the presence of hydrolyzable ATP and soluble factors from the post ribosomal supernatant. Ricin has been shown to have a $K_m = 0.1$ mM for rabbit reticulocyte ribosomes and a kcat = 1500/min.[27] Similar parameters are measured for *Artemia salina* ribosomes. Km = 0.4 µM and kcat = 1000/min in the presence of hydrolyzable ATP and an enzyme-containing supernatant fraction. In the absence of either ATP or the postribosomal fraction, the Km rises to 1.3 µM while kcat drops to about 350/min.[28] Presumably ribosomes have altered conformations, depending on their environment, which have slightly differing susceptibility to RIP attack.

Ricin A chain has been shown to be an N-glycosidase, removing a specific adenine base from a very conservative region of the 28 S rRNA.[29,30] Ricin is generally much more active against animal than plant ribosomes. Intact bacterial ribosomes are not susceptible to ricin intoxication, although their rRNA possesses the conserved base sequence.[8]

Perhaps the most thoroughly studied type 1 RIP is pokeweed antiviral protein (PAP). Its enzymology is very similar to that of RTA. Like RTA, it functions by enzymatically attacking the 60S subunit of eucaryotic ribosomes. The kcat is 400 moles/mole/min, and the Km for ribosomes is 0.2 mM.[31] Electron microscopy has been used to show that PAP is sequestered in the cell wall matrix.[32] It was hypothesized that this is consistent with an antiviral role for the protein. PAP is sequestered away from host cell ribosomes until the wall is breached, allowing virus entry, and also allowing PAP entry to the cytoplasm. It can then inhibit host cell ribosomes, retarding viral replication. An analysis of the amino terminal sequence of PAP shows that it arose from a common ancestor with ricin A chain.[11]

It was observed that the type 1 RIP mirabilis anti-viral protein (MAP) is active against bacterial ribosomes.[33] MAP had an IC50 of 2 nM against reticulocyte ribosomes, roughly 30-fold higher than for RTA. However, MAP had an IC50 of 1 µM against *E. coli* ribosomes, whereas RTA shows no detectable action against them. It has subsequently been shown that several other type 1 RIPs, including PAP, are active against both eucaryotic and bacterial ribosomes in contrast to the type 2 RIPs ricin and abrin.[34] Presumably, structural differences between the type 1 RIPs and the A chain of the type 2 RIPs must allow them to attack bacterial ribosomes.

X-RAY STRUCTURE OF RIPS

The first RIP for which a high resolution X-ray structure was solved was the 2.8 Å model of ricin.[35] That model was refined crystallographically at 2.5 Å resolution.[36] The refinement allowed a detailed description to be made of the A chain,[37] and of the B chain.[38] RTA, expressed from a clone in *E. coli*, was examined crystallographically at 2.3 Å and compared with the A chain as seen in the heterodimeric ricin structure.[39] More recently, recombinant RTA was crystallized in a novel space group and solved to 1.8 Å resolution.[40] Because the RTA amino acid sequence is homologous to the type 1 RIPs, it was predicted that the RTA structure would define a protein fold that would represent the other members of the family;[37] this was shown to be correct as other related structures were elucidated.

Trichosanthin is a type 1 RIP isolated from *Trichosanthes kirilowii*; it has been used in Chinese folk medicine as an abortifacient for centuries. A 4 Å electron density map allowed the backbone of the protein to be traced[41] but problems with the amino sequence precluded an accurate atomic model.[42] The molecular structure of trichosanthin was subsequently solved by several laboratories, in several different space groups. These include models at 2.7 Å resolution,[43] 1.88 Å resolution,[44] and 1.73 Å resolution.[45]

PAP, another type 1 RIP, was solved at 2.5 Å resolution and a detailed comparison made with RTA.[46] α-Momorcharin, an RIP closely related in amino acid sequence to trichosanthin, was solved to 2.0 Å.[47] Momordin, isolated from *Momordica charantia* seeds, was solved to 2.1 Å resolution by molecular replacement methods using RTA as a model.[48]

Figure 13.1 shows an amino acid sequence alignment of the five RIPs for which high resolution X-ray structures are presently available. It confirms that the proteins are homologs and shows that a description of the ricin A chain fold will pertain, in general, to all the other members of this class. Ricin is the only type 2 RIP for which a published structure is available, and so it provides the only view of the B chain.

Figure 13.2 is a backbone drawing of the ricin heterodimer. In native ricin, RTA and RTB are connected by a disulfide bridge between Cys 259 of RTA and Cys 4 of RTB. The thermodynamic parameters of the association between ricin chains have been evaluated using analytical ultracentrifuge data.[18] The dissociation constant was found to measure 1.72×10^{-6} M and the values for entropy and enthalpy were determined to be positive suggesting that hydrophobic interactions are responsible for chain association. The interface between RTA and RTB contains aromatic rings and aliphatic side chains which interact across the intermolecular boundary. For example, the RTB side chain of Tyr 262 interacts with RTA residues Phe 140 and Phe 240. In addition a few polar contacts are made between the chains. At least 16 water molecules are trapped between the two ricin chains.[38]

RICIN A CHAIN

Figure 13.3 shows a drawing of the RTA backbone. Major elements of secondary structure are labeled, helices with capital letters and β-strands with small letters. Table 13.1 shows the residues included in those secondary structural elements.[37,39,40] This basic fold is seen in all the other RIP enzymes. As an example, Figure 13.4 shows a least squares superposition of RTA with trichosanthin. The r.m.s. deviation of 243 Cα atoms between the two proteins is 2.26 Å. Figure 13.1 shows that there are 23 nonpolar, 12 polar, and 4 glycines

```
                                                         20
RTA        IFPKQYPI-----------------INFTTAGATVQSYTNFIRAVRGRL
PAP        V----------------------NTIIYNVGSTTISKYATFLNDLRNEA
TRICH      -----------------------DVSFRLSGATSSSYGVFISNLRKAL
MOMRDN     -----------------------DVSFRLSGADPRSYGMFIKDLRNAL
α-MMC      M--SRFSVLSFLILAIFLGGSIVKGDVSFRLSGADPRSYGMFIKDLRNAL
                                      .  . ...    * *.  .*

           40              60              80
RTA        TTGADVRHEIPVLPNRVGLPINQRFILVELSNHAELSVTLALDVTNAYVV
PAP        KDPSLKCYGIPSLPNTNTNP---KYVLVELQGSNKKTITLMLRRNNLYVM
TRICH      PNERKL-YDIPLL--RSSLPGSQRYALIHLTNYADETISVAIDVTNVYIM
MOMRDN     PFREKV-YNIPLL--LPSVSGAGRYLLMHLFNYDGKTITVAVDVTNVYIM
α-MMC      PFREKV-YNIPLL--LPSVSGAGRYLLMHLFNYDGKTITVAVDVTNVYIM
                    ** *       .    .. *..*      ....   * *..

                           100             120
RTA        GYRAG-----NSAYFFHPDNQEDAEAITHLF-TDVQNRYT--FAFGGNYD
PAP        GYSDPFETNKCRTHIFNDISGTERQDVETTLCPNANSRVSKNINFDSRYP
TRICH      GYRAG-----DTSYFFNEASATEAAK--YVF-KDAMRKVT--LPYSGNYE
MOMRDN     GYLAD-----TTSYFFNQPAAELASQ--YVF-RDAR-KIT--LPYSGNYE
α-MMC      GYLAD-----TTSYFFNEPAAELASQ--YVF-RDARRKIT--LPYSGNYE
           **        . .*.            .   .  . .. .         *

                           140             160
RTA        RLEQLAG-NLRENIELGNGPLEEAI---SALYYYSTGGTQLPTLARSFII
PAP        TLESKAGVKSRSQVQLGIQILDSNIGKISGVMSFTEKTE-----AEFLLV
TRICH      RLQTAAG-KIRENIPLGLPALDSAI---TTLFYYNANSA-----ASALMV
MOMRDN     RLQIAAG-KPREKLPIGLPAIDSAI---STLLHYDSTAA-----AGALLV
α-MMC      RLQIAAG-KPREKIPIGLPALDSAI---STLLHYDSTAA-----AGALLV
           *.  **  .  *  .     ..*       *             *   ...

                   180             200
RTA        CIQMISEAARFQYIEGEMRTRIRYNRRSAPDPSVITLENS-WGRLSTAIQ
PAP        AIQMVSEAARFKYIENQV--KTNFNRAFNPNPKVLNLQET-WGKISTAIH
TRICH      LIQSTSEAARYKFIEQQIGKRV--DKTFLPSLAIISLENS-WSALSKQIQ
MOMRDN     LIQTTAEAARFKYIEQQIERA--YRDEVPSIATLSLENSLWSGLSKQIQ
α-MMC      LIQTTAEAARFKYIEQQIERA--YRDEVPSLATISLENS-WSGLSKQIQ
           **  .****...**  ..     .     *    . *... *. .* *.

           220             240
RTA        --ESNQGAFASPIQLQRRNGSKFSVYDVS--ILIPIIALMVYR-------
PAP        --DAKNGVLPKPLELVDASGAKWIVLRVDE--IKPDVALLNYV-------
TRICH      IASTNNGQFETPVVLINAQNQRVMITNVDAGVVTSNIALLLNRNNMA---
MOMRDN     LAQGNNGIFRTPIVLVDNKGNRVQITNVTSKVVTSNIQLLLVTRNIAEGD
α-MMC      LAQGNNGIFRTPIVLVDNKGNRVQITNVTSKVVTSNIQLLLNTRNIAEGD
             ..* . *. *     .      *    . . .*.

           260
RTA        ----CAPPPSSQF
PAP        -GGSCQTT-----
TRICH      -------------
MOMRDN     -------------
α-MMC      NGDVSTTHGFSSY
```

*Fig. 13.1. Amino acid alignment for RIPs of known X-ray structure. The proteins are ricin A chain, RA; pokeweed antiviral protein, PAP;[63] trichosanthin, trich;[42] momordin, momrn;[64] α-momorcharin, α-MMC.[65] The alignment was carried out using CLUSTAL.[66] At the top of each block are numbers corresponding to RTA residues; at the bottom an * denotes an invariant position and the · denotes conservative changes among the five proteins.*

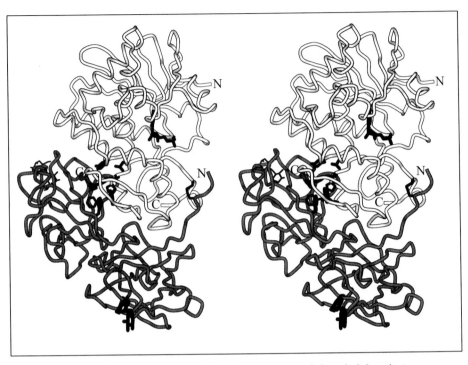

Fig. 13.2. A stereo ribbon drawing of ricin. RTB is at the lower left in dark bonds; it appears as a dumbbell shaped protein. RTA is in the upper right with open bonds. The N- and C-termini of each chain are labeled. Several chemical groups are shown in black bonds. These include the hydrophobic residues which interact between the two chains, the two lactose moieties binding to the extreme ends of the B chain, and the FMP substrate analog bound in the active site of the A chain.

Table 13.1 Secondary structure in RTA

α Helix	Residues	β Strand	Residues
A	18-32	a	7-13
B	99-104	b	56-64
C	122-127	c	68-76
D	141-152	d	79-86
E	161-180	e	88-93
F	184-192	f	113-117
G	202-210	g	230-234
H	211-219	h	237-242

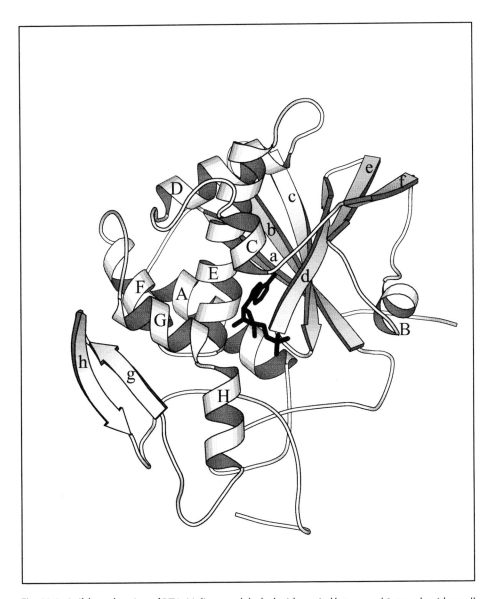

Fig. 13.3. A ribbon drawing of RTA. Helices are labeled with capital letters and β-strands with small letters. FMP is shown in black bonds and marks the active site.

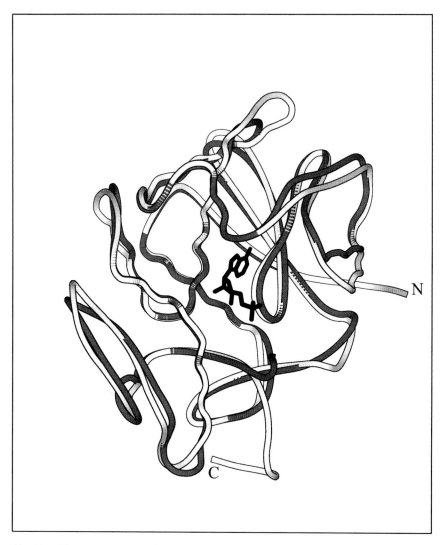

Fig. 13.4. The superposition of RTA with trichosanthin. The Cα trace of RTA is shown as a light coil, while trichosanthin is dark. FMP, in black bond, marks the active site of the two toxins.

which are identical in all five RIPs with X-ray structures. In addition there are many more conservative positions. The polar residues tend to lie in or near the active site cleft and tend to be involved in substrate binding and catalysis. The conserved hydrophobic residues form the core of the RIP fold and dictate its overall folding pattern.

It is known that the RIPs act by depurinating a specific adenine base (A4324 in rat) within a highly conserved stem and loop structure of the 28 S rRNA.[29,30] A nonhydrolyzable analog of AMP, formycin monophosphate (FMP), was diffused into ricin crystals and observed by difference Fourier methods.[49] The RTA FMP interactions at the catalytic center are shown in Figure 13.5. It is likely that RTA makes many more interactions with the intact rRNA or ribosome, evident from the fact that RTA has a Km ∼1 µM for ribosomes, but has a Kd for FMP near 5 mM. Figure 13.5 shows that the substrate purine ring is roughly sandwiched between Tyr 80 and Tyr 123. The binding requires a rotation of Tyr 80 to accommodate the purine. The FMP (adenine) 6 amino group donates a hydrogen bond to the carbonyl of Val 81, while N1 receives a bond from the amido of Val 81. The carbonyl oxygen of Gly 121 receives hydrogen bonds from N7 and the 6-amino group of FMP. These help bind the substrate and preclude binding of guanine. A strong hydrogen bond is seen between Arg 180 and N3 of the substrate analog; Glu 177 is seen to be near the ribose ring. In addition to the nonhydrolyzable analog, the dinucleotide ApG was also bound to ricin. The interactions with adenosine were essentially those described for FMP, with only minor differences. For example, because FMP is protonated at N7 and adenine is not, that position cannot donate a hydrogen bond to the carbonyl of Gly 121.[49]

FMP has also been bound to PAP, and shows very similar interactions to those seen for RTA.[46] The analog of Tyr 80 (Tyr 72) does not move much upon substrate binding. Even in the apoenzyme that side chain is positioned like Tyr 80 in liganded RTA. FMP has also been bound to α-momorcharin and shows the same general pattern as seen previously for RTA and PAP.[47] NADPH has been observed to bind to trichosanthin.[50] The adenosine half of the ligand binds in a similar fashion to that seen in the other RIPs while the nicotinamide nucleotide makes few specific contacts with the enzyme.

Most of the active site residues of RTA have been subjected to site-directed mutagenesis and the effect on enzyme activity observed. These kinetic data, together with the X-ray models of substrate interaction, allow plausible models of the catalytic mechanism to be made and tested. Table 13.2 gives a synopsis of important mutagenic experiments carried out on RTA. It is evident that converting either Tyr 80 or Tyr 123 to Phe has a small effect; converting them to Ser is much more damaging. This probably arises from two effects, (1) the inability to form the nonpolar ring interactions of the wild-type and (2) conformational changes in the protein, apparent from the greater instability of the Ser mutants.

Glu 177 is clearly an important residue. Conversion to Asp inhibits activity nearly 100-fold, and conversion to Gln reduces activity at least 200-fold. Conversion to Ala seems to have only a small effect, but this is misleading. Removal of the Glu 177 side group allows that of Glu 208 to move into place and rescue partial activity.[51,52] Of greatest

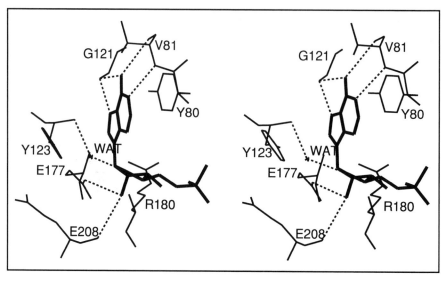

Fig. 13.5. The active site of RTA. The local backbone and key protein side chains are shown in narrow bonds. FMP is in heavier bonds and hydrogen bonds are shown as dashed lines. A water molecule, the likely ultimate nucleophile in the depurination reaction, is shown in a position observed in a PAP FMP complex (pH 7.5) but not seen in RTA FMP (pH 4.7).

Table 13.2 Kinetics of RTA mutants

Protein	Relative Activity	Reference
Wild-type	100.	
Y80F	7.	28
Y80S	0.6	53
Y123F	15.	28
Y123S	1.4	53
E177D	1.3	67
E177Q	0.6	28
E177A	5.	67
R180K	38.	51
R180Q	0.04	53

importance is Arg 180. Conversion to Lys decreases activity about 3-fold, but conversion to the neutral Gln reduces activity 2500-fold, based on kcat/Km.[53] It is important to note that kinetic parameters show alterations of Glu 177[51,54] and Arg 180[53] affect kcat and have little or no effect on Km. This suggests they are involved in catalysis, that is in transition state stabilization, and not in substrate binding. This contrasts with mutations of the active site tyrosines, or with Asn 209 which have mixed effects or primarily affect Km.[28] Site-directed mutations of trichosanthin have recently confirmed the importance of the Glu 177 analog (Glu 160) in that RIP. Conversion to Ala, together with removal of the rescuing residue (Glu 189 is the analog of Glu 208 in RTA), reduces enzyme activity 1800-fold.[55]

The RIPs are N-glycosidases hydrolyzing the C1'-N9 bond of adenine. Knowledge of N-glycosidation reactions is rather sparse, compared with the substantial body of knowledge about the chemistry of O-glycosidases.[56] It is evident, however, that an N-glycosidase must arrange water as the ultimate nucleophile in the reaction and protonate the leaving adenine group.

Schramm and coworkers have used kinetic isotope effects to investigate the mechanism of a nucleosidase which hydrolyzes the adenine base from AMP.[57] Their results are best explained by a transition state in which there is considerable carbocation (oxycarbonium) character in the ribose ring, protonation of the leaving adenine, partial bonding between the ribose and the base, and participation of a water nucleophile attacking at C1' of ribose.

Water must attack the ribose at C1' and an attacking water has been suggested as part of the transition state structure for AMP nucleosidase.[57] A water molecule bonding between Glu 177 and Arg 180 is seen in the ricin A chain and RTA structures and it was suggested that this might move upon substrate binding to become the attacking water.[39] No such water is seen in the FMP complex, which is not a true transition state analog, but modeling can readily accommodate a water between Arg 180 and Glu 177 as the attacking group on the underside of ribose.[49] Attack by a water on this side of the ribose would result in inversion of the product hand. In the PAP FMP complex[46] electron density is seen for a water molecule which is above the plane of the ribose, bonded to the ribose ring oxygen (O4'), to O5', and to N8 of the formycin ring; this water is seen in only one of the two molecules of the PAP asymmetric unit, however. It would attack the ribose from above and result in a retention of hand. There is also space for a water, bonded between the Arg 180 analog (Arg 179) and the Glu 177 analog (Glu 176). At this time we do not know if the RIP N-glycosidase proceeds with retention or inversion of the ribose anomer. Kinetic measurements must be made to further define mechanistic details of the RIP family.

A complex between α-momorcharin and FMP shows two waters in the active site which may represent the attacking water. One is above the ribose, bonded to N8 of formycin, to the amido group of the Tyr 123 analog (Tyr 111), and to the phosphate of FMP. The second water is below the ribose plane, bonded to O3' and nearer to the Glu 177 analog than to the Arg 180 analog. The authors feel either water may be the ultimate nucleophile but favor the attack from above the ring, perhaps with polarization by the phosphate group.[47] The crystal structure of a trichosanthin NADPH complex shows a water molecule, above the ribose plane, bonded between the analog of Tyr 80 (Tyr 70) and the ribose ring oxygen (O4') which the authors believe is the likely attacking agent.[50] That water is on the opposite side of the substrate from the Glu 177 analog but might be polarized by the Arg 180 analog which is nearby.

Figure 13.6 shows one plausible mechanism for RTA, and RIPs in general.[49] In a concerted mechanism, C-N bond cleavage develops oxycarbonium character on the ribose which is stabilized by the negative charge of Glu 177. The negative charge in the adenine ring must be neutralized by partial or complete protonation by the enzyme. Some

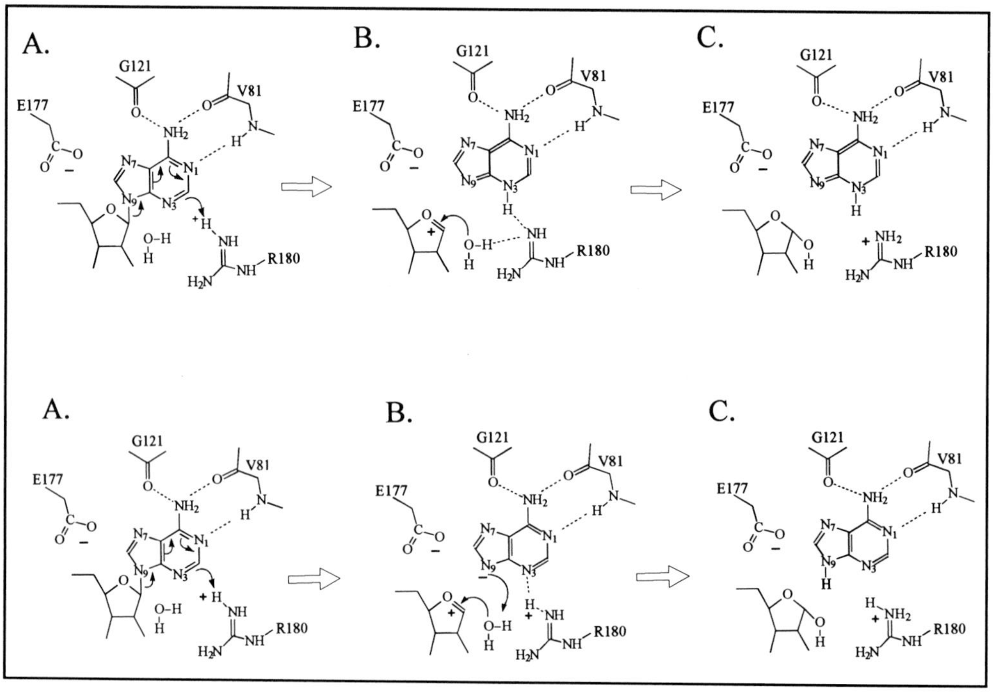

Fig. 13.6. *Two variations of a plausible mechanism of action for RTA and the RIP family. The top series shows: (A) The target adenosine is stacked with the aromatic side chains of tyrosines 80 and 123. They lie above and below the adenine plane but for clarity are not shown here. Specific hydrogen bonds are formed to the adenine ring as indicated by the dotted lines. (B) Bond cleavage between adenine N9 and C1' of ribose is facilitated by protonation at N3 by Arg 180. As Arg 180 transfers a proton to adenine, it attracts one from water and thereby activates the solvent as it develops hydroxide character. Cationic development as an oxycarbonium ion on ribose is stabilized by ion pairing to Glu 177. (C) Activated water attacks the oxycarbonium, releasing the adenine leaving group. The bottom series is the same except that in (B) hydrogen bonding serves as partial protonation of the leaving adenine and N9 helps polarize the water. In (C) the product adenine is in the aromatic resonance form without proton rearrangement.*

stabilization probably arises from hydrogen bonding between the backbone amido of Val 81 and N1, but at least partial protonation at N3 by Arg 180 is probably the most crucial factor. The attacking water is shown on the underside of the ribose plane, bound to Arg 180 although it is also close to Glu 177. Protonation of N3 by Arg 180 increases its basicity toward water, polarizing it with the second function group of the guanidine. The polarized water attacks the backside of the sugar inverting the anomer. It may also be that the attacking water is above the ribose plane and attacks the oxycarbonium ion directly or is polarized by some other group, like the phosphate. In this case the ribose hand would be retained.

THE RICIN B CHAIN

As can be seen in Figure 13.2, RTB is made of two globular domains, each with one galactose binding site.[35] Domain 1 consists of residues 1-135, and domain 2 is residues 136-262. Each domain is composed, in turn, of four peptides designated α, β, γ and λ. These subdomains are each approximately 40 residues in length. The λ peptides are linkers, 1λ contains Cys 4 and is involved in linking RTB to RTA in the heterodimer. Peptide 2λ spans domain 2 to link it to domain 1. The two λ peptides are homologs reflecting the fact that RTB arose as a gene duplication.[58]

The α, β, and γ peptides are all homologous to one another.[36,59] In each domain one copy of the three peptides assembles around a pseudo 3-fold axis to create a globular domain. Three fold symmetric hydrophobic interactions secure the core structure of each domain. The core is composed mainly of conserved Trp and Ile residues.

Only one of the three potential galactose binding sites retains binding affinity in each domain; these are the α peptide of domain 1 and the γ peptide of domain 2. Both active sugar binding sites are relatively shallow pockets.[59] The binding of lactose is illustrated in Figure 13.7. The top of the pocket is simply the side chain of an aromatic residue, Trp 37 and Tyr 248 respectively in domains 1 and 2. The bottom of the pocket is formed by a three residue kink in the polypeptide chain residues 24-26 and 236-238. The galactosyl moiety of lactose is seen to be oriented and secured by hydrogen bonds to the side chains of several polar residues while the glucosyl moiety extends freely into solvent. The C3 and C4 hydroxyls of the bound galactose hydrogen bond to Asp 22 in site 1 and Asp 234 in site 2. These carboxylates are anchored into their correct positions by a hydrogen bond from Gln 47 in site 1 and Gln 256 in site 2. The C3 hydroxyl of galactose hydrogen bonds with Asn 46 in site 1 and Asn 255 in site 2. Site directed mutagenesis of these last two residues has confirmed that they are important in galactose binding and loss of the side group may reduce binding by several orders of magnitude.[60-62]

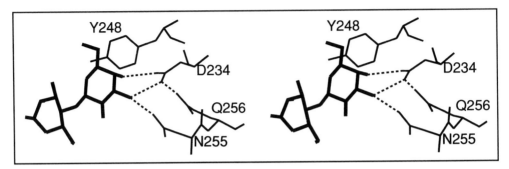

Fig. 13.7. Lactose binding the ricin B chain. The local backbone and key protein side chains are shown in narrow bonds. Lactose is in heavier bonds and hydrogen bonds are shown as dashed lines. The binding site shown is from the C terminal domain of RTB; an equivalent arrangement is seen in the N terminal domain.

Equilibrium dialysis revealed two binding sites for lactose with association constants of 3000 M[-1] and 19,000 M[-1] at 25 C.[23] A Tyr residue has been implicated in the stronger binding site,[25] while chemical modifications suggested a Trp residue was important to sugar binding in the lower affinity site.[26] It appears that the binding site in domain 1 is the weak site, while that in domain 2 is stronger.

ACKNOWLEDGMENTS

This work was supported by grants GM 30048 and GM35989 from the National Institutes of Health and by grants from the Foundation for Research and the Welch Foundation.

REFERENCES

1. Albersheim P, Valent BS. Host-pathogen interactions in plants. J Cell Biol 1978; 78:627-43.
2. Walker-Simmons M, Hadwider L, Ryan CA. Chitosans and pectic polysaccharides both induce the accummulation of the antifungal phytoalexin pisatin in pea pods and antinutrient proteinase inhibitors in tomato leaves. Biochem Biophys Res Commun 1983; 110:194-99.
3. Roberts WK, Selitrennikoff CP. Zeamatin, an antifungal protein from maize with membrane-permeabilizing activity. J Gen Microbiol 1990; 136:1771-78.
4. Abeles FB, Bosshart RP, Forrence LE et al. Preparation and purification of glucanase and chitinase from bean leaves. Plant Physiol 1970; 47:129-34.
5. Boller T, Gehri A, Mauch F et al. Chitinase in bean leaves: induction by ethylene, purification, properties, and possible function. Planta 1983; 157:22-31.
6. Olsnes S, Pihl A. Toxic lectins and related proteins. In: Cohen P, Van Heynigen S, eds. The molecular action of toxins and viruses. New York: Elsevier Biomedical Press, 1982:52-105.

7. Lord JM, Roberts LM, Robertus JD. Ricin: structure, mode of action, and some current applications. FASEB Journal 1994; 8:201-08.

8. Stirpe F, Barbieri L. Ribosome-inactivating proteins up to date. FEBS Letters 1985; 195:1-8.

9. Lodge JK, Kaniewski WK, Tumer NE. Broad-spectrum virus resistance in transgenic plants expressing pokeweed antiviral protein. Proc Natl Acad Sci USA 1993; 90:7089-93.

10. Logemann J, Jach G, Leah R et al. Expression of a ribosome inhibiting protien (RIP) leads to fungal resisitance in transgenic tobacco plants. Mol Biol of Plant Growth and Dev 1991; 3:1363.

11. Ready M, Wilson K, Piatak M et al. Ricin-like plant toxins are evolutionarily related to single-chain ribosome-inhibiting proteins from Phytolacca. J Biol Chem 1984; 259:15252-56.

12. Frankel AE. Immunotoxins. Boston: Kluwer Academic Publishers, 1988.

13. Funatsu G, Kimura M, Funatsu M. Primary Structure of Ala Chain of Ricin D. Agric Biol Chem 1979; 43:2221-24.

14. Funatsu G, Yoshitake S, Funatsu M. Primary Structure of Ile Chain of Ricin D. Agric Biol Chem 1978; 42:501-03.

15. Lamb FI, Roberts LM, Lord JM. Nucleotide sequence of cloned cRNA coding for preproricin. Eur J Biochem 1985; 148:265-70.

16. Halling KC, Halling AC, Murray EE et al. Genomic cloning and characterization of a ricin gene from *Ricinus communis*. Nucl Acids Res 1985; 13:8019-33.

17. Kimura Y, Hase S, Kobayashi Y et al. Structures of sugar chains of ricin D. J Biochem 1988; 103:944-49.

18. Lewis MS, Youle RJ. Ricin subunit association: thermodynamics and the role of the disulfide bond in toxicity. J Biol Chem 1986; 261:11571-77.

19. Nicolson GL, Blaustein J. The interaction of *Ricinus communis* agglutinin with normal and tumor cell surfaces. Biochem Biophys Acta 1972; 266:543-47.

20. Baenziger JU, Fiete D. Structural determinants of Ricin communis agglutinin and toxin specificity for oligosaccharides. J Biol Chem 1979; 254:9795-99.

21. Nicolson GL, Lacorbiere M, Hunter TR. Mechanism of cell entry and toxicity of an affinity-purified lectin from *Ricinus communis* and its differential effects on normal and virus-transformed fibroblasts. Cancer Res 1975; 35:144-55.

22. Simmons BM, Stahl PD, Russell JH. Mannose receptor-mediated uptake of ricin toxin and ricin A chain by macrophages: multiple intracellular pathways for A chain translocation. J Biol Chem 1986; 261:7912-20.

23. Zentz C, Frenoy JP, Bourrillon R. Binding of galactose and lactose to ricin: equilibrium studies. Biochim Biophys Acta 1978; 536:18-26.

24. Houston LL, Dooley TP. Binding of two molecules of 4-methylumbelliferyl galactose or 4-methylumbelliferyl N-acetylgalactosamine to the B chains of ricin and *Ricinus communis* agglutinin and to purfied ricin B chain. J Biol Chem 1982; 257:4147-51.

25. Mise T, Shimoda T, Funatsu G. Identification of tyrosyl residue in the high-affinity saccharide-binding site of ricin D. Agric Biol Chem 1986; 50:151-55.

26. Hatakeyama T, Yamasaki N, Funatsu G. Evidence for involvement of tryptophan residue in the low-affinity saccharide binding site of ricin D. J Biochem 1986; 99:1049-56.

27. Olsnes S, Fernandez-Puentes C, Carrasco L et al. Ribosome inactivation by the toxic lectins abrin and ricin: kinetics of the enzymic activity of the toxin A-chains. Eur J Biochem 1975; 60:281-88.

28. Ready MP, Kim YS, Robertus JD. Directed alteration of active site residues in ricin A chain and implications for the mechanism of action. Proteins 1991; 10:270-78.

29. Endo Y, Tsurugi K. RNA N-glycosidase activity of ricin A-chain. J Biol Chem 1987; 262:8128-30.

30. Endo Y, Chan YL, Lin A et al. The cytotoxins α-sarcin and ricin retain their specificity when tested on a synthetic oligoribonucleotide (35-mer) that mimics a region of the 28 S ribosomal ribonucleic acid. J Biol Chem 1988; 263:7917-20.

31. Ready M, Bird S, Rothe G et al. Requirements for antiribosomal activity of pokeweed antiviral protein. Biochim Biophys Acta 1983; 740:19-28.

32. Ready MP, Brown DT, Robertus JD. Extracellular localization of pokeweed antiviral protein. Proc Nat Acad Sci 1986; 83:5053-56.

33. Habuka N, Akiyama K, Tsuge H et al. Expression and secretion of Mirabolis antiviral protein by *Eschericia coli*, and its inhibition of in vitro eukaryotic and prokaryotic protein synthesis. J Biol Chem 1990; 264:6629-37.

34. Hartley MR, Legname G, Osborn R et al. Single-chain ribosome inactivating proteins from plants depurinate *Escherichia coli* 23S ribosomal RNA. FEBS Lett 1991; 290:65-68.

35. Montfort W, Villafranca JE, Monzingo AF et al. The three-dimensional structure of ricin at 2.8 Å. J Biol Chem 1987; 262:5398-403.

36. Rutenber E, Katzin BJ, Collins EJ et al. The crystallographic refinement of ricin at 2.5 Å resolution. Proteins 1991; 10:240-50.

37. Katzin BJ, Collins EJ, Robertus JD. The structure of ricin A chain at 2.5 Å. Proteins 1991; 10:251-59.

38. Rutenber E, Robertus JD. The structure of ricin B chain at 2.5 Å resolution. Proteins 1991; 10:260-69.

39. Mlsna D, Monzingo AF, Katzin BJ et al. The structure of recombinant ricin A chain at 2.3 Å. Prot Sci 1993; 2:429-35.

40. Weston SA, Tucker AD, Thatcher DR et al. X-ray structure of recombinant ricin A-chain at 1.8 Å resolution. J Mol Biol 1994; 244:410-22.

41. Pan K, Zhang Y, Lin Y et al. The course of the polypeptide chain of trichosanthin molecule. Scientia Sinica 1986; 29:26-32.

42. Collins EJ, Robertus JD, LoPresti M et al. Primary amino acid sequence of α-trichosanthin and molecular models for abrin A-chain and α-trichosanthin. J Biol Chem 1990; 265:8665-69.

43. Xia Z, Shang L, Zhang Z et al. The three-dimensional structure of trichosanthin refined at 2.7 Å resolution. Chinese J Chem 1993: 11:280-88.

44. Zhou K, Fu Z, Chen M et al. Structure of trichosanthin at 1.88 Å resolution. Proteins 1994; 19:4-13.

45. Gao B, Ma X, Wang Y et al. Refined structure of trichosanthin at 1.73 Å resolution. Sci in China Series B Chem 1994; 37:59-73.

46. Monzingo AF, Collins EJ, Ernst SR et al. The 2.5 Å structure of pokeweed antiviral protein. J Mol Biol 1993; 233:705-15.

47. Ren J, Wang Y, Dong Y et al. The N-glycosidase mechanism of ribosome-inactivating proteins implied by crystal structures of alpha-momorcharin. Structure 1994; 2:7-16.

48. Husain J, Tickle IJ, Wood SP. Crystal structure of momordin, a type I ribosome inactivating protein from the seeds of *Momordica charantia*. FEBS Letters 1994; 342:154-58.

49. Monzingo AF, Robertus JD. X-ray analysis of substrate analogs in the ricin A-chain active site. J Mol Biol 1992; 227:1136-45.

50. Xiong J, Xia Z, Wang Y. Crystal structure of trichosanthin-NADPH complex at 1.7 Å resolution reveals active-site architecture. Nature Struct Biol 1994; 1:695-700.

51. Frankel A, Welsh P, Richardson J et al. The role of arginine 180 and glutamic acid 177 of ricin toxin A chain in the enzymatic inactivation of ribosomes. Mol Cell Biol 1990; 10: 6257-63.

52. Kim Y, Mlsna D, Monzingo AF et al. The structure of a ricin mutant showing rescue of activity by a noncatalytic residue. Biochemistry 1992; 31:3294-96.

53. Kim YS, Robertus JD. Analysis of several key active site residues of ricin A chain by mutagenesis and X-ray crystallography. Protein Engineering 1992; 5:775-79.

54. Chaddock JA, Roberts LM. Mutagenesis and kinetic analysis of the active site Glu177 of ricin A-chain. Prot Engin 1993; 6:425-31.

55. Wong KB, Ye YB, Dong YC et al. Structure/function relationship study of Gln 156, Glu 160 and Glu 189 in the active site of trichosanthin. Eur J Biochem 1994; 221:787-91.

56. Sinnot ML. Catalytic mechanisms of enzymic glycosyl transfer. Chem Rev 1990; 90:1171-1202.

57. Mentch F, Parkin DW, Schramm VL. Transition-state structures for N-glycoside hydrolysis of AMP by acid and by AMP nucleosidase in the presence and absence of allosteric activator. Biochemistry 1987; 26:921-30.

58. Villafranca JE, Robertus JD. Ricin B-chain is a product of gene duplication. J Biol Chem 1981; 256:554-56.

59. Rutenber E, Ready M, Robertus JD. Structure and evolution of ricin B chain. Nature 1987; 326:624-26.

60. Vitetta ES, Yen N. Expression and functional properties of genetically engineered ricin B chain lacking galactose binding activity. Biochim Biophys Acta 1990; 1049:151-57.

61. Swimmer C, Lehar SM, McCafferty J et al. Phage display of ricin B chain and its single binding domains: system for screening galactose-binding mutants. Proc Natl Acad Sci USA 1992; 89:3756-60.

62. Newton DL, Wales R, Richardson PT et al. Cell surface and intracellular functions for ricin galactose binding. J Biol Chem 1992; 267:11917-22.

63. Lin Q, Chen ZC, Antoniw JF et al. Isolation and characterization of a cDNA clone encoding the anti-viral protein from *Phytolacca americana*. Plant Mol Biol 1991; 17:609-14.

64. Minami Y, Funatsu G. The complete amino acid sequence of momordin-a, a ribosome- inactivating protein from the seeds of bitter gourd, *Momordica charantia*. Biosci Biotech Biochem 1993; 57:1141-44.

65. Ho WKK, Liu SC, Shaw PC et al. Cloning of the cDNA of α-momorcharin: a ribosome inactivating protein. Biochim Biophys Acta 1991; 1088:311-14.

66. Higgins DG, Sharp PM. Fast and sensitive multiple sequence alignments on a microcomputer. CABIOS 1989; 5:152-253.

67. Schlossman D, Withers D, Welsh P et al. Expression and characterization of mutants of ricin toxin A chain in *Escherichia coli*. Mol Cell Biol 1989; 9:5012-21.

CHAPTER 14

THE STRUCTURES AND EVOLUTION OF SNAKE TOXINS OF THE THREE-FINGER FOLDING TYPE

Dominique Housset and Juan C. Fontecilla-Camps

Because of their impact on human health, snake venoms have been generally the subject of a great deal of attention and several recent reviews describe the properties of their components.[1-3] Many of these have enzymatic activity and are either related to digestive proteins such as phospholipases, proteases and nucleases[4] or to trypsin inhibitor polypeptides such as the dendrotoxins.[5] This has led to the hypothesis that the venom glands of snakes may have been shaped after the pancreas, secreting at first enzymes that later turned into toxic substances.[4] The most studied toxins are those found in Elapidae and Hydrophidae snake venoms and several hundred amino acid sequences as well as tens of three-dimensional structures are currently known. One of the best characterized group is the so-called three-finger snake toxins. These proteins, with molecular weights close to 7,000 Da, share a three-dimensional fold first described for erabutoxin b.[6,7] The best known members of this group are the short (59-62 amino acids, 4 disulfide bridges), and the long (66-80 amino acids, 4 or 5 disulfide bridges) α-neurotoxins. These proteins bind to the acetylcholine receptor located on the post-synaptic membrane thereby blocking the transformation of the chemical signal transmitted by acetylcholine into a depolarizing one.[1] All Elapidae and Hydrophidae venoms seem to contain long neurotoxins and most of them have short neurotoxins (Table 14.1). Other three-finger molecules, devoid of post-synaptic neurotoxic activity, have been purified from cobras and mambas.[8-12] One group of

Protein Toxin Structure, edited by Michael W. Parker. © 1996 R.G. Landes Company.

Table 14.1. Toxin contents of snake venoms

	SNT	LNT	CTX	FAS/CAL
Sea Kraits	+	+	–	–
Sea Snakes	+	+	–	–
Australian elapids	+	+	–	–
Cobras and Ringhals	+	+	+	–
Mambas	+	+	–	+
Kraits	–	+	–	–

SNT = short neurotoxins, LNT = long neurotoxins, CTX = cardiotoxins, FAS = fasciculins and CAL = calciseptine and FS_2

such molecules, the cardiotoxins, also called cytotoxins, seem to interact with cell membranes, although the exact nature of their receptor is still not fully elucidated.[13-14] Sea-snake Krait venoms appear atypical in that they are devoid of short neurotoxins (Table 14.1). Mamba snake venoms are the only ones known to posses three-finger type inhibitors of acetylcholine esterases (fasciculins) and Ca^{2+} channels (calciseptins). Since the first crystallographic structure determination of erabutoxin b in 1976,[6] numerous structural studies have been undertaken by both X-ray crystallography and NMR. Today, not less than 27 structures are available from the Protein Data Bank, corresponding to 21 different proteins (Table 14.2).

The three-finger family of toxins constitute an excellent example of a protein motif that has served as template to generate multiple functions during evolution.[15] Here, we will analyze structural similarities, as well as basic differences, among these proteins and will provide an insight into the protein folding, structure/function relationships, and evolution of these proteins.

STRUCTURAL COMPARISON OF THREE-FINGER SNAKE TOXINS

Typically, the three-finger snake toxins are relatively flat proteins, made of a dense core cross-linked by four to five disulfide bridges, out of which emerge three loops, disposed as the central fingers of a hand (Fig. 14.1). The secondary structure consists of two antiparallel β-sheets: a two-stranded one in loop I and a three-stranded one involving loop II and one segment of loop III. In some cases, the two β-sheets may be connected through main chain: main chain hydrogen bonds, forming a single five-stranded β-sheet. Schematic drawings of this fold are shown in Figure 14.1. Although the core of the protein is usually very well conserved, the three loops exhibit significant structural

Table 14.2. 3D-structures of snake toxins available at the Protein Data Bank

name	source	PDB code	method	resolution (Å)*	number of amino acids	reference
erabutoxin a	*Laticauda semifasciata*	5ebx	X–ray	2.0	62	34
erabutoxin b	*Laticauda semifasciata*	3ebx	X–ray	1.4	62	35
erabutoxin b	*Laticauda semifasciata*	6ebx	X–ray	1.7	62	16
erabutoxin b	*Laticauda semifasciata*	1era	nmr	710 – 38 0.66	62	36
neurotoxin b	*Laticauda semifasciata*	1nxb	X–ray	1.4	62	7
α–neurotoxin	*Dendroaspis p. polylepis*	1ntx	nmr	656 – 93 0.66	62	37, 38
toxin α	*Naja nigricollis*	1nea	nmr	409 – 73 0.51	61	39
neurotoxin I	*Naja naja oxiana*	1ntn	X–ray	1.9	72	40
neurotoxin II	*Naja naja oxiana*	1nor	nmr	489 – 140 0.53	61	41
cobrotoxin	*Naja naja atra*	1cod	nmr	461 – 29 1.67	62	42, 43
cardiotoxin CTX I	*Naja naja atra*	2cdx	nmr	715 – 27 1.01	60	44
cardiotoxin II	*Naja naja atra*	1cre	nmr	0.79	60	45
cardiotoxin CTX IIb	*Naja m. mossambica*	2ccx	nmr	597 – 135 0.72	60	46, 47
cardiotoxin III	*Naja naja atra*	2crt	nmr	0.81	60	48
cardiotoxin V	*Naja naja atra*	1cvo	nmr		62	49
cardiotoxin V II4	*Naja m. mossambica*	1cdt	X–ray	2.5	60	50, 51
γ–cardiotoxin	*Naja nigricollis*	1cxn	nmr	489 – 81 1.18	60	52
γ–cardiotoxin	*Naja nigricollis*	1tgx	X–ray	1.55	60	53
α–cobratoxin	*Naja naja siamensis*	2ctx	X–ray	2.4	71	54
κ–bungarotoxin	*Bungarus multicinctus*	1kba	X–ray	2.3	66	55
κ–bungarotoxin	*Bungarus multicinctus*	1nbt	nmr	582 – 27	66	56, 57

* for nmr models: number of distance and angle restraints and rms deviation from average model in Å.

Table 14.2 continues on following page.

Table 14.2 continues from previous page.

Table 14.2. 3D-structures of snake toxins available at the Protein Data Bank

name	source	PDB code	method	resolution (Å)*	number of amino acids	reference
α–bungarotoxin	*Bungarus multicinctus*	2abx	X-ray	2.5	74	58
α–bungarotoxin	*Bungarus multicinctus*	1abt	nmr	389 – 0	74	59
dendroaspin	*Dendroaspis j amesoni kaimose*	1drs	nmr	466 – 16 1.2	59	15
toxin FS2	*Dendroaspis p. polylepis*	1tfs	nmr	600 – 55	60	18
fasciculin 1	*Dendroaspis augusticeps*	1fas	X-ray	1.9	61	19
fasciculin 2	*Dendroaspis augusticeps*	1fsc	X-ray	2.0	61	17

* for nmr models: number of distance and angle restraints and rms deviation from average model in Å.

variations, concerning both their orientation (relative to the core) and their internal conformation. To analyze these variations in detail, we have performed a series of superpositions of the available three-dimensional models on erabutoxin b (3ebx), the best determined structure. The comparison of crystallographic and NMR models brings out the question of reliability of the three-dimensional structures. Although high resolution X-ray crystallography provides the most accurate picture of a protein, the backbone derived from NMR models seems to be accurate enough to allow for the comparison of toxin structures obtained by these two methods. For the majority of the structures, over 30 Cαs could be superposed to corresponding atoms in erabutoxin b. In some cases, however, the fit was poor: 2abx, 1abt, 1nbt, 1drs and 2cdx (PDB codes, Table 14.2). These structures present significant differences in either the usually conserved loop III and/or loop II or some of the cysteine Cα atoms positions. The crystallographic model of α-bungarotoxin (2abx) has relatively poor refinement statistics (R-factor = 0.24, 0.032 Å for bond length r.m.s. deviations from ideality) which suggests that there may be some errors in the model. The corresponding NMR structure (1abt) neither matches 2abx nor 3ebx. 2cdx differs significantly from all other cardiotoxins, especially in loop III; this may represent a real difference since the structure was determined using a rather large number of NMR restrains (Table 14.2). 1drs exhibits unique conformations for loop II and loop III which are most likely due to its function,[15] but might also arise from the relatively

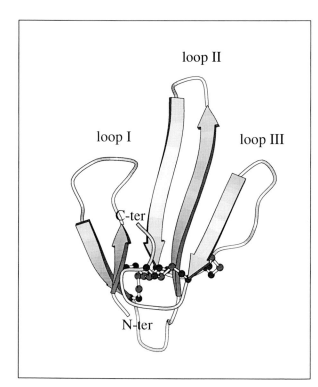

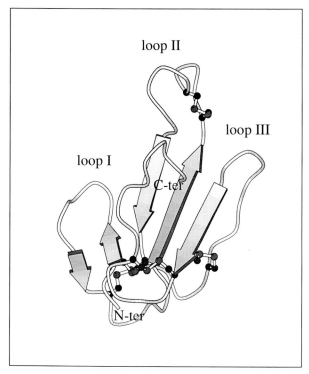

Fig. 14.1. Schematic drawing of six representative snake toxins backbone: β-strands are shown with arrows, disulfide bridges with ball-and-stick. These figures were generated with MOLSCRIPT.[60] (a, shown left) erabutoxin b (3ebx); (b, shown below) cobratoxin(2ctx); (c) κ-bungarotoxin (1kba); (d) cardiotoxin g (1tgx); (e) toxin FS$_2$ (1tfs); (f) fasciculin 1 (1fas).

Fig. 14.1c.

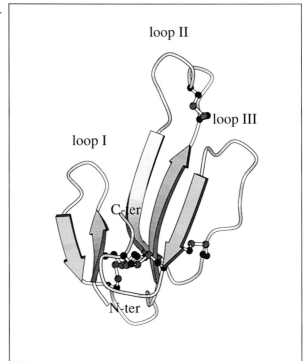

Fig. 14.1d.

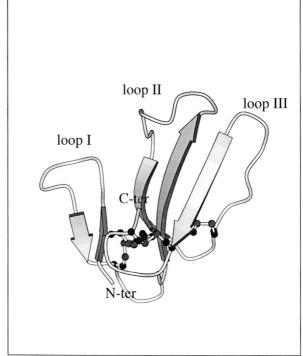

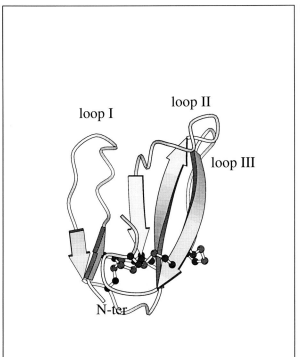

Fig. 14.1e.

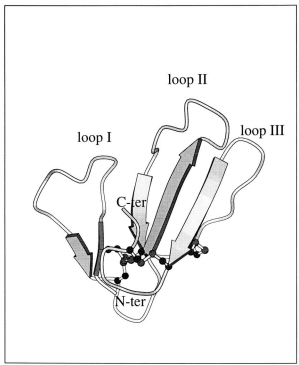

Fig. 14.1f.

low number of NMR restraints used. These structures will be excluded
from some aspects of the following structural analysis which will be
based on the most reliable models from each group of snake toxins.

The superposition of 20 structures indicates that 30 residues are
topologically conserved among snake toxins, with a r.m.s. difference
of 0.89 Å from the average position of the α-carbons of these 30 amino-
acids. This fact shows that the core structure, including disulfide bridges
and the three-stranded β-sheet, is remarkably conserved and that, as
may be expected, major structural differences are restricted to loops
(Fig. 14.2). One may expect that loop conformation could be a function of experimental conditions such as crystal packing interactions or
the action of chemicals present in the crystallization solution, or it
could be poorly determined due to a local lack of restraints in the
case of NMR structures. In fact, crystal packing forces do not necessarily alter the loop structure: comparison of erabutoxin b structures
obtained from different crystal forms and NMR analyses indicates essentially identical loop conformations in all cases.[16] This would suggest that the observed differences in loop conformations among the
different functional groups of toxins are real features of these molecules. There is, however, one known exception to this: comparison
of fasciculins 1 and 2 has shown that the presence of detergent in the
crystallization medium might have been responsible for a drastic change
in the conformation of loop I.[17]

Figure 14.2 shows the superposition of 23 different snake toxins:
conserved and variable regions are clearly distinguishable. In general,
the β-sheet structures are topologically equivalent and superpose well
whereas loops (especially loops I & II) can be oriented in different
directions. Most of the compared snake toxins share a common conformation for loop III except for toxin FS_2, where the plane of this
loop is tilted orthogonally relative to the average plane of the molecule and dendroaspin where the loop diverge from the β-sheet structure.[15,18]

Specific Structural Features

From a purely structural point of view, the usual classification of
toxins is not the most adequate. Some specific features appear more
significant. These are the deletion of three residues at the loop I end
(long neurotoxins including κ-bungarotoxin, and cardiotoxins), the
deletion of one or two residues between cysteines 17 and 24 (long
neurotoxins and fasciculins), the presence of a fifth disulfide bridge at
the loop II extremity (long neurotoxins including κ-bungarotoxin) and
the insertion of two residues between cysteines 41 and 43 (3ebx), which
corresponds to residues 38-42 in cardiotoxins (long neurotoxins including κ-bungarotoxin, and cardiotoxins).

The structural consequence of the deletion at loop I will be discussed in the next paragraph. The deletion of one or two residues

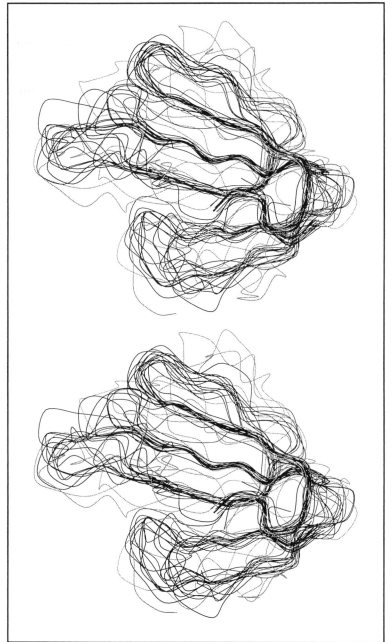

Fig. 14.2. Stereoscopic view of the backbone superposition of 23 snake toxins. All the models were superposed on erabutoxin b, using the program Align.[61] Gray lines represent structures with a poor structural fit (1nbt, 2abx, 1abt, 1drs, 2cdx).

between cysteine 17 and 24 modifies the type II turn emerging from the core at the bottom of the molecule (Fig. 14.1). In fasciculins, the turn connecting loop I and loop II is spatially shifted and of type I. The presence of the fifth disulfide bridge, together with an insertion of four residues, located at the extremity of loop II, is responsible for a turn which is very similar in α-cobratoxin, κ-bungarotoxin and neurotoxin I. The nonstandard turn between cysteines 38 and 42 of cardiotoxin is also shared by long neurotoxins and κ-bungarotoxin. Thus, Cys 42 is not superposed to Cys 44 of erabutoxin (both Cα and Sγ have clearly distinct positions). The proline next to it is more or less superposed for short and long neurotoxins, while κ-bungarotoxin and cardiotoxins present a significantly shifted position. Finally, the C-terminal residues of the long toxins seem to be very flexible, sharing no common features within the group. Thus, the length of the amino-acid chain does not appear as the most significant structural determinant.

Loop Conformation

The conformational variability of loop I seems to be essentially due to its flexibility, as suggested by several NMR structures and the fasciculins. The deletion of three residues in cardiotoxins and in long toxins adds to this variability. Three groups can be distinguished as a function of loop I conformation: an extended conformation made of two 4-residue-β-strands is shared by erabutoxin, cobrotoxin, dendroaspin and fasciculin 2, even though the orientation of the loop is quite different in each case. A folded conformation, pointing towards the C-terminal end is found in cardiotoxins and in fasciculin 1. κ-Bungarotoxin and α-cobratoxin have conformations similar to the one in cardiotoxin, but with the extremity pointing to the other side of the protein. Three basic conformations can be described for loop II: an extended one, found in erabutoxin and cobrotoxin; a partially extended one, with a bend at the extremity which confers a different convexity to the protein, for both fasciculins and cardiotoxins; and one that is specific to the toxins with five disulfide bridges, with an extra turn stabilized by the additional bridge. The conformation of Loop III is the most conserved one, especially for the strand belonging to the three-stranded β-sheet. The first part of the loop is more flexible, and can have a very specific conformation in some cases (toxin FS_2).

It is not always easy to detect the sequence-structure relationships that could explain the structural variability. Moreover, although as indicated above it does not seem to be a general rule, in certain cases the crystallization conditions may have lead to changes in loop conformation, as observed in fasciculins.[17] One intriguing feature is the convexity of the molecule, originating from the orientation of loop II. The presence of a classic β-bulge in fasciculins (X = 25, 1 = 35, 2 = 36) and cardiotoxins (X = 24, 1 = 34, 2 = 35) forces the extremity of loop

II to bend toward the C-terminal region of the molecule. This conformational change, which can be illustrated by the value of main chain torsion angles for residues 37-38, is facilitated by the fact that the side chains of residues 36 and 37 of erabutoxin are inverted by the bent conformation and superpose to residues 35 and 34, respectively, in the fasciculins. Furthermore, this conformation generates a main chain: main chain hydrogen bond between loop II and loop III (N 28-O 47 in fasciculins, N 27-O 48 in cardiotoxins).[19] However, no specific sequence seems to be related to the presence of the bulge.

THE EVOLUTION OF THREE-FINGER SNAKE TOXINS

ORIGIN OF THE THREE-FINGER TOXINS

Several authors have addressed the problem of the origin of three-finger toxins. Most of the early work was performed through amino acid sequence analyses.[20-22] Erickson[20] has compared sequences from snake, scorpion and bee toxins and has suggested that the first two groups evolved from an ancestor molecule about half the length of present toxins. Other authors favor phospholipase-ribonuclease[21] and proinsulin molecules[22] as ancestors of snake toxins. Phospholipase has been considered a likely candidate because it is the most widespread molecule in snake venoms in general.[4] More recently, additional sequence and structural data have shed new light on this problem. Van Tilbeurgh et al[23] have remarked on the general similarities between the three-dimensional structures of colipase, a pancreatic protein involved in the digestion of fatty acids, and the three-finger toxins. Colipase, which like the toxins is a flattened molecule rich in disulfide bridges, binds to lipase and anchors the enzyme to substrate:bile-salt micelles. This is a most intriguing observation given the general similarities between digestive enzymes and snake venom components (see above). However, a detailed topological comparison reveals fundamental differences between the two types of molecules casting a doubt about their relationship (not shown). If colipases are nevertheless related to snake toxins this would argue for a membrane-associated ancestral activity for the latter. In fact, the evolutionary relation: cardiotoxin → long, short neurotoxins has been often postulated.[21,24] More recently, the NMR structure of the human complement regulatory protein CD59[25,26] has shown that, as guessed from the limited amino acid sequence homologies,[27] it is also a member of the three-finger fold proteins. It is difficult to establish an evolutionary relationship between CD59, a cell-surface glycoprotein that protects host cells from complement-mediated lysis, and snake toxins. The same can be said for dendroaspin, an inhibitor of platelet aggregation and platelet adhesion.[15] Perhaps a more illuminating similarity is the one found between amino acid sequences of skin secretion polypeptides from the frog *Xenopus laevis*, the xenoxins, and three-finger snake proteins[28] The strict sequence homology between xenoxins and snake venom proteins is quite high. It is comparable to the one found between a neurotoxin and a cytotoxin by Kolbe et al[28] (20 identities) and is highest between xenoxin-1 and a cytotoxin from *Naja naja siamensis* (21 identities). Xenoxins do not appear to be highly toxic to mice. They display neither antibacterial nor anti-coagulant activity. The authors conclude that a common ancestor of xenoxins and snake three-finger proteins must have existed in the early phases of vertebrate evolution. They also consider that the ancestral polypeptide must have had low toxicity, the highly active neurotoxins being a later product of evolution. We will come back to this below.

In addition to CD59 and xenoxins, other proteins have been found to have amino acid sequences that suggest a three-finger topology. One example is the squid glycoprotein Sgp-2 of unknown function.[29]

Phylogenetic Trees and Dendrograms Based on Amino Acid Sequences

The phylogenetic relationships of snake toxins having the three-finger fold have been extensively reviewed.[2,30,31] In spite of these efforts no clear picture has yet emerged to how the various toxin types are related to each other. One of the major problems has been to reconcile structural and functional aspects. Thus, the long and short neurotoxins have similar functional residues but, in terms of inter-cystine loop lengths, the latter are more similar to cardiotoxins. On the other hand, cardiotoxins and long α-neurotoxins do share a three-residue loop between the 6th and 7th cysteine (short toxin notation) that is only one-residue long in the short neurotoxins. Given this situation, it is not surprising that the various authors have, sometimes, reached radically different conclusions. Strydom[21] has concluded that no clear-cut choice of snake toxin groupings can be reached by using the amino acid sequences alone. He has suggested, instead, that the key to snake toxin classification is found in the weak sequence homologies between venom phospholipase A and the snake toxins. In that author's opinion, an ancestral "cardiotoxin-phospholipase" led to cardiotoxins, on the one hand, and to long and short neurotoxins on the other. The transition from a short neurotoxic ancestor to the long neurotoxins would have been provoked by an unequal crossing-over event. Both Strydom[21] and Tamiya and Yagi[30] have concluded that there is no clear correlation between toxin types and snakes species. The latter authors have even suggested that horizontal gene transfer took place during snake evolution and that any attempt at classification of snakes based on their venom components cannot succeed. Dufton[31] and Dufton and Harvey[32] have used an inter-cystine loop length analysis to divide the three-finger snake toxins into groups and to establish a dendrogram carrying evolutionary implications. Because of the fact that, as mentioned above, short neurotoxins and cardiotoxins have very similar inter-cystine loop lengths but very different functions

Dufton[31] has concluded that cardiotoxicity must have developed from neurotoxicity, unless strong convergent evolution has taken place between short and long neurotoxins. Thus, cardiotoxins have been suggested to be both ancestors and descendants of neurotoxins.

The position of "angusticeps types" such as fasciculins (and calciseptins) seems to be better established according to these authors who place these toxins as closely related to short neurotoxins.[21,32]

THREE-DIMENSIONAL STRUCTURES AS BASIS
FOR EVOLUTIONARY RELATIONSHIPS

The availability of NMR or X-ray structures for the basic structural types: long neurotoxins, short neurotoxins, cytotoxins, fasciculins and calciseptin (or FS_2) opens the road to structure-based evolutionary analysis that can be freed from considerations other than the conservation of three-dimensional features. It is a well established fact that three-dimensional structure is much more conserved that amino acid sequences.[33]

Thirty-two positions are structurally conserved among all the toxins (Table 14.3). Their high degree of conservation implies that they are key elements in the three-finger type toxin folding. By the same token, they may also be relatively independent of function. However, when comparing the structurally conserved segments of cardiotoxins and short neurotoxins, 13 positions appear to differ systematically between the two types of toxins (Table 14.4). Assuming that cardiotoxins and short neurotoxins represent extreme structural and functional types, these differences may be good descriptors of evolutionary distance between the two molecules for residues with a fundamentally structural role. Based on these ideas, we have compared the residues in these thirteen positions among the six classes: long and short neurotoxins, κ-bungarotoxin, cardiotoxins, fasciculins and calciseptin (FS_2). The following conclusions can be reached: (1) the degree of similarity to cardiotoxins decreases as follows: long neurotoxin > κ-bungarotoxin > fasciculin > FS_2; (2) the degree of similarity to short neurotoxins is long neurotoxin > fasciculin > κ-bungarotoxin = FS_2. In general, the most dissimilar toxin is cardiotoxin and the most similar to the others is the short neurotoxin. In between the two extremes, the long neurotoxins (including κ-bungarotoxin) appear as intermediates whereas the fasciculins and FS_2 seem to be closer related to the short neurotoxins.

Taken at face value, these results suggest that cardiotoxins are distantly related to the other classes of toxins except for the long neurotoxins. They also indicate that fasciculins, short neurotoxins, and to a lesser extend, FS_2, are more closely related to each other and to long neurotoxins than to cardiotoxins. The intermediate position of the long neurotoxins is also observed for the overall three-dimensional structure with features in common with both cardiotoxins and short neurotoxins (Fig. 14.1). It is not clear, however, that they are ancestral

Table 14.3. Superposition of snake toxins Cα on the erabutoxin b structure (3ebx)

PDB code	r.m.s. for selected pairs (Å)	number of pairs	number of Ca with d < 1.5Å
5ebx	0.19	60	62
1era	1.09	54	45
1nxb	0.43	59	59
1ntx	0.71	51	51
1nea	1.03	47	42
1ntn	0.79	48	46
1nor	1.02	49	47
1cod	1.00	51	47
2cdx	1.04	27	24
1cre	0.95	27	25
2ccx	0.86	35	34
2crt	1.04	37	33
1cvo	0.97	36	31
1cdt	0.77	37	36
1cxn	1.02	35	31
1tgx	0.69	35	34
2ctx	0.82	42	39
1kba	0.85	37	34
1nbt	1.18	16	15
2abx	1.10	17	18
1abt	1.18	18	18
1drs	0.97	25	23
1tfs	0.89	37	35
1fas	0.86	38	37
1fsc	0.89	35	34

If both X-ray and nmr models were available, the X-ray model was chosen. The program ALIGN[61] was used to superimpose the Cα atoms, using the eight conserved cysteines as a starting point. Then, superposition proceeded iteratively, superposing Cαs which were distant by less than 2 Å (or less than 3 times the r.m.s. if < 2 Å).

to short neurotoxins, that is, whether their similar modes of action are the results of either convergent or divergent evolution.

Cardiotoxins, fasciculin and, to a lesser extent, calciseptine (FS_2) display the same concavity, a feature that does not appear to be dependent on any particular amino acid and have a hydrogen bond between residues 28 and 47. This is a most intriguing similarity since it is not related to the mode of action of the proteins and is found in mamba (Africa) and cobra (Asia) venoms. Neither is a close relationship between cardiotoxins and fasciculins especially supported by our sequence comparisons. The conserved concavity is likely to be an ancestral

Table 14.4. Structural alignment of snake toxins based on the superposition of PDB models on erabutoxin b (3ebx)

		5	10	15	20 25	30	35	40
3ebx	RICFNHQSSQPQTTKTCSPGESSCYHKQWSDFRGTIIERGCGC
5ebx	RICFnH		TTkTC		SCYNKqw		IERGCGc
1ntx	RICY H		TT SC		SCYKK		IERGCG
1nea	LECH Q		TT TC		NCYKK		IERGCG
1nor	LECH Q		TT TC		NCYKK		IERGCG
1cod	LECH Q		TT GC		NCYKK		TERGCG
1cdt	LKCNkL		AYkTC		LCYKMml		PKRGCIc
1tgx	LKCNqL		FWkTC		LCYKMtm		PKRGCIc
1cre	LKCN L		FY TC		LCYKM		VKRGCI
2crt	LKCN L		FY TC		LCYKM		PKRGCI
1cxn	LKCN L		FW TC		LCYKM		VKRGCI
2ccx	LKCN L		FW TC		LCYKM		VKRGCI
1cvo	LKCH T		IY TC		LCFKA		PKRGCA
2cdx	LKCN K		AS TC		LCYKM		PKRGCI
1ntn	ITCYkT		TSeTC		LCYTKtw		IELGCAc
2ctx	IRCFiT		ISkDC		VCYTKtw		VDLGCAc
1kba	RTCLiS		TPqTC		ICFLKaq		IEQGCVc
fasc	TMCYsH		ILtNC		SCYRKsr		VGRGCGc
fasc	TMCYsA		ILtNC		SCYRKsr		VGRGCGc
1tfs	RICY R		AT TC		TCYKM		SERGCG

	45	50 55	60
3ebx	PTVKPGIKLSCCESEVCNN
5ebx	pgIkLSCCESEVCn
1ntx	V IHCCQSDKC
1nea	I LNCCTTDKC
1nor	V LNCCRTDRC
1cod	I INCCTTDRC
1cdt	slVkYVCCSTDRCn
1tgx	slIkYMCCNTDKCn
1cre	V YVCCNTDRC
2crt	V YVCCNTDRC
1cxn	I YMCCNTDKC
2ccx	I YMCCNTNKC
1cvo	L YVCCSTDKC
2cdx	M YVCCNTDRC
1ntn	syQdIKCCSTDNCn
2ctx	tgVdIQCCSTDNCn
1kba	rnYsLLCCTTDNCn
fasc	dyLeVKCCTSDKCn
fasc	dnLeVKCCTSDKCn
1tfs	Q TECCKGDRC

For considering pairs of Cα superposed, no distance criteria is applied. Cα^i of 3ebx is structurally conserved if for each structure, the closest Cα of Cα^i, has Cα^i as the closest 3ebx's Cα. Superposed residues for the 20 selected structures are shown with uppercase (30 Cα, r.m.s.: 0.89 Å). Lower cases represent extra residues superposed among the 9 available crystallographic structures (39 Cα, r.m.s.: 0.78 Å). Numbering is that of erabutoxin b, straight lines above numbers are for β-strands.

feature. This argues for a direct link between a "cardiotoxin-like" ancestor on one hand, and modern cardiotoxins and fasciculins, on the other. Since short neurotoxins are most similar to fasciculins and calciseptine they too may be the result of direct evolution from "cardiotoxin-like" ancestors.

From these comparisons it appears that the long neurotoxins are the consequence of an unusual genetic event and that they cannot be considered as a mainstream feature in snake toxin evolution. When did the split from the central evolutionary trend take place? Here, a short digression will be necessary. From the recent data on xenoxins' amino acid sequences, one can conclude that three-finger polypeptides with substantial sequence homology to snake venom molecules were present long before the emergence of venomous snakes. Also, as stated by Kolbe et al,[28] it is unlikely that these ancestral molecules had neurotoxic activity since such a trait would not be easily lost. One must conclude then that the ancestor of these molecules was not neurotoxic and that postulates that place a "cardiotoxin-like" molecule at their origin are most likely correct. Thus if the similarities between long neurotoxins and cardiotoxins are meaningful and if a "cardiotoxin-like" molecule was ancestral to the snake toxins, then the divergence of long neurotoxin-like molecules may have been a relatively early event in toxin evolution. This is reinforced by the fact that long neurotoxins are a constant feature of both Elapidae and the evolutionary more primitive Hydrophidae venoms. However, there are many similarities between the long neurotoxins and short neurotoxins, fasciculins and FS_2. This suggests that many of the similarities existed already at the time of the split of the former from the common branch.

In this chapter, we have tried to profit from the availability of several three-dimensional structures to establish structural and evolutionary relationships among snake toxins of the three-finger type. Superposition of several of the structures has allowed us to define 32 α-carbon positions which appear to constitute the basic core of the three-finger folding. Although many of the residues present in this core may play functional roles, it is likely that functional variety depends greatly in the sequence and conformation of the extra-core loops. In spite of the high degree of conservation of the main chain atoms of the 32 core residues, only some of them are strictly conserved or conservely changed in terms of amino acid sequence. Differences in the core residue sequences may be interpreted as representing evolutionary distance, if the assumption that core residues play mostly a structural role is correct. Such an approach favors an ancestral "cardiotoxin-like" molecule at the origin of all three-finger type toxins. It also suggests that long neurotoxins branched off from the main sequence at an early stage in the toxin evolution.

REFERENCES

1. Endo T, Tamiya N. Structure-function relationships of postsynaptic neurotoxins from snake venom. In: Harvey AL, ed. Snake toxins. Pergamon Press, 1991:165-222.
2. Dufton MJ, Hider RC. The structure and pharmacology of elapid cytotoxins. In: Harvey AL, ed. Snake toxins. Pergamon Press, 1991:259-302.
3. Rees B, Bilwes A. Three-dimensional structures of neurotoxins and cardiotoxins. Chem Res Toxicol 1993; 6:385-406.
4. Kochva E. The origin of snakes and evolution of the venom apparatus. Toxicon 1987; 25:65-106.
5. Harvey AL, Anderson AJ. Dentrotoxins: snake toxins that block potassium channels and facilitate neurotransmitter release. In: Harvey AL, ed. Snake Toxins. Pergamon Press, 1991:131-64.
6. Low B, Preston HS, Sato A et al. Three dimensional structure of erabutoxin b neurotoxic protein: inhibitor of acetylcholine receptor. Proc Natl Acad Sci USA 1976; 73:2991-94.
7. Tsernoglou D, Petsko GA. The crystal structure of a postsynaptic neurotoxin from sea snake at 2.2 Å resolution. FEBS Letters 1976; 68:1-4.
8. Joubert F, Taljaard N. The complete primary structure of toxin C from *Dendroaspis polylepis polylepis* (black mamba) venom. S Afr J Chem 1978; 31:107-10.
9. Rodriguez-Ithurralde D, Silveira L, Barbeito L et al. Fasciculin, powerful anticholinesterase polypeptide from *Dendroaspis augusticeps* venom. Neurochem Int 1983; 5:267-74.
10. Karlsson E, Mbuga PM, Rodriguez-Ithuralde D. Fasciculins, anticholinesterase toxins from the venom of the green mamba *Dendroaspis augusticeps*. J Physiol 1984; 79:232-40.
11. Cerveñanski C, Engström A, Karlsson E. Study of structure-activity relationship of fasciculin I. Modification of amino groups. Toxicon 1991; 29:1163.
12. Watanabe TX, Itahara Y, Kuroda H et al. Smooth muscle relaxing and hypotensive activities of synthetic calciseptine and homologous snake venom peptide FS2. Jap J Pharmacol 1995; 68:305-13.
13. Bougis P, Rochat H, Piéroni G et al. Penetration of phospholipid monolayers by cardiotoxins. Biochemistry 1981; 20:4915-20.
14. Harvey AL. Cardiotoxins from cobra venoms: possible mechanisms of action. J Toxicol—Toxin Reviews 1985; 4:41-49.
15. Sutcliffe MJ, Jaseja M, Hyde EI et al. Three-dimensional structure of the RGD-containing neurotoxin homologue, dendroaspin. Nature Struct Biol 1994; 1:802-07.
16. Saludjian P, Prange T, Navaza J et al. Structure determination of a dimeric form of erabutoxin-b, crystallized from a thiocyanate solution. Acta Crystallogr 1992; B48:520-31.
17. leDu MH, Housset D, Marchot P et al. Crystal structure of fasciculin 2 from green mamba snake venom: evidence for unusual loop flexibility. Acta Crystallogr D51; in press.

18. Albrand JP, Blackledge MJ, Pascaud F et al. Nmr and restrained molecular dynamics study of the three-dimensional solution structure of toxin fs2, a specific blocker of the l-type calcium channel, isolated from black mamba venom. Biochemistry 1995; 34:5923-37.
19. le Du MH, Marchot P, Bougis PE et al. 1.9 Å resolution structure of fasciculin 1, an anti-acetylcholinesterase toxin from green mamba snake venom. J Biol Chem 1992; 267:22122-30.
20. Erickson BW. Sequence homology of snake, scorpion, and bee toxins. In: Rosenberg P, ed. Toxins—animal, plant and microbial. Proceedings of the 5th Internat. Symp. on Animal Plant and Microbial Toxins—San José, Costa Rica. Pergamon Press 1976: 1071-86.
21. Strydom DJ. The Evolution of Toxins Found in Snake Venoms. In: Lee CV ed. Snake Venoms, Handbook of Experimental Pharmacology. Berlin: Springer-Verlag, 1979: 52, 258-75.
22. Ivanov Ch P, Ivanov OCH. The evolution and ancestors of toxic proteins. Toxicon 1979; 17:205-20.
23. van Tilbeurgh H, Sarda L, Verger R et al. Structure of the pancreatic lipase-procolipase complex. Nature 1992; 359:159-62.
24. Yang CC. Chemistry and evolution of toxins in snake venoms. Toxicon 1974; 12: 1-43.
25. Fletcher CM, Harrison RA, Lachmann PJ et al. Structure of a soluble, glycosylated form of the human complement regulatory protein CD59. Structure 1994; 2:185-99.
26. Kieffer B, Driscoll PC, Campbell ID et al. Three-dimensional solution structure of the extracellular region of the complement regulatory protein CD59, a new cell-surface protein domain related to snake venom neurotoxins. Biochemistry 1994; 33:4474-82.
27. Fletcher CM, Harrison RA, Lachmann PJ et al. Sequence-specific ^{1}H-NMR assignments and folding topology of human CD59. Protein Sci 1993; 2: 2015-27.
28. Kolbe HVJ, Huber A, Cordier P et al. Xenoxins, a family of peptides from dorsal gland secretion of *Xenopus laevis* related to snake venom cytotoxins and neurotoxins. J Biol Chem 1993; 268:16458-64.
29. Williams AF, Tse AGD, Gagnon J. Squid glycoproteins with structural similarities to Thy-1 and Ly-6 antigens. Immunogenetics 1988; 27:265-72.
30. Tamiya N, Yagi T. Nondivergence theory of evolution: sequence comparison of some proteins from snakes and bacteria. J Biochem 1985; 98:289-303.
31. Dufton MJ. Classification of elapid snake neurotoxins and cytotoxins according to chain length: evolutionary implications. J Mol Evol 1984; 20:128-34.
32. Dufton MJ, Harvey AL. The long and the short of snake toxins. TIPS 1989; 10:258-59.
33. Johnson MS, Sali A, Blundell TL. Phylogenies from structural comparisons. Methods in Enzymology 1988; 183:670-89.
34. Corfield PWR, Lee T-J, Low BW. The crystal structure of erabutoxin a at 2.0 angstroms resolution J Biol Chem 1989; 264:9239-42.

35. Smith JL, Corfield PWR, Hendrickson WA et al. Refinement at 1.4 angstroms resolution of a model of erabutoxin b. Treatment of ordered solvent and discrete disorder. Acta Crystallog 1988; A44:357-68.

36. Hatanaka H, Oka M, Kohda D et al. Tertiary structure of erabutoxin b in aqueous solution elucidated by two-dimensional nuclear magnetic resonance. J Mol Biol 1994 240:155-66.

37. Labhardt AM, Hunziker-Kwik EH, Wüthrich K. Secondary structure determination for α-neurotoxin from *Dendroaspis polylepis polylepis* based on sequence-specific [1]H-nuclear-magnetic-resonance assignments. Eur J Biochem 1988; 177:295-305.

38. Brown LR, Wüthrich K. Nuclear magnetic resonance solution structure of the α-neurotoxin from the black mamba (*Dendroaspis polylepis polylepis*). J Mol Biol 1992; 227:1118-35.

39. Zinn-Justin S, Roumestand C, Gilquin B et al. Three-dimensional solution structure of a curaremimetic toxin from *Naja nigricollis* venom: a proton NMR and molecular modeling study. Biochemistry 1992; 31:11335-47.

40. Nickitenko AV, Michailov AM, Betzel C et al. Three-dimensional structure of neurotoxin-1 from *Naja naja oxiana* venom at 1.9 Å resolution. FEBS 1993; 320:111-17.

41. Golovanov P, Lomize AL, Arseniev AS et al. Two-dimensional ^{1}H-NMR study of the spatial structure of neurotoxin II from *Naja naja oxiana*. Eur J Biochem 1993; 213:1213-23.

42. Yu C, Lee C-S, Chuang, LC et al. Two-dimensional NMR studies and secondary structure of cobrotoxin in aqueous solution. Eur J Biochem 1990; 193:789-99.

43. Yu C, Bhaskaran R, Chuang LC et al. Solution conformation of cobrotoxin: a nuclear magnetic resonance and hybrid distance geometry-dynamical simulated annealing study. Biochemistry 1993; 32:2131.

44. Jahnke W, Mierke DF, Beress L et al. Structure of cobra cardiotoxin CTXI as derived from nuclear magnetic resonance spectroscopy and distance geometry calculations. J Mol Biol 1994; 240:445.

45. Bhaskaran R, Huang CC, Tsai YC et al. Cardiotoxin II from taiwan cobra venom, *Naja naja atra*: structure in solution and comparision among homologous cardiotoxins. J Biol Chem 1994; 269:23500-08.

46. Steinmetz WE, Bougis PE, Rochat H et al. [1]H nuclear-magnetic-resonance studies of the three-dimensional structure of the cardiotoxin CTX IIb from *Naja mossambica mossambica* in aqueous solution and comparison with the crystal structures of homologous toxins. Eur J Biochem 1988; 172:101-16.

47. O'Connell JF, Bougis PE, Wüthrich K. Determination of the nuclear magnetic resonance solution structure of cardiotoxin CTX IIb from *Naja mossambica mossambica*. Eur J Biochem 1993; 213:891-900.

48. Bhaskaran R., Huang CC, Chang KD, Yu C. Cardiotoxin III from taiwan cobra (*Naja naja atra*): determination of structure in solution and comparison with short neurotoxins. J Mol Biol 1994; 235:1291-301.

49. Singhal AK, Chien K-Y, Wu W-G et al. The solution structure of cardiotoxin V from *Naja naja atra*. Biochemistry 1993; 32:8036-44.

50. Rees B, Samama JP, Thierry JC et al. Crystal structure of a snake venom cardiotoxin. Proc Natl Acad Sci USA 1987; 84:3132-38.

51. Rees B, Bilwes A, Samama JP et al. Cardiotoxin V^4$_{II}$ from *Naja mossambica mossambica*: the refined crystal structure. J Mol Biol 1990 214:281-97.

52. Gilquin B, Roumestand C, Zinn-Justin S et al. Refined three-dimensional solution structure of a snake cardiotoxin: analysis of the side-chain organisation suggests the existence of a possible phospholipid binding site. Biopolymers 1993; 33:1659-75.

53. Bilwes A, Rees B, Moras D et al. X-ray structure at 1.55 Å of toxin gamma, a cardiotoxin from *Naja nigricollis* venom. Crystal packing reveals a model for insertion into membranes. J Mol Biol 1994; 239:122-36.

54. Betzel C, Lange G, Pal GP et al. The refined crystal structure of α-cobratoxin from *Naja naja siamensis* at 2.4 Å resolution. J Biol Chem 1991; 266:21530-36.

55. Dewan JC, Grant GA, Sacchettini JC. Crystal structure of kappa-bungarotoxin at 2.3-angstrom resolution. Biochemistry 1994; 33:13147.

56. Oswald RE, Sutcliffe MJ, Bamberger M et al. Solution structure of neuronal bungarotoxin determined by two-dimensional NMR spectroscopy: sequence-specific assignments, secondary structure, and dimer formation. Biochemistry 1991; 30:4901-09.

57. Sutcliffe MJ, Dobson CM, Oswald RE. Solution structure of neuronal bungarotoxin determined by two-dimensional nmr spectroscopy: calculation of tertiary structure using systematic homologous model building, dynamical simulated annealing, and restrained molecular dynamics. Biochemistry 1992; 31:2962-70.

58. Love RA, Stroud RM. The crystal structure of α-bungarotoxin at 2.5 Å resolution: relation to solution structure and binding to acetylcholine receptor. Protein Engineering, 1986; 1:37-46.

59. Basus VJ, Song G, Hawrot E. Nmr solution structure of an alpha-bungarotoxin(slash)nicotinic receptor peptide complex. Biochemistry 1993; 32:12290-298.

60. Kraulis PJ. Molscript: a program to produce both detailed and schematic plots of protein structures. J Appl Crystallogr 1991; 24:946-50.

61. Satow Y, Cohen GH, Padlan EA et al. Phosphocholine binding immunoglobulin Fab McPC603: An x-ray diffraction study at 2.7 Å. J Mol Biol 1986; 190:593-604.

CHAPTER 15

THE STRUCTURE OF A VIRALLY ENCODED FUNGAL TOXIN FROM *USTILAGO MAYDIS* THAT INHIBITS FUNGAL AND MAMMALIAN CALCIUM CHANNELS

Fei Gu, Anis Khimani, Stan Rane, William H. Flurkey,
Robert F. Bozarth and Thomas J. Smith

Several strains of *Ustilago maydis* (corn smut fungus) and *Saccharomyces cerevisiae* (yeast) have been classified as "killer strains" by their ability to kill similar strains of fungi in culture. This killer phenotype is due to secretion of toxins produced by dsRNA virions that persistently infect the host cell. The fungal toxins do not affect the host cell strain but do kill similar strains of fungi in that locale. These fungal viruses are unusual in that they are not expressed externally but instead are transmitted vertically through basidiospores or horizontally through anastomosis.[1,2] Therefore, unlike most nonfungal viruses, the killer dsRNA fungal viruses have a symbiotic relationship with their host. The viruses are solely dependent upon host cell survival for replication and lend the host a selective advantage by encoding small toxins that are secreted by the host cell.

The multi-segmented viral genome is encased by unusual capsids composed of 120 copies of gag protein.[3-6] The structures of two of these viruses have been determined using cryo-electron microscopy and image reconstruction techniques (Fig. 15.1).[4] With two identical copies

---

Protein Toxin Structure, edited by Michael W. Parker. © 1996 R.G. Landes Company.

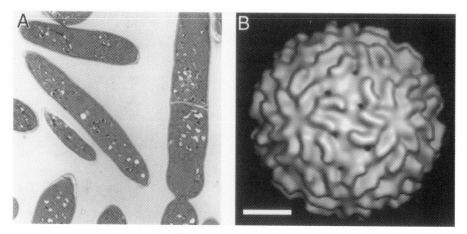

Fig. 15.1. (a) Electron micrograph of cross-sections of several Ustilago maydis *cells (photo courtesy of Ms. Deborah Sherman, Purdue University). (b) Image reconstruction of the UMV4 capsid.[4] The white arrows denote the large holes in the capsid. Fig. 15.1b reprinted from: J Mol Biol 1995: 244:255-258.*

of gag in each icosahedral asymmetric unit, these capsids were found to contradict Casper and Klug rules[7] for icosahedral equivalency without any topological evidence of acceptable pseudo-symmetry such as P = 4. Since the virus is not transmitted via expulsion into the extracellular milieu, it is not obvious why the viral capsids are produced. In the case of LA-1, it was shown that a small number of the capsid protein, gag, are fused with the RNA polymerase, pol.[6] In addition, holes were found in the LA-1 and UMV capsids that are large enough for the uptake of nucleotides and extrusion of nascent viral RNA.[4] Perhaps most interestingly, these capsids resemble the cores of reoviruses[8] rotaviruses,[9] and bacteriophage φ6[10] by the active role that they play in genome replication. In addition, all of these unusual capsids contain 120 copies of at least one protein subunit (these are the only capsids found to have such stoichiometry). Therefore, these fungal capsids are probably not merely protective carapaces for the viral genome but instead are cytoplasmic compartments for genome replication. Furthermore, the notable similarity between these dsRNA virus cores suggest that they may have evolved from a common ancestral virus.[4]

Over the past two decades, a great deal of work has characterized the killing mechanisms of the toxins. In *Saccharomyces cerevisiae*, the two best studied classes of toxins are K1 and K2. The members of one class are immune to the toxins of other members of the same class but sensitive to the killing action of the other class.[11] K1 is composed of two subunits linked together via a disulfide bond.[12-15] One subunit is predicted to have a hydrophobic, α-helical secondary structure typical of membrane spanning proteins. The other domain is hydrophilic and thought to be involved in protecting the host from toxic effects.[16] It was proposed that the K1 toxin first binds to the cell wall, targets secondary receptors at the cell membrane, and then either acts as a protonophore or alters existing ionophores.[17-19] Subsequent experiments demonstrated that the K1 toxin forms ion channels in sensitive yeast sphereoplasts and in artificial membranes.[20] In addition, a toxin with physiological properties similar to K1 from the yeast *Pichia kluyveri* has also been shown to have channel forming activity in synthetic membranes.[21]

In *Ustilago maydis*, three killer toxins have been described; KP1, KP4, and KP6.[22-24] KP1 and KP6 are both composed of two non-covalently linked subunits with each subunit being composed of ~80 amino acids. As with the yeast toxins, KP6 receptors are thought to reside in the cell wall. However, unlike the yeast toxins, both subunits of KP6 are thought to be involved in killing.[25-27] The lack of spheroplast sensitivity to KP6 suggests that there is not a secondary cell membrane receptor.[26] Preliminary evidence suggests that KP6, using both subunits, may also form channels in planar phospholipid membranes,[28] and that these subunits form mostly β-sheets in the presence of phospholipids.[29]

KP4 is unlike other previously described toxins from killer strains of *Saccharomyces cerevisiae* (yeast; K1, K2;[11]) and *Ustilago maydis* (corn smut; KP1, KP6;[22-24]). While these other toxins are composed of two polypeptides and thought to form ion channels in the cell membrane[17-19] KP4 is a single polypeptide with a previously undefined mechanism of action.[22] The yeast toxins are acidic proteins,[30] the other *Ustilago* toxins are neutral, whereas KP4 is very basic with a pI greater than 9.[32]

KP4 STRUCTURE

The structure of KP4 was determined to 1.9 Å resolution using a single isomorphous derivative and phases were improved using real space averaging (Fig. 15.2).[33] KP4 belongs to the α/β sandwich family of proteins (Fig. 15.3) and has a double split β-α-β motif.[34] The toxin has a total of seven β-strands and three α-helices with the major secondary structure elements consisting of a five anti-parallel stranded β-sheet (β1, β3, β4, β6, β7) with two anti-parallel α-helices (α2, α3) lying at approximately 45° to the strands in the β-sheet.

The first β-strand, β1, is only two residues long and is terminated by the disulfide bond between residues 5 and 78. A short, four residue helix (α1) lies between β1 and the first main helix, α2. This helix contains a disulfide bond connecting it to β-strand, β6 (residues 11 and 81, respectively). α2 is a ten residue helix that transverses the β-sheet via a left-handed β-α-β connection and is connected via a disulfide bond to the last alpha helix, α3 (residues 27 and 67, respectively). The carboxyl end of α2 is stabilized by the short β-strand, β2,

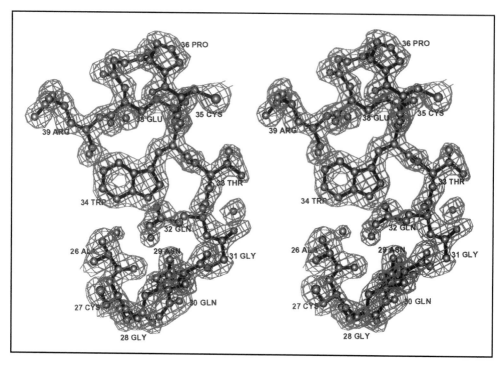

Fig. 15.2. Typical example of the 2Fo-Fc electron density calculated to 1.9 Å resolution using model phases. The atomic model (residues 26-39) is represented by a ball and stick. For clarity, the program MolView[44] was used to trim away electron density more than 2.0 Å away from any atom within this loop. It should be noted, however, that no unassigned density was removed during this process. Reprinted with permission from: Structure 1995:805-814. © 1995 Current Biology Limted.

Fig. 15.3. (Figure is on opposite page) Topology and structure of KP4. (a) Stereo, depth-cued, C-α backbone of KP4 with every tenth residue labeled. Disulfide connections are represented by dashed lines. (b) Ribbon drawing of KP4. The orientation of KP4 is identical in (a) and (b). (c) Topology diagram of KP4 drawn in a style used in a recent review of α/β folds.[34] The triangles represent β-strands and the circles, α-helices. The smaller elements that are only comprised of a few residues are represented by smaller symbols. Note that the main secondary elements form a topology represented by item (j) of Figure 9 from ref. 34. Reprinted with permission from: Structure 1995:805-814. © 1995 Current Biology Limted.

and then by a disulfide bond between residues 35 and 60. The β3-β4 turn forms a very large protuberance at the edge of the sheet that is stabilized by the carboxyl terminus via β-sheet H-bonds and a disulfide bond (residues 44 and 105). α3 transverses back across the β-sheet towards the amino terminus and lies anti-parallel to α2. As with the α2 helix, α3 is part of a left-handed β-α-β cross-over. Similar to the β3-β4 turn, the β6-β7 turn forms a large extension from the β-sheet. Unlike the β3-β4 turn, the β6-β7 protuberance is not stabilized by a disulfide bond and is not as extended as the β3-β4 turn. While the

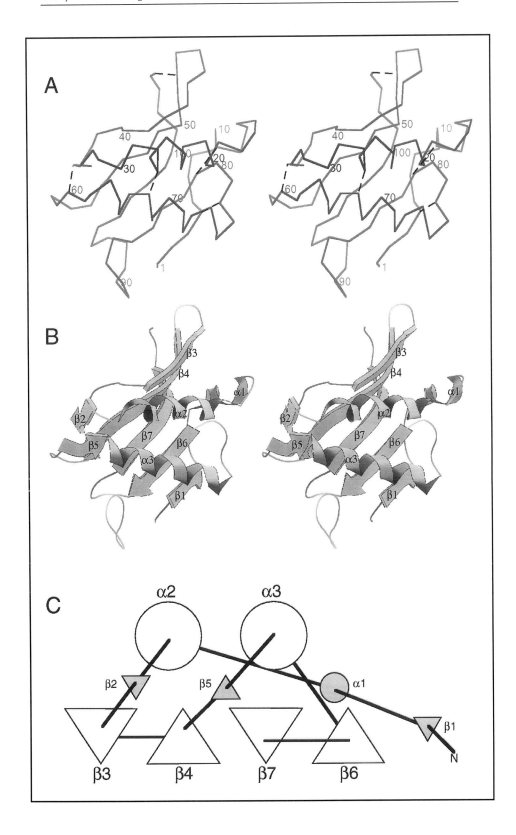

two left-handed β-α-β cross-overs are unusual, it is not clear whether they have a special functional role in KP4.

The toxin is very stable; resisting organic solvents, elevated temperatures, and digestion by proteolytic enzymes (Gu and Smith, unpublished results). The reasons for this stability is clearly evident in the structure. Almost all of the hydrophobic residues are buried between the amphipathic helices and a hydrophobic patch of the β-sheet. The protein is further stabilized by the disulfide bonds that link the major helices together at their ends and middles as well as disulfide bonds that affix their amino and carboxyl termini to the major β-sheet. Finally, the protein is stabilized by having very short random coil elements. Therefore, it seems highly unlikely that the toxin is capable of undergoing the kinds of conformational changes required to expose the amphipathic helices and form ion channels (as has been proposed for the other fungal toxins).

MECHANISMS OF KP4 ACTION SUGGESTED BY ITS STRUCTURE

Since ~80% of the proteins in the α/β sandwich family function by binding to other proteins (e.g., protein G,[35,36] ubiquitin[37] and scorpion toxin[38,39]) it seemed plausible that KP4 acts by binding to cell membrane proteins. Regions of the toxin that might be involved in such interactions are suggested by the extremely asymmetric distribution of charge about the toxin surface. The toxin has a large patch of positive charge covering about two-thirds of the surface surrounding the large protrusion formed by the β3-β4 loop. In contrast, the exposed face of the β-sheet is "cup shaped" and covered by hydrophilic residues. Therefore, it was speculated that the β3-β4 may be part of the "active" site of the toxin and that the positive charge about it allows it to interact with either the phospholipid surface or the target membrane protein.

The scorpion toxins are a family of neurotoxins with some interesting, albeit tenuous, similarities to KP4. There are two classes of these single polypeptide chain neurotoxins: short toxins with ~37 amino acids (e.g., charybdotoxin), and long toxins containing ~60-70 amino acids (e.g., *Androctonus australis* Hector II, AaHII). The long neurotoxins, which are of the α/β sandwich family, can be further subdivided on the basis of their effects on excitable membranes. Albeit about half the size of KP4, the long toxins are similar to KP4 in that they are all highly basic proteins that are stabilized by an extensive network of disulfide bonds. The core motif of the α and β neurotoxins are quite similar and the functional differences between them may lie in the length and orientation of protruding loops extending from this core region.[40] In AaHII, Lys58 is a highly reactive moiety that, upon chemical modification, renders the toxin inactive.[41] This lysine is near the C-terminus of AaHII that is covalently linked via a disulfide bond to a protrusion formed by the turn between the first α-helix and the

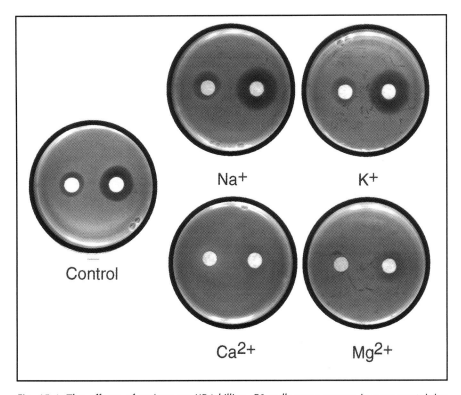

Fig. 15.4. The effects of cations on KP4 killing. P2 cells were grown in agar containing complete media³² with the addition of 0.1M of KCl, NaCl, MgCl₂, CaCl₂ and without any additional salts. For each condition, two filter blots were placed on top of the agar. The left filter blot contained a ~0.8 µM solution of KP4 and the right filter blot contained ~8.0 µM solution of KP4. KP4 killing is shown by a clearing about the filter blot after incubation overnight at room temperature. Reprinted with permission from: Structure 1995:805-814. © 1995 Current Biology Limted.

first β-strand.³⁹ Similarly, the β3-β4 loop in KP4 has several basic groups along the sides and base of the protrusion (Fig. 15.4) and is stabilized by a disulfide bond between it and the carboxyl terminal cysteine (Fig. 15.2). One significant difference between KP4 and AaHII, other than size and topology, is that AaHII has an unusual cluster of four tyrosines on the exposed β-sheet face whereas KP4 is much more polar at the corresponding face with only a single tyrosine protruding into solvent. It was speculated that this hydrophobic patch, conserved amongst the long toxins, might play a role in channel binding.³⁸

BIOCHEMICAL STUDIES ON KP4 MECHANISM OF ACTION

In light of this tenuous similarity to the scorpion toxins, KP4 killing efficacy was tested on P2 cells (sensitive to KP4) in the presence of additional KCl, NaCl, $MgCl_2$ and $CaCl_2$ (Fig. 15.3). The addition of up to 0.2M K^+ and 0.2M Na^+ did not have a significant effect on KP4 killing, whereas as little as 40mM Ca^{2+} and 80mM Mg^{2+} completely rescued P2 cells. It should be noted that the growth of P2 cells away from the test area appeared to be equivalent under all test conditions. These results suggest that KP4 inhibits Ca^{2+} channels and high external Ca^{2+} concentrations rescue the P2 cells by forcing Ca^{2+} through the remaining channel activity. The rescue by Mg^{2+} may be due to competition between KP4 and divalent cation binding or the targeted cationic channels may not be entirely Ca^{2+} specific.

These rescue experiment results strongly suggested that KP4 affects Ca^{2+} channels. Therefore, the effect of KP4 was tested on mammalian cells lines where more is known about the types and kinetics of cationic channels. Standard whole-cell patch clamp techniques were used to examine the effects of KP4 on voltage-activated Na^+, K^+ and Ca^{2+} currents in PC12, GH_3 and adrenal chromaffin cells. Focal application of 8.0 μM KP4 for up to 2 minutes had no effect on activation or inactivation time-courses of Na^+ or K^+ currents, and caused amplitude variations of less than 5% (Na^+ currents, n = 9 chromaffin cells and 3 PC12 cells; K^+ currents, n = 5 chromaffin cells and 4 PC12 cells). KP4 did, however, modulate Ca^{2+} currents. In GH_3 cells KP4, at 0.8 and 8.0 μM, inhibited high-threshold voltage-activated Ca^{2+} current[42] by an average of 30 ± 5% (n = 3) and 48 ± 2% (n = 7), respectively (Fig. 15.5).

In chromaffin cells, KP4 application caused an initial enhancement of current similar to that observed in response to a depolarizing prepulse (Fig. 6b of ref. 43). Indeed, in the presence of KP4 (0.5 min KP4, at ~0.8μM) this current appears identical to the facilitated current, that is KP4 mimics prepulse facilitation. This effect persists with continued application of KP4 (1.5 min KP4), but is overlaid on a more slowly developing inhibition of steady-state inhibition persists (1.5 min post KP4). Similar results were observed in PC12 cells.

With continued application of KP4 (8μM), steady-state current amplitude was decreased by an average of 28 ± 3% in chromaffin cells (n = 5), and 25 ± 5% in PC12 cells (n = 7). In these mammalian neuronal cells, therefore, the prevalent action of KP4 is inhibition of voltage-activated Ca^{2+} channel currents. Since mammalian cells do not contain Mg^{2+} channels, the Mg^{2+} dependent rescue of P2 fungal cells cannot be analyzed.

CONCLUSIONS

Some rather tenuous similarities between the structures of KP4 and the scorpion toxins suggest that the active site on KP4 lies about the $\beta3$-$\beta4$ loop. This hypothesis will be tested using site-directed mutagenesis. The KP4 motif is unique and future studies may determine whether some of the more unusual topological characteristics are related to function.

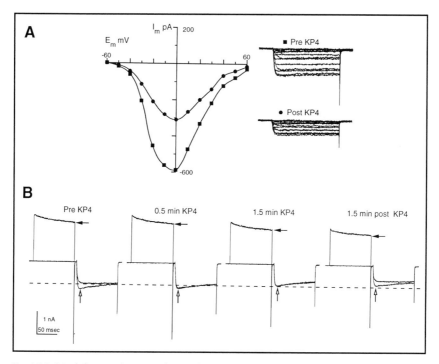

Fig. 15.5. Modulation of neuronal Ca²⁺ channels by KP4. (a) Current-voltage plots and raw current traces show inhibition of voltage-activated Ca²⁺channel currents in a GH₃ cell after application of ~8 μM KP4. Currents were evoked by 50 msec command steps given from -90 mV, and KP4 was focally applied from a micropipette at a distance of less than 2 cell diameters. (b) In bovine adrenal chromaffin cells, application of a depolarizing prepulse causes Ca²⁺ current facilitation. Pre-KP4, a large outward current (filled arrow) is observed during the prepulse (125 mV, 100 msec), and is followed by facilitation (open arrow) of the inward current activated by a step to 0 mV. The starting amplitude of the current evoked at 0 mV in the absence of a prepulse is indicated by the dashed line. Scale bars pertain to (b) only. Reprinted with permission from: Structure 1995:805-814. © 1995 Current Biology Limted.

The effects of KP4 on mammalian Ca²⁺ channels, in conjunction with the fungal rescue experiments, strongly support the hypothesis that KP4 affects target fungal cells via Ca²⁺ (and/or possible Mg²⁺) channels. It is tempting to speculate that, since Ca²⁺ plays such a major role in cell growth and regulation, KP4 may have been originally expressed by the virus to amplify virus production by controlling host replication. This concurs with our preliminary evidence that KP4 may cause cessation of fungal growth rather than the large morphological changes (e.g., loss in cell mass) caused by the other types of fungal toxins.

Perhaps the most surprising result was that fungal and mammalian Ca²⁺ channels are conserved to such a degree that both appear to be affected by the same protein inhibitor. In addition, these results suggest that Ca^{2+} channel blockers might serve as lead compounds for new anti-fungal agents. Further whole-cell patch clamp studies are underway to define the types of mammalian Ca^{2+} channels affected by KP4. In addition, we will attempt to perform patch clamp analysis on fungal cells. Because of KP4's effect on mammalian cells, it seems likely that KP4 has a broader spectrum of activity than suggested by the killing assay alone. Preliminary evidence suggests that KP4 may be affecting cardiac Ca^{2+} channels. Therefore, the effects of intravenous application of KP4 on mammalian heart rate and blood pressure are also being examined. Finally, immuno-electron microscopy are underway to determine the distribution of the toxin within treated target cells. It is not clear whether KP4 is transported into the cytoplasm of the target cell, or if its effects are exerted only via the outer plasma membrane surface.

ACKNOWLEDGMENTS

We would like to thank Ms. Tiffany Sullivan for producing some of the KP4 used in this study. The program MOLVIEW[44] was used to create Figures 15.1, 15.2a and 15.2b, 15.3, and 15.5. This work was supported by grants from the National Institutes of Health (GM10704 to T.J.S., R01GM43462 to S. R., and GM422182 to W.F.F.), from the Lucille P. Markey Charitable Trust (Purdue Structural Biology Center), the American Heart Association to S. R., the Indiana Affiliate Grant to S. R., the Indiana State Faculty Research Grant to R.J.B, the SCi Research Society to A. K., and the Indiana Academy of Science to A.K.

REFERENCES

1. Bevan EA, Mankower M. The physiological basis of the killer character in yeast. Proc Int Congr Genet 1963; XI:1202-03.

2. Wood HA, Bozarth RF. Heterokaryon transfer of viruslike particles and a cytoplasmically inherited determinant in *Ustilago maydis*. Phytopathology 1973; 63:1019-21.

3. Bozarth RF, Koltin Y, Weissman MB et al. The molecular weight and packaging of dsRNAs in the mycovirus from *Ustilago maydis* killer strains. Virology 1981; 113:492-502.

4. Cheng RH, Caston JR, Wang G et al. Fungal virus capsids: cytoplasmic compartments for the replication of double-stranded RNA formed as icosahedral shells of asymmetric Gag dimers. J Mol Biol 1994; 244:255-58.

5. Esteban R, Wickner RB. Three different M_1 RNA-containing viruslike particle types in *Saccharomyces cerevisiae*: in-vitro M_1 double stranded RNA synthesis. Mol Cell Biol 1986; 6:1552-61.

6. Fujimura T, Ribas JC, Makhov AM et al. Pol of gag-pol fusion protein required for encapsidation of viral RNA of yeast L-A virus. Nature 1992; 359:746-49.

7. Caspar DLD, Klug A. Physical principles in the construction of regular viruses. Cold Spring Harbor Symp Quant Biol 1962; 27:1-24.

8. Dryden KA, Wang G-J, Yeager MA et al. Early steps in reovirus infection are associated with dramatic changes in supramolecular structure and protein conformation: Analysis of virions and subviral particles by cryoelectron microscopy and three-dimensional image reconstruction. J Cell Biol 1993; 122:1023-41.

9. Mansell EA, Patton JT. Rotavirus RNA replication: VP2, but not VP6, is necessary for viral replicase activity. J Virol 1990; 64:4988-96.

10. Mindich M, Bamford DH. In Calendar R, ed. The Bacteriophages. Plenum Press, New York, 1988:475-520.

11. Young TW, Yagiu M. A comparison of the killer character in different yeasts and its classification. Antonie van Leeuwenhoek 1978; 44:59-77.

12. Bostian KA, Elliot Q, Bussey H et al. Sequence of the preprotoxin dsRNA gene of type 1 killer yeast: multiple processing produces a two component toxin. Cell 1984; 36:741-51.

13. Bussey H, Boone C, Zhu H et al. Genetic and molecular approaches to synthesis and action of the yeast killer toxin. Experientia 1990; 46:193-200.

14. Bussey H, Saville D, Greene DJ et al. Secretion of *Saccaromyces cerevisiae* killer toxin: processing of the glycosylated precursor. Mol Cell Biol 1983; 3:1362-70.

15. Dimochowska A, Dignard D, Henning D et al. Yeast KEX1 gene encodes a putative protease with a carboxypeptidase B-like function involved in the killer toxin and α-factor precursor processing. Cell 1987; 50:573-84.

16. Tipper DJ, Bostian KA. Double-stranded ribonucleic acid killer systems in yeast. Microbiol Rev 1984; 48:125-56.

17. Bussey H. Physiology of killer factor in yeast. Adv Microb Physiol 1981; 22:93-122.

18. De La Pena P, Barros F, Gascon S et al. Effect of yeast killer toxin on sensitive cells of *Saccharomyces cerevisiae*. J Biol Chem 1981; 256:10420-25.

19. Middelbeek EJ, Von De Larr HHAM et al. Physiological conditions affecting the stability of *Saccharonyces cerevisiae* killer toxin and energy reqirement for toxin action. Antonie van Leeuwenhoek 1980; 46:483-97.

20. Martinac B, Zhu H, Kubalski A et al. Yeast K1 killer toxin forms ion channels in sensitive yeast spheroplasts and in artificial liposomes. Proc Natl Acad Sci USA 1990; 87:6228-32.

21. Kagan B. Mode of action of yeast killer toxins: channel formation in lipid bilayer membranes. Nature 1983; 302:709-11.

22. Koltin Y. The killer system of *Ustilago maydis*: secreted polypeptides encoded by viruses. In: Koltin Y, Leibowitz M, eds. Viruses of Fungi and simple eukaryotes. New York: Marcel Dekker, 1988; 209-42.

23. Koltin Y, Day PR. Specifity of *Ustilago maydis* killer proteins. Applied Microbiol 1975; 30:694-96.

24. Puhalla JE. Compatibility reactions on solid medium and interstrain inhibition in *Ustilago maydis*. Genetics 1968; 60:461-74.

25. Peery T, Shabat-Brand T, Steinlauf R et al. The virus encoded toxin of *Ustilago maydis*—two polypeptides are essential for activity. Mol Cell Biol 1987; 7:470-77.

26. Steinlauf R, Peery T, Koltin Y et al. The *Ustilago maydis* virus encoded toxin—effect of KP6 on cells and spheroplasts. Exp Mycol 1988; 12:264-74.

27. Tao J, Ginzberg I, Banerjee N et al. *Ustilago maydis* KP6 toxin: structure, expression in *Saccharonyces cerevisiae*, and relationship to other cellular toxins. Mol Cell Biol 1990; 10:1373-81.

28. Zizi M, Finkler A, Koltin Y. Association of both subunits of *Ustilago maydis* toxin, KP6, forms large voltage-independent channels. In: Bloomfield V, ed. Biophysical Society Meeting. San Francisco: Biophysical Journal, 1985:A203.

29. Duax WL, Ghosh D, Langs D et al. Crystallization of the α polypeptide for the KP6 killer toxin. In: Bloomfield V, ed. Biophysical Society Meeting. Biophysical Journal, San Francisco, California, 1995; A203.

30. Bussey H. Effects of yeast killer factor on sensitive cells. Nature New Biology 1972; 235:73-5.

31. Levine R, Koltin Y, Kandel J. Nuclease activity associated with the *Ustilago maydis* virus induced killer proteins. Nucleic Acids Research 1979; 6:3717-32.

32. Ganesa C, Chang Y-H, Flurkey WH et al. Purification and molecular properties of the toxin coded by *Ustilago maydis* virus P4. Biophys Biochem Res Commun 1989; 162:651-57.

33. Gu F, Khimani A, Rane SG et al. Structure and function of a virally encoded fungal toxin from *Ustilago maydis*: a fungal and mammalian Ca^{2+} channel inhibitor. Structure 1995; 3:805-14.

34. Orengo CA, Thornton JM. Alpha plus beta folds revisited: some favoured motifs. Structure 1993; 1:105-20.

35. Achari A, Hale SP, Howard AJ et al. 1.67 Å X-ray structure of the B2 immunoglobulin domain of streptococcal protein G and comparison to the NMR structure of the B1 domain. Biochemistry 1992; 31:10449-57.

36. Derrick JP, Wigley DB. Crystal structure of a streptococcal protein G domain bound to an Fab fragment. Nature 1992; 359:752-54.

37. Vijay-Kumar S, Bugg CE, Wilkinson KD et al. Comparison of the three-dimensional structures of human, yeast, and oat ubiquitin. J Biol Chem 1987; 262:6396-99.

38. Fontecilla-Camps JC, Almassy RJ, Ealick SE et al. Architecture of scorpion neurotoxins: a class of membrane-binding proteins. Trends Biochem Soc 1981; 6:291-96.

39. Housset D, Habersetzer-Rochat C, Astier J-P et al. Crystal structure of Toxin II from scorpion *Androctonus australis* hector refined at 1.3 Å resolution. J Mol Biol 1994; 238:88-104.

40. Fontecilla-Camps J-C. Three-dimensional model of the insect-directed scorpion toxin from *Androctonus australis* Hector and its implication for the evolution of scorpion toxins in general. J Mol Evol 1989; 29:63-67.

41. Sampieri F, Habersetzer-Rochat C. Structure-function relationships in scorpion neurotoxins: identification of the superreactive lysine in toxin I of *Androctonus australis* Hector. Biochim Biophys Acta 1978; 535:100-09.

42. Matteson DR, Armstrong CM. Properties of two types of calcium channels in clonal pituitary cells. J Gen Physiol 1986; 87:161-82.

43. Artalejo CR, Mogul DJ, Perlman RL et al. Three types of bovine chromaffin cell Ca²⁺ channels: facilitation increases the opening probability of a 27 pS channel. J Physiol 1991; 444:213-40.

44. Smith TJ. MolView: a program to analyze and display atomic structures on the Macintosh personal computer. J Molecular Graphics 1995; 13:122-25.

INDEX

Page numbers in italics denote figures (f) or tables (t).